D1190131

RADIO SYSTEM DESIGN FOR TELECOMMUNICATIONS (1–100 GHz)

RADIO SYSTEM DESIGN FOR TELECOMMUNICATIONS (1–100 GHz)

ROGER L. FREEMAN

A Wiley-Interscience Publication

JOHN WILEY & SONS

New York · Chichester · Brisbane · Toronto · Singapore

The dust jacket shows a digital tropospheric scatter terminal for military communications—The AN/TRC–170. Picture courtesy of the Raytheon Company, Lexington, MA.

Copyright © 1987 by Roger L. Freeman

Published by John Wiley & Sons, Inc.

All rights reserved. Published simultaneously in Canada.

Library of Congress Cataloging in Publication Data:

Freeman, Roger L.
    Radio system design for telecommunications
(1–100 GHz)

    Bibliography: p.
    Includes index.
    1. Radio relay systems—Design and construction.
I. Title.
TK6553.F7254   1987   621.3841'56   86-28934
ISBN 0-471-81236-6

Printed in the United States of America

10 9 8 7 6 5 4 3 2 1

To my mother, Mary Alice Bing

PREFACE

OBJECTIVE

The purpose of this book is to provide essential design techniques for radiolinks in the point-to-point service operating in the range of 1–100 GHz. It treats general propagation in this frequency range, the design of line-of-sight (LOS) microwave/millimeter links, troposcatter/diffraction, and satellite systems, both analog and digital. The book has been prepared with both the student and working engineer in mind.

BACKGROUND

Line-of-sight radiolinks began to be widely implemented in the 1950s using essentially World War II technology. Application has accelerated since then, being driven by the demand of toll-telephone and television relay for broadcasters. A typical (LOS) link (hop) extends from under 10 miles (16 km) to over 40 miles (64 km), the distance from the near-end transmitter to the far-end receiver; the range is limited by intervening terrain, antenna height, and performance requirements. Some 100 such links in tandem can (and do) provide U.S. transcontinental broadband telecommunication service.

Diffraction/troposcatter is an extension of microwave line-of-sight beyond the horizon with a typical link or hop extending from below 100 miles (160 km) to more than 300 miles (480 km). Such systems require much higher power transmitters, larger antennas, and diversity of operation when compared to their line-of-sight counterparts. The primary demand driver for these systems was, and still is, the armed forces. Troposcatter links operate generally from 900 MHz to over 5 GHz and presently can support no more than about 240 FDM (frequency division multiplex) telephone channels in an analog operational mode, and in a digital mode about 72 PCM (pulse code modulation), or 128 CVSD (continuous variable slope delta modulation) channels.

Geostationary satellites extended the working range of radiolinks dramatically. Considering a link in this case as a single satellite system from near-end terminal transmitter through a satellite repeater to a far-end terminal receiver, a broadband signal can be extended about one third of the way around the earth or about 8000 miles (12,000 km). Three properly spaced geostationary satellites can provide worldwide access for all the world's major population centers. Domestic/regional and special service satellite communication systems have added still another dimension to this technology.

Such systems provide telephone trunking on a national or local regional basis both for common carrier and private service, direct-to-user service, which bypasses local telephone interconnects, TV relay for broadcasters and cable TV and direct TV broadcast. Although rather conceptually simple repeater-type satellites will continue solid growth in the future, the more complex digital processing satellites will show an accelerated implementation. These latter satellites will have up to 1 Gbps or more capacity, provide baseband and/or IF (intermediate frequency) switching and forward error correction on some or all services.

The vast implementation of radiolink systems since the 1950s and of satellite systems since the 1970s has brought about a severe frequency spectrum congestion forcing two direct results: more bandwidth-conservative systems and the utilization of the higher frequencies—those available bands above 10 GHz.

SCOPE

The aim of this text is to describe how radiolinks operate, how to size or dimension terminals and ancillary subsystems, and how to select the necessary performance parameters and equipment specifications to meet the needs of various customers. The seven chapters are organized so that each forms a background for the subsequent chapters. Chapter 1, Radio Propagation 1–100 GHz, contains a general review of the mechanisms of radio propagation in free space, over obstacles, and in the earth's atmosphere. It serves as a basis for all subsequent chapters. If there were no atmosphere nor obstacles in or near a radio transmission path, the propagation problem would become less complex and our concern would only be that of free space propagation in the presence of ground and earth curvature. With the earth's atmosphere included, the propagation problem becomes more complex, resulting in additional signal absorption and fading. Obstacles in or near the transmission path can cause both back and forward scatter of the radio wave leading to a discussion of the Fresnel zone structure of the signal around the area of the obstacle.

The entire discussion in Chapter 1 impacts Chapter 2, Line-of-Sight Radiolinks (1–10 GHz). This chapter covers the essential design principles of broadband radio communication links where the transmitting antenna on one end and the receiving antenna on the other are within "line-of-sight" of each other. LOS (line-of-sight) radiolinks have the broadest application for point-to-point multichannel communications and video transmission and are the most widely implemented of all the radio disciplines covered in this book.

Chapter 3 provides practical design techniques for over-the-horizon radiolinks (e.g., non-LOS paths). The chapter stresses tropospheric scatter. It also covers single obstacle and smooth earth diffraction paths. The material in this chapter relies heavily on a background built in the two previous chapters.

Chapters 4 and 5 are extensions of Chapter 2. They cover the design and sizing of satellite earth stations that interoperate with geostationary satellites. The chapters emphasize propagation, system impairments, especially noise, as well as link design and equipment sizing. Chapter 4 treats the general theme, analog system design with repeater-type satellites, whereas Chapter 5 deals with digital systems both for repeater-type satellites and processing satellites.

Propagation and system design peculiar to the operation of systems above 10 GHz are analyzed in Chapter 6. Topics treated include attenuation and noise due to rainfall, depolarization, attenuation due to gaseous absorption, as well as mitigation techniques available to meet performance requirements, such as spatial or path diversity.

Chapter 7 treats system design from the hardware standpoint. It describes how to develop hardware configurations for LOS radiolinks, both analog and digital, LOS repeaters, over-the-horizon radiolinks and satellite terminals, from small single voice channel facilities to large INTELSAT Type "A" terminals. The chapter takes the performance criteria and sizing parameters developed in Chapters 2 through 6 and details how to configure equipment that will meet these requirements showing tradeoffs to optimize the systems from the standpoint of costs.

NOTES TO THE READER

The text richly utilizes tables, figures, and equations. A double number system is used for identifying each element. The first digit identifies the chapter and the second digit or digits identifies the table, figure, or equation sequentially inside each chapter. Section numbering is also sequential inside each chapter. Acronyms are defined on first usage and then used freely thereafter. The acronyms used are common to the Industry.

The reader is expected to have a working knowledge of electrical communication, algebra, trigonometry, logarithms, and time distributions. The more difficult communication system concepts are referenced to other texts for further reading or clarification.

Nearly every key formula is followed by at least one worked example. Every chapter is followed by a set of review questions and problems. However, the answer to every problem can be found in the text or in a similar example problem worked in the chapter in question. References are identified sequentially in each chapter by a digit which identifies that reference listed at the end of the chapter.

ACKNOWLEDGMENTS

I am very grateful to three of my Raytheon colleagues for their arduous efforts of manuscript review. Dan Odom of Raytheon's Equipment Development Laboratories System Engineering Group provided an excellent critique of Chapter 1. I consider Dan one of the leading propagation engineers in the United States. Don Hastings was kind enough to go through the entire text. Don is from Raytheon's Equipment Division, Communication Systems Directorate and has more than 20 years background in LOS radiolink and tropo system implementation. He kept me on the right track from a practical viewpoint. Dr. Jim Mullen of Raytheon's Research Division painstakingly reviewed the entire book. Jim is one of the most positive people I have ever met. His critique was very constructive. The comments and suggestions offered by these reviewers have greatly enhanced the book. My heartiest thanks to the three of you. Mark Kiryelejza helped resolve some sticky problems. Jack Dicks, INTELSAT's director of system engineering, kindly supported me with INTELSAT reports and standards. His help proved invaluable. My son, Bob, assembled the index on our PC. Bob is always right there when I need him.

ROGER L. FREEMAN

Sudbury, Massachusetts
February 1987

CONTENTS

4. BASIC PRINCIPLES OF SATELLITE COMMUNICATIONS— ANALOG SYSTEMS   221

6. SYSTEM DESIGN ABOVE 10 GHz   397

RADIO SYSTEM DESIGN FOR TELECOMMUNICATIONS (1–100 GHz)

RADIO PROPAGATION
1–100 GHz

1.1 INTRODUCTION

The purpose of this book is to describe methods for the design of broadband radio systems that operate in the 1–100-GHz frequency band. These systems include line-of-sight (LOS) radiolinks, diffraction/scatter, and satellite links. Such transmission links are commonly part of a larger system, the telephone network, for example, and this aspect must not be lost sight of. The three classes of radiolinks also have certain issues and phenomena of propagation in common. Such common aspects of propagation are described in this chapter including free space or spreading loss, the effects of obstacles on propagation, and quiet atmospheric effects as well as the mechanisms that cause fading.

The objective of this chapter is to prepare the reader with the necessary background information for the general propagation problem impacting those three classes of transmission. However, those propagation issues peculiar to each of the transmission types are dealt with in the appropriate chapter.

An arbitrary division of the spectrum of interest has been made at 10 GHz. In general, for those systems operating below 10 GHz, we can say that atmospheric absorption and precipitation play a less important role than for frequencies above 10 GHz. Chapter 6, "System Design above 10 GHz," treats the problems of propagation, such as rainfall and gaseous absorption, for these higher frequencies.

The following discussion uses the isotropic as a reference antenna. Isotropic antennas are not physically possible but they are the simplest reference to use as an intermediate numerical reference to which the actual antenna characteristics are added later.

1.2 LOSS IN FREE SPACE

Our first step is to define the loss between a transmitting and receiving antenna separated by a distance d with the transmission medium assumed to

1

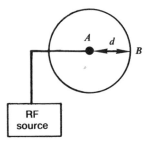

be a vacuum. The antenna at each end of the link is assumed to be an isotropic.

We now can say that point A is an isotropic source, and by definition, an isotropic source radiates uniformly in all directions (i.e., has a gain of 1 or 0 dB). Let the total power in watts radiated by the source be P_T. The envelope containing the radiation around the source can be considered to be an expanding sphere of radius d. The net power flow through the surface of a sphere at its center point must also be P_T, whence it follows that the power flow per unit area through any portion of the sphere's surface must be

$$P_{av} = \frac{P_T}{4\pi d^2} \tag{1.1}$$

where d is the distance in meters from the center to the surface of the sphere.

In the preceding figure, point A, the isotropic source, is at the center of the sphere, and point B, the receiving antenna, is on the sphere's surface.

Conventionally, an isotropic antenna may serve either as a transmitting antenna or a receiving antenna. In the receive function, it absorbs power from the radiation field in which it is situated. The amount of power that the receiving antenna absorbs in relation to the RF power density of the field is determined by its effective aperture, which is defined as the area of the incident wave front that has a power flux equal to the power dissipated in the load connected to the receive antenna output terminals. For an isotropic antenna, the effective area is $\lambda^2/4\pi$, where λ is the wavelength of the incident radiation field. From equation (1.1) it then follows that an isotropic antenna situated in a radiation field with a power density of P_{av} will deliver into its load a power P_R given by

$$P_R = P_T\left(\frac{\lambda}{4\pi r}\right)^2 \tag{1.2}$$

where r is the radius of the sphere or the distance d_t (i.e., $r = d_t$).

The transmission loss between transmit and receive antennas is defined conventionally as

$$L_{dB} = 10\log_{10}\frac{P_T}{P_R} \tag{1.3}$$

Combining equations (1.2) and (1.3), the transmission loss becomes

$$L_{dB} = 21.98 + 20\log_{10}\left(\frac{r}{\lambda}\right) \tag{1.4}$$

Equation (1.4) can now be restated in a more useful form:

$$L_{dB} = 32.4 + 20\log d_{km} + 20\log F_{MHz} \tag{1.5}$$

where d $(d = r)$ is the distance in kilometers between transmit and receive antennas and F is the frequency of the emitted radio field in megahertz (i.e., the source at A). F is obtained from the familiar equation

$$\lambda = \frac{c \times 10^{-6}}{F} \tag{1.6}$$

where $c = 2.998 \times 10^8$ m/sec, the velocity of light in free space. If d is measured in statute miles, then equation (1.4) becomes

$$L_{dB} = 36.58 + 20\log d_{sm} + 20\log F_{MHz} \tag{1.7a}$$

and if d is measured in nautical miles, equation (1.4) becomes

$$L_{dB} = 37.80 + 20\log d_{nm} + 20\log F_{MHz} \tag{1.7b}$$

If d is measured in feet, the constant is -37.87, and in meters, -27.55.

1.3 ATMOSPHERIC EFFECTS ON PROPAGATION

1.3.1 Introduction

If a radio beam is propagated in free space, where there is no atmosphere (by definition), the path followed by the beam is a straight line. The transmission loss in free space was derived in Section 1.2.

However, a radio ray propagated through the earth's atmosphere encounters variations in the atmospheric refractivity index along its trajectory that causes the ray path to become curved. Atmospheric gases will absorb and scatter the radio path energy, the amount of absorption and scattering being a function of frequency and altitude above sea level. Absorption and scattering do become serious contributors to transmission loss above 10 GHz and are discussed in Chapter 6. The principal concern in this section is the effect of the atmospheric refractive index on propagation. Refractivity of the atmosphere will affect not only the curvature of the ray path (expressed by a factor K) but will also give some insight into the fading phenomenon.

1.3.2 Refractive Effects on Curvature of Ray Beam

1.3.2.1 K-Factor

The K-factor is a scaling factor (actually assumed as a constant for a particular path) that helps quantify curvature of an emitted ray path. Common radiolinks, that are described as line-of-sight, incorrectly suggest that effective communications is limited by the optical horizon (i.e., $K = 1$). In most cases radiolinks are not restricted to line-of-sight propagation. In fact we often can achieve communications beyond the optical horizon by some 15% (i.e., $K = 1.33$). Figure 1.1 shows this concept in a simplified fashion, and Figure 1.2 shows the effects of various K-factors on the bending of the radio ray beam. This bending is due to angular refraction.

Angular refraction through the atmosphere occurs because radio waves travel with differing velocities in different parts of a medium of varying dielectric constant. In free space the group velocity is maximum, but in the nonionized atmosphere, where the dielectric constant is slightly greater due to the presence of gas and water molecules, the radio wave travels more slowly. In what radiometeoroligists have defined as a standard atmosphere, the pressure, temperature, and water vapor content (humidity) all decrease with increasing altitude. The dielectric constant, being a single parameter combining the resultant effect of these three meteorological properties, also decreases with altitude. Since electromagnetic waves travel faster in a medium of lower dielectric constant, the upper part of a wavefront tends to travel with a greater velocity than the lower part, causing a downward deflection of the beam. In a horizontally homogeneous atmosphere where the vertical change of dielectric

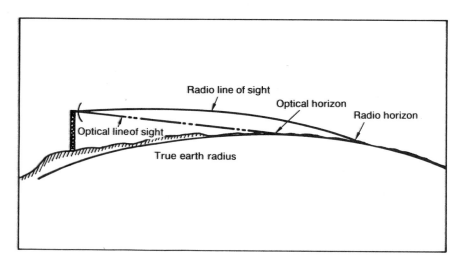

FIGURE 1.1 Optical line-of-sight versus radio line-of-sight.

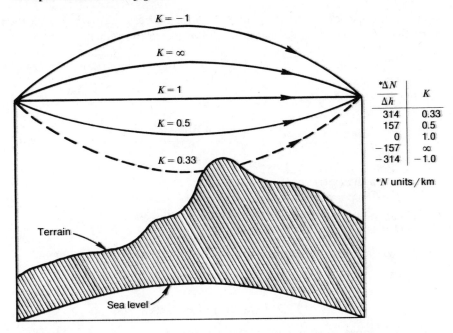

FIGURE 1.2 Ray beam bending for various K-factors (linear refractivity gradients assumed).

constant is gradual, the bending or refraction is continuous, so that the ray is slowly bent away from the thinner density air toward the thicker, thus making the beam tend to follow the earth's curvature. This bending can be directly related to the radii of spheres. The first sphere, of course, is the earth itself (i.e., radius = 6370 km) and the second sphere is that formed by the curvature of the ray beam with its center coinciding with the center of the earth. The K-factor can now be defined as the ratio of the radius, r, of the ray beam curvature to the true radius of the earth, r_0, or

$$K \approx \frac{r}{r_0} \qquad (1.8)$$

where K is often called the effective earth radius factor and r is the effective earth radius.

1.3.2.2 Refractivity

The radio refractive index is defined as the ratio of the velocity of propagation of a radio wave in free space to the velocity in a specified medium. At standard atmosphere conditions near the earth's surface, the radio refractive index (n) has a value of approximately 1.0003. However, in the design of radio systems, the use of a scaled-up unit is more desirable. This is called the refractivity (N), which is defined in the following relationships:

$$N \approx (n - 1)10^6 \qquad (1.9)$$

For the earth's atmosphere,

$$N = 77.6/T\left[P + \frac{4810e_s(\text{RH})}{T}\right] \qquad (1.10)$$

where P = atmospheric pressure in millibars
T = temperature in Kelvins
e_s = saturation water vapor pressure in millibars
RH = relative humidity expressed as a fraction.

Under standard atmospheric conditions, the refractivity is about 300 ± 20 N units.

1.3.3 Refractivity Gradients

Probably of more direct interest to the radiolink design engineer is refractivity gradients. If we assume that the refractive index, n, of air varies linearly with the height h for the first few tenths of a kilometer above the earth's surface and does not vary in the horizontal direction, then we can restate the K-factor in terms of the gradient $\Delta n/\Delta h$ by (Ref. 5):

$$\frac{r}{r_0} = K \approx \left[1 + \frac{r_0\Delta n}{\Delta h}\right]^{-1} \qquad (1.11)$$

again where $r_0 \approx 6370$ km and h is the height above earth's surface.
   As in equation (1.9), $N \approx (n - 1)\,10^6$, so that

$$\frac{\Delta n}{\Delta h} = \frac{\Delta N}{\Delta h}(10^{-6})\ N \text{ units/km} \qquad (1.12)^*$$

and

$$K \approx \left[1 + \left(\frac{\Delta N}{\Delta h}\right)\bigg/157\right]^{-1} \qquad (1.13)$$

Return to Figure 1.2 where several values of K and $\Delta N/\Delta h$ are listed to the right. The figure also illustrates subrefractive conditions, $0 < K < 1.0$, where the refractivity gradients are positive. The worst case in Figure 1.2 is where the ray beam is interrupted by the surface, where $K = 0.33$, placing the receiving terminal out of normal propagation range.

The more commonly encountered situation is where $\infty \geq K \geq 1.0$ or $-157 \leq \Delta N/\Delta h \leq 0$. In this case the ray is bent toward the earth. When $\Delta N/\Delta h = -157$, the ray has the same curvature as the earth and the ray path acts like straight line propagation over a flat earth.

As we can now see, the bending of a radio ray beam passing through the atmosphere is controlled by the gradient refractive index. For most purposes the horizontal gradient is so small that it can be neglected. The vertical change under standard atmospheric conditions is approximately -40 N units/km, which approximates the value at noon on a clear, summer day at sea level, in the temperature zone with a well mixed atmosphere. However, in the very lowest levels of the atmosphere, the vertical gradient may vary between extreme values as large as $+500$ to -1000 N units/km over height intervals of several hundred feet. The variance is a function of climate, season, time of day, and/or transient weather conditions. It is also affected by terrain, vegetation, radiational conditions, and atmospheric stratification. The more extreme stratification tends to occur in layers less than 100 m in thickness, and which at times extend over long distances. When averaged over 500–1000 m heights above ground, the radio refractivity gradient is likely to vary between 0 and -300 N units/km. However, during a large percentage of the year, the standard value of -40 N units/km is more likely.

Commonly we will encounter two refractivity parameters used in estimating radio propagation effects; these are surface refractivity N_s and surface refractivity reduced to sea level N_0. Figures 1.3–1.5 give data on mean N_0 values for the world. Figure 1.3 provides worldwide mean values of N_0 for February; Figure 1.4 gives similar information for August; and Figure 1.5 provides data on mean monthly values of N_0 in excess of 350 N units. High values of N are usually due to the wet term in the refractivity equation.

For long-term median estimates, an empirical relation has been established between the average mean refractivity gradient $\Delta N/\Delta h$ for the first kilometer above the surface and the value of the average monthly mean refractivity N_s at the surface. For the continental United States the relationship is

$$\frac{\Delta N}{\Delta h} = -7.32 \exp(0.005577\overline{N}_s) \tag{1.14}$$

where $\Delta N/\Delta h$ is in N-units/km and $\overline{N}_s$ is in N-units. For the Federal Republic of Germany

$$\frac{\Delta \overline{N}}{\Delta h} = -9.30 \exp(0.004565\overline{N}_s) \tag{1.15}$$

and for the United Kingdom

$$\frac{\Delta \overline{N}}{\Delta h} = -3.95 \exp(0.0072\overline{N}_s) \tag{1.16}$$

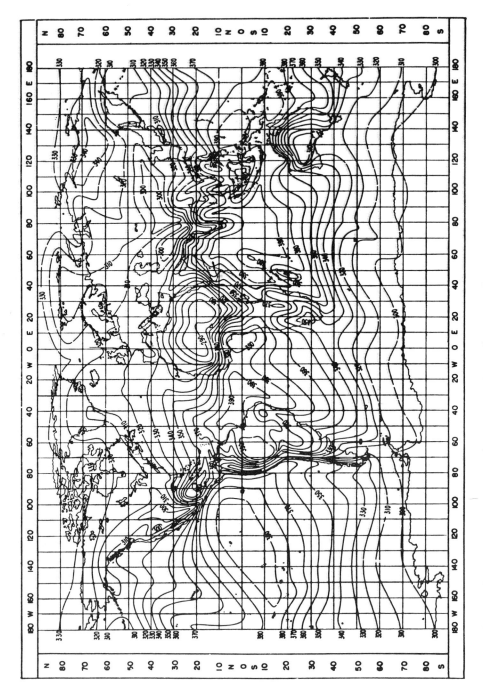

FIGURE 1.3 Worldwide mean value of N_0: February. (From CCIR Rep. 563-1, Courtesy of ITU-CCIR, Geneva.)

8

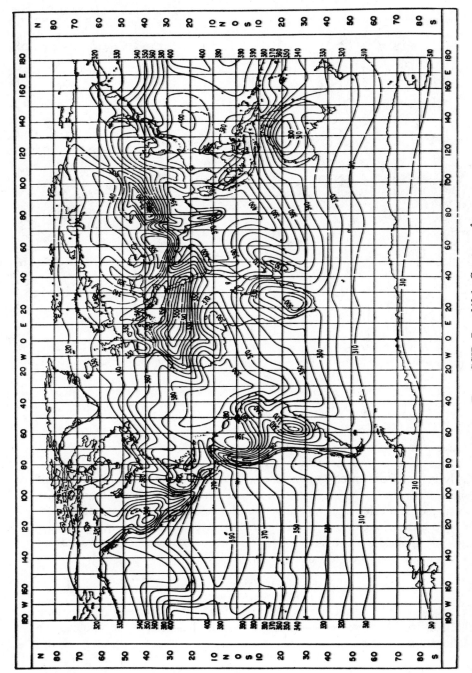

FIGURE 1.4 Worldwide mean value of N_0; August. (From CCIR Rep 563-1. Courtesy of ITU-CCIR, Geneva.)

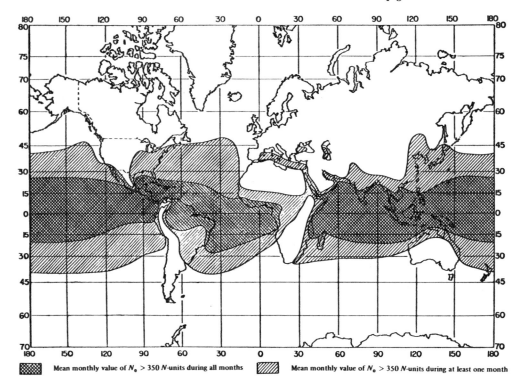

FIGURE 1.5 Mean monthly values of N_0 in excess of 350 N units.

These relationships are valid for $250 \leq \overline{N}_s \leq 400$ N units and are only applicable to average negative gradients close to the surface. Often, particularly in transhorizon communication, it is more convenient to use values of surface refractivity N_s because information for this parameter is more readily available. The value of N_s is a function of temperature, pressure, and humidity, and, therefore, decreases on the average with elevation. For a particular link the applicable values are read for $\overline{N}_0$ from a map (such as Figures 1.3 and 1.4) and are converted to values of N_s by

$$N_s = N_0 \exp(-0.1057 h_s) \qquad \text{(Ref. 5)} \qquad (1.17a)$$

where h_s is the height above mean sea level (in kilometers) of the radio horizon in the direction of the far end of the link. For h_s in kilofeet, the following expression may be used:

$$N_s = N_0 \exp(-0.03222 h_s) \qquad \text{(Ref. 7)} \qquad (1.17b)$$

Figure 1.6 is a nomogram to convert N_0 to N_s for values of h_s in thousands of feet (kilofeet).

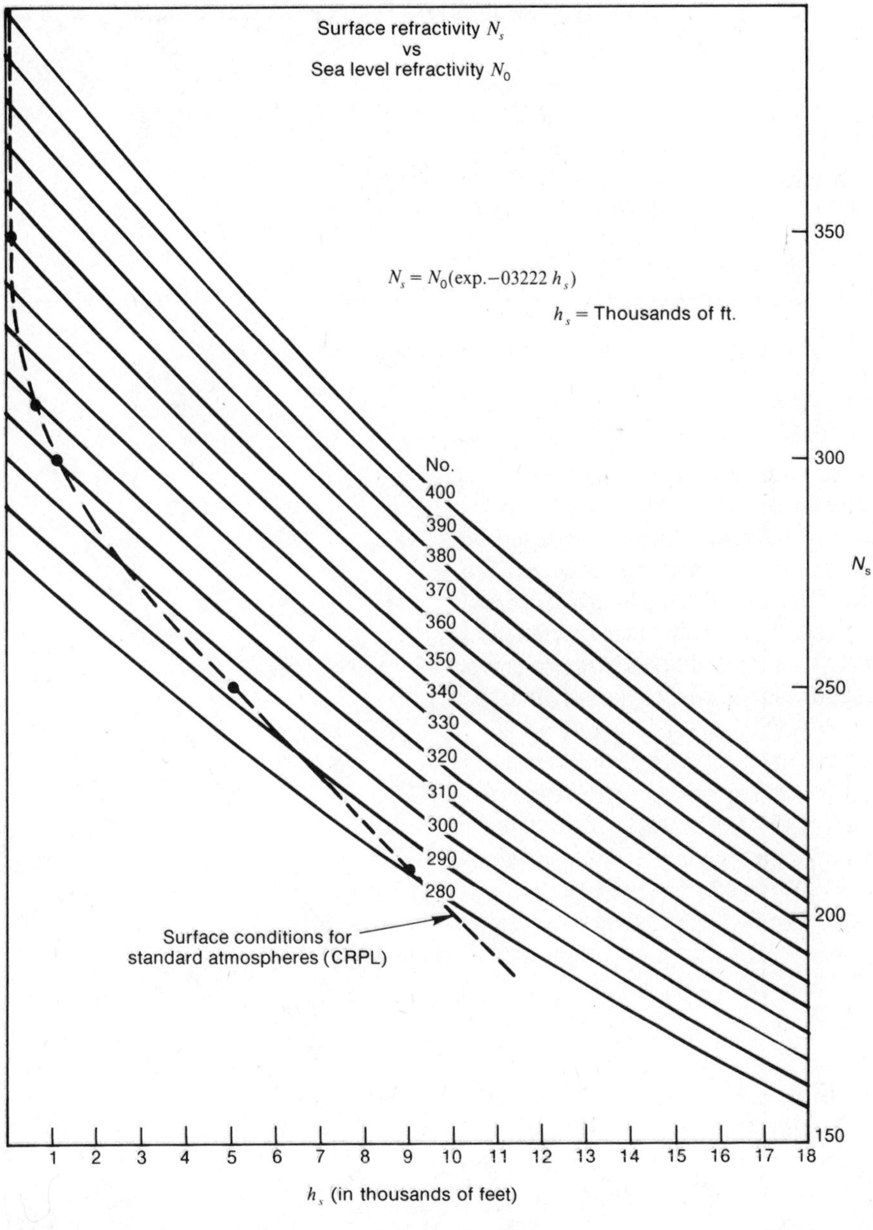

Surface refractivity N_s
vs
Sea level refractivity N_0
$N_s = N_0(\text{exp.}-03222\ h_s)$
h_s = Thousands of ft.
No.
400
390
380
370
360
350
340
330
320
310
300
290
280
Surface conditions for
standard atmospheres (CRPL)
350
300
250
200
150
N_s
1 2 3 4 5 6 7 8 9 10 11 12 13 14 15 16 17 18
h_s (in thousands of feet)

FIGURE 1.6 Surface refractivity N_s versus sea level refractivity N_0, based on equation (1.17b) (From Navelex 0101, 112, Ref. 7).

The effective earth radius (Section 1.3.2) can be calculated from N_s with the following formula

$$r = r_0\left[1 - 0.04665\exp(0.005577\overline{N}_s)\right]^{-1} \qquad (\text{Ref. 9}) \qquad (1.18)$$

where $r_0 = 6370$ km (as in Section 1.3.2).

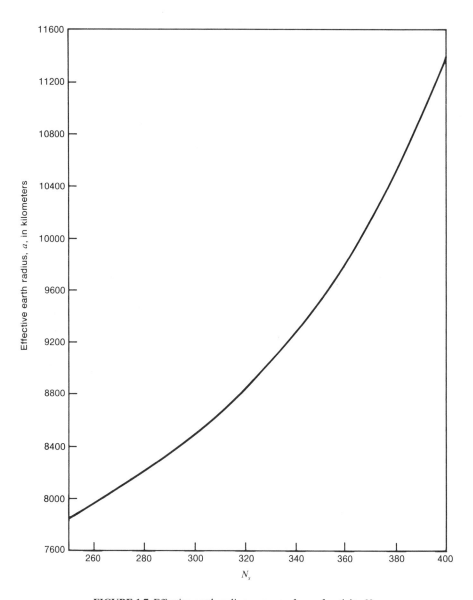

FIGURE 1.7 Effective earth radius versus surface refractivity N_s.

Figure 1.7 is a curve to derive effective earth radius from surface refractivity N_s, where the earth radius is given in kilometers.

1.4 DIFFRACTION EFFECTS—THE FRESNEL ZONE PROBLEM

Diffraction of a radio wave front occurs when the wave front encounters an obstacle that is large when compared to the wavelength of the ray. Below about 1000 MHz there is diffraction or bending from an obstacle with increasing attenuation as a function of obstacle obstruction. Above about 1000 MHz, with increasing obstruction of an obstacle, the attenuation increases even more rapidly such that the path may become unusable by normal transmission means than of the lower frequencies (see Chapter 3). The actual amount of obstruction loss is dependent on the area of the beam obstructed in relation to the total frontal area of the energy propagated and to the diffraction properties of the obstruction.

Under normal transmission conditions (i.e., nondiffraction paths, Chapters 2 and 3), the objective for the system designer is to provide sufficient clearance of the obstacle without appreciable transmission loss due to the obstacle. To calculate the necessary clearance we must turn to wave physics, Huygen's principle, and the theory developed by Fresnel. When dealing with obstacle diffraction, we will assume that the space volume is small enough that gradient effects can be neglected so that the diffraction discussion can proceed as though in an homogeneous medium.

Consider Figure 1.8. The Huygens–Fresnel wave theory states that the electromagnetic field at a point S_2 is due to the summation of the fields caused by reradiation from small incremental areas over a closed surface about a point source S_1, provided that S_1 is the only source of radiation. The field at a constant distance r_1 from S_1, which is a spherical surface, has the same phase over the entire surface since the electromagnetic wave travels at a constant phase velocity in all directions in free space. The constant phase surface is

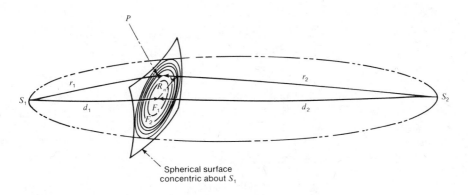

Spherical surface
concentric about S_1

FIGURE 1.8 Fresnel zone geometry.

called a wave front. If the distances r_2 from the various points on the wave front to S_2 are considered, the contributions to the field at S_2 are seen to be made up of components that will add vectorially in accordance with their relative phase differences. Where the various values of r_2 differ by half-wavelength ($\lambda/2$), the strongest cancellation occurs. Fresnel zones distinguish between the areas on a closed surface about S_1 whose components add in phase.

Let us consider a moving point P_1 in the region about the terminal antenna locations S_1 and S_2 such that the sum of the distances r_1 and r_2 from the antennas to P is constant. Such a point, then, will generate an ellipsoid with S_1 and S_2 as its foci. We now can define a set of concentric ellipsoidal shells so that the sum of the distances r_1 and r_2 differs by multiples of half-wavelength ($\lambda/2$). The intersection of these ellipsoids defines Fresnels zones on the surface as shown in Figure 1.8. Thus, on the surface of the wave front, a first Fresnel zone F_1 is defined as bounded by the intersection with the sum of the straight line segments r_1 and r_2 equal to the distance d plus one-half wavelength ($\lambda/2$). Now the second Fresnel zone F_2 is defined as the region where $r_1 + r_2$ is greater than $d + \lambda/2$ and less than $d + 2(\lambda/2)$. Thus the general case may now be defined where F_n is the region where $r_1 + r_2$ is greater than $d + (n - 1)\lambda/2$ but less than $d + n\lambda/2$. Field components from even Fresnel zones tend to cancel those from odd zones since the second, third, fourth, and fifth zones (etc.) are approximately of equal area.

Fresnel zone application to path obstacles may only be used in the far field. The minimum distance d_F where the Fresnel zone is applicable may be roughly determined by $d_F > 2D^2/\lambda$, where D is the antenna aperture measured in the same units as λ.

To calculate the radius of the nth Fresnel zone R_n on a surface perpendicular to the propagation path, the following equation provides a good approximation:

$$R_n \simeq \sqrt{n\lambda\left(\frac{d_1 d_2}{d_1 + d_2}\right)} \qquad (1.19a)$$

where R_n and d are in the same units, or

$$R_n \simeq 17.3\sqrt{\frac{n}{F_{GHz}}\left(\frac{d_1 d_2}{d_1 + d_2}\right)} \qquad (1.19b)$$

where d_1 is the distance to the near end antenna and d_2 is the distance to the far end antenna from the obstacle, and in equation (1.19b) all distances are in kilometers, the frequency of the emitted signal is in gigahertz and R_n is in meters.

If R_1 is the first Fresnel zone, then

$$R_n = R_1\sqrt{n} \qquad (\text{Ref. 4}) \qquad (1.20)$$

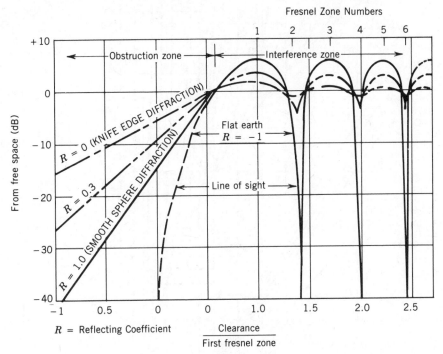

FIGURE 1.9 Path attenuation versus path clearance (From Navelex 0101, 112, Ref. 7.)

To calculate the radius of the first Fresnel zone in feet where d is measured in statute miles,

$$R_1 = 72.1 \sqrt{\frac{d_1 d_2}{F_{\text{GHz}}(d_1 + d_2)}} \qquad \text{(Ref. 4)} \qquad (1.21)$$

Conventionally we require 0.6 Fresnel zone clearance of the beam edge (3-dB point) due to obstacles in the path. Figure 1.9 shows path attenuation versus path clearance. Providing 0.6 Fresnel zone clearance usually is sufficient to ensure that attenuation due to an obstacle in or near the ray beam path is negligible. Figure 1.10 is a nomogram to calculate 0.6 Fresnel zone clearance for some of the more common line-of-sight radiolink frequencies.

Reference 4 gives some practical Fresnel zone clearance guidelines related to K-factor:

Although there are some variations, there are two basic sets of clearance criteria which are in common use in microwave communications systems. One is the

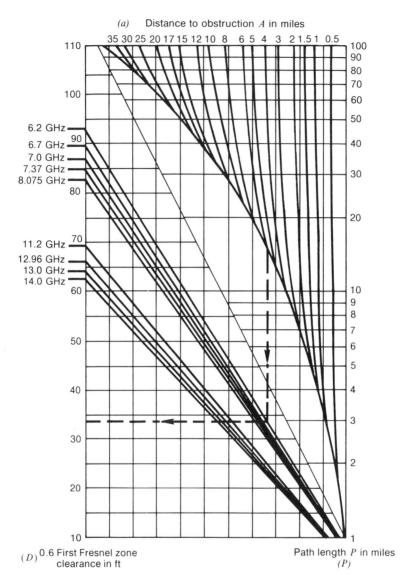

(a) Distance to obstruction A in miles

(D) 0.6 First Fresnel zone
clearance in ft

Path length P in miles
(P)

Locate path length on P scale (20). Locate distance to obstruction on the A scale curve (5). Where they intersect, drop vertical line to freq. (6.2 GHz). From the intersection of the vertical line and frequency line read clearance on D scale (33.5)

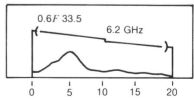

FIGURE 1.10 Nomogram for 0.6 first Fresnel zone clearance. From *Telecommunications Engineering Information* (1969). (Courtesy of Rockwell International, Collins Transmission Division, Dallas, TX.)

"heavy route" set used for those systems with the most stringent reliability requirements, the other a "light route" set used for systems where some slight relaxation of requirements can be made. The following are typical clearance criteria:

For "heavy route," highest-reliability systems: At least $0.3F_1$ at $K = \frac{2}{3}$, and $1.0F_1$ at $K = \frac{4}{3}$, whichever is greater. In areas of very difficult propagation, it may be necessary to ensure a clearance of at least grazing at $K = \frac{1}{2}$ (for 2 GHz paths above 36 miles, substitute $0.6F_1$ at $K = 1.0$).

Note that the evaluation should be carried out along the entire path and not just at the center. Earth bulge and Fresnel zone radii vary in a different way along the path, and it often happens that one criterion is controlling for obstacles near the center of the path and the other is controlling for obstacles near one end of the path.

For "light-route" systems with slightly less stringent reliability requirements at least $0.6F_1 + 10$ feet at $K = 1.0$. At points quite near the ends of the paths, the Fresnel zones and earth bulge become vanishingly small, but it is still necessary to maintain some minimum of perhaps 15 to 20 feet above obstacles to avoid near field obstructions.

1.5 GROUND REFLECTION

When a radio wave is incident upon the earth's surface, it is not actually reflected from a point on the surface, but from a sizable area. The area of reflection may be large enough to encompass several Fresnel zones or it may have a small cross-sectional area such as a ridge or peak encompassing only part of a Fresnel zone.

The significance of ground-reflected Fresnel zones is similar to free-space Fresnel zones. However, radio waves reflected from the earth's surface are generally changed in phase depending on the polarization of the signal and the angle of incidence. Horizontally polarized waves in our band of interest are reflected from the earth's surface and are shifted in phase very nearly 180°, effectively changing the electrical path length by approximately one-half wavelength ($\lambda/2$). For vertically polarized waves, on the other hand, the phase shift varies between 0 and 180° depending on the angle of incidence and the reflection coefficient, which depends largely on ground conditions. For the horizontally polarized case, if the reflecting surface is large enough to encompass the total area of any odd-numbered Fresnel zones, the resulting reflections will arrive at the receiving antenna out of phase with the direct wave causing fading. In some cases similar phenomenon has been observed for vertically polarized signals.

To mitigate ground reflections on line-of-sight paths, tower heights can be adjusted (i.e., low–high technique) to effectively move the reflection point to a portion of the intervening path that is on rough terrain where the reflected signal will be broken up. Methods of adjusting the reflection point are discussed in Chapter 2.

1.6 FADING *

1.6.1 Introduction

Fading is defined as any time varying of phase, polarization, and/or level of a received signal. The most basic definitions of fading are in terms of the propagation mechanisms involved: refraction, reflection, diffraction, scattering, attenuation, and guiding (ducting) of radio waves. These are basic because they determine the statistical behavior with time of measurable field parameters including amplitude (level), phase, and polarization, as well as frequency and spatial selectivity of the fading. Once these mechanisms are understood, remedies can be developed to avoid or mitigate the effects.

Fading is caused by certain terrain geometry and meterological conditions that are not necessarily mutually exclusive. Basic background information has been established on these conditions in previous sections of this chapter. All radio transmission systems in the 1–100-GHz frequency range can suffer fading including satellite earth terminals operating at low elevation angles and/or in heavy precipitation.

1.6.2 Multipath Fading

Multipath fading is the most common type of fading encountered, particularly on line-of-sight radiolinks. It is the principal cause of dispersion, which is particularly troublesome on digital troposcatter and high-bit-rate line-of-sight links.

For an explanation of atmospheric multipath fading, we must turn to the refractive index gradient (Section 1.3.3). As the gradient varies, multipath fading results owing to the interference between direct rays (Figure 1.11) and the specular component of a ground-reflected wave; the nonspecular component of the ground-reflected wave; partial reflections from atmospheric sheets or elevated layers; or additional direct wave paths (i.e., nonreflected paths).

Of interest to the radiolink design engineer is the fading rate, meaning the number of fades per unit of time and the fading depth, meaning how much the signal intensity at the receiver varies from its free space value, generally expressed in decibels.

The four multipath fading mechanisms previously listed can operate individually or concurrently. Fade depths can exceed 20 dB, particularly on longer line-of-sight paths and more than 30 dB on longer troposcatter paths. Fade durations of up to several minutes or more can be expected.

---

*Section 1.6 has been adapted from Ref. 3.

MULTIPATH FADING MECHANISMS

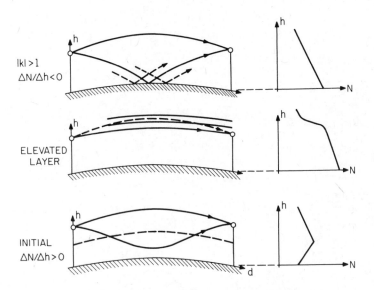

FIGURE 1.11 Mechanisms of multipath fading.

Often multipath fading is frequency selective and the best technique for mitigation is frequency diversity. For effective operation of frequency diversity, sufficient frequency separation is required between the two transmit frequencies to provide sufficient decorrelation. On most systems a 5% frequency separation is desirable. However, on many installations such a wide separation may not be feasible owing to frequency congestion and local regulations. In such cases it has been found that a 2% separation is acceptable. Frequency diversity design is described further in Chapters 2, 3, and 7.

1.6.3 Power Fading

Power fading results from a shift of the beam from the receiving antenna due to one or several of the following:

☐ intrusion of the earth's surface or atmospheric layers into the propagation path

☐ antenna decoupling due to variation of the refractive index gradient (variation of K-factor)

☐ partial reflection from elevated layers that have been interpositioned in the ray beam path

ATTENUATION   FADING   MECHANISMS

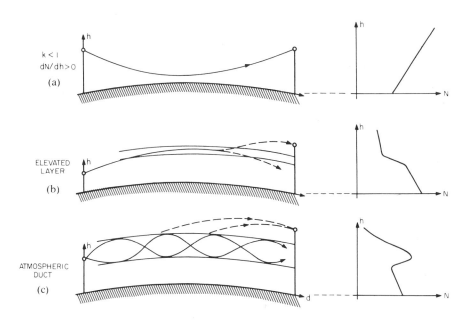

FIGURE 1.12 The mechanisms of power or attenuation fading. (From MIL-HDBK-416, Ref. 5.)

□ where one of the terminal antennas is in a ducting formation
□ precipitation in the propagation path (discussed in Chapter 6).

Examples of the mechanisms of power fading are given in Figure 1.12.

1.6.3.1 Fading due to Earth Bulge

When there is a positive gradient (subrefractive) of the refractive index, power fading may be expected owing to diffraction by the earth's surface, as shown in Figure 1.12a. Fade depths of 20–30 dB can be expected with fade durations lasting for several hours or more. This type of fading may not be normally mitigated by frequency diversity, but may be reduced or completely avoided by the proper adjustment of antenna tower heights.

The guidelines for Fresnel zone clearance of Section 1.4 must be modified on those microwave line-of-sight paths where subrefractive index gradients occur (i.e., where K is less than 1). Clearances greater than one Fresnel zone are recommended particularly where the intervening path approaches smooth earth. In mountainous regions where antennas are mounted on dominating ridges or peaks, a single Fresnel zone clearance would be sufficient. Similar guidelines may be used where a limited range of refractive index gradients is encountered.

1.6.3.2 Duct and Layer Fading

Fades of about 20 dB or more can occur due to atmospheric ducts and elevated layers. These fades can persist for hours or days and are more prevalent during darkness. Neither space nor frequency diversity mitigate this type of fading.

An elevated duct is often characterized as a combination of superrefractive layer above a subrefractive layer. Such a condition has the effect of guiding or focusing the signal along the duct. The reverse condition, namely, a subrefractive layer above a superrefractive layer, will tend to defocus signal energy introduced with the layer combination. The defocusing effect produces power fading.

One obvious cure for this type of fading is repositioning one or both antennas. Another would be to select other sites.

1.6.4 K-Factor Fading

This type of fading involves either multipath fading from direct ray and ground reflections or diffraction power fading that depends on the K-factor value. These two types of fading can supplement one another and cause fading throughout a wide range of refractive index gradients (values of K-factor). K-factor fading can be expected when the intervening terrain is comparatively smooth such as over-water paths, maritime terrain, or gently rolling terrain.

Figure 1.13, from Ref. 3, illustrates the resulting signal variations, the spherical earth transmission loss versus the refractive index gradient. The figure also shows the effect of terrain roughness, expressed by σ/λ and the divergence–convergence factor under the dynamic influence of the refractive index gradient. Reference 3 describes the ratio σ/λ, where σ is the standard deviation of the surface irregularities about a median spherical surface and λ is the transmission wavelength. Smooth earth is then defined as $\sigma/\lambda = 0$. For the smooth earth case, the fading, marked by nulls of interference between the direct wave and the specularly reflected wave, is serious only over a limited range of refractive index gradients. For the path parameters of Fig. 1.13 as an example, the fades due to interference nulls can exceed 20 dB only within the range of -115 to -195 N units/km and for gradients in excess of 300 N units/km, as the surface roughness is increased from the smooth earth case or where $\sigma/\lambda = 0$, the critical region of gradients shifts to more negative values. In Figure 1.13, the critical region for negative gradients shifts to the range of

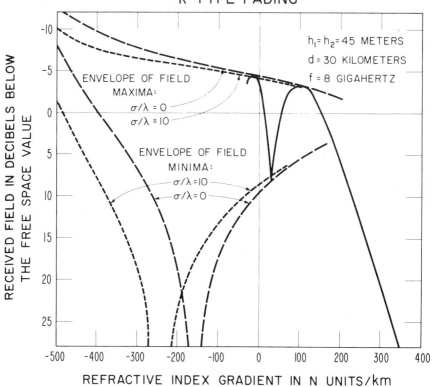

ILLUSTRATION OF THE VARIATION OF FIELD
STRENGTH WITH REFRACTIVE INDEX GRADIENT,
k-TYPE FADING

$h_1 = h_2 = 45$ METERS
$d = 30$ KILOMETERS
$f = 8$ GIGAHERTZ

ENVELOPE OF FIELD
MAXIMA:
$\sigma/\lambda = 0$
$\sigma/\lambda = 10$

ENVELOPE OF FIELD
MINIMA:
$\sigma/\lambda = 10$
$\sigma/\lambda = 0$

RECEIVED FIELD IN DECIBELS BELOW THE FREE SPACE VALUE

REFRACTIVE INDEX GRADIENT IN N UNITS/km

FIGURE 1.13 *K*-type fading, illustration of variation of field strength with refractive index gradient. (From Ref. 3.)

-180 to -290 N units/km for $\sigma/\lambda = 10$. Irregularities or roughness that would cause the median terrain surface to depart slightly from a sphere could also shift the range of critical negative gradients in either direction. These critical ranges, as well as those due to the diffraction fade (at values greater than 300 N units/km), depend on specific link parameters such as transmission frequency, antenna heights, and path lengths. Figure 1.13 clearly shows that unless the terrain roughness is sufficient to shift the critical range of negative gradients outside the range of refractive index gradients expected to occur at a particular location, reflections from the terrain surface cannot be neglected. Similarly, high points of the terrain cannot be considered to eliminate terrain reflection unless they also partially obstruct the reflected wave over the critical range of refractive index gradients.

The effects of *K*-type fading can be reduced by:

☐ increasing the terminal antenna heights to provide adequate protection against the diffraction fading for the expected extreme positive gradients of refractive index and

☐ diversity reception that effectively reduces the attenuation due to multipath out of the expected extreme negative gradients of refractive index.

1.6.5 Surface Duct Fading on Over-Water Paths

Long line-of-sight paths over water can encounter a special type of fading due to the presence of surface ducts. Such surface ducts can be a semipermanent condition, particularly in high-pressure regions such as the Bermuda High, which is in the Atlantic Ocean between 10 and 30°N latitude. In this case, the ducts are formed less than 2 km from the shoreline and extend along the sea surface up to heights from 7 to 20 m for wind velocities from 15 to 55 km/hr. They persist during fair weather and generally reform after rain showers and squalls. The resulting fading is due to a combination of multipath fading caused by sea reflections and power fading in the presence of the surface duct. Figure 1.14, from Ref. 3, illustrates two typical situations.

Because of the continual disturbance of the sea surface, a reflected wave consists of a diffuse or randomly distributed component superimposed upon a specular component. This time distribution of the reflected wave is a Beckmann distribution (a constant plus a Hoyt distribution). This constitutes the received field of Figure 1.14*a*. In Figure 1.14*a* an addition of the direct wave produces an enhanced or reduced constant component due to phase inter-

SURFACE DUCT FADING MECHANISM

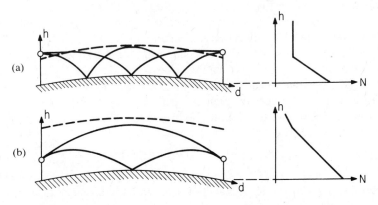

FIGURE 1.14 The fading mechanisms of surface ducts. (From Ref. 3.)

EXAMPLES OF SURFACE DUCT PROPAGATION
FOR EFFECTIVE EARTH RADIUS AND TRUE EARTH RADIUS

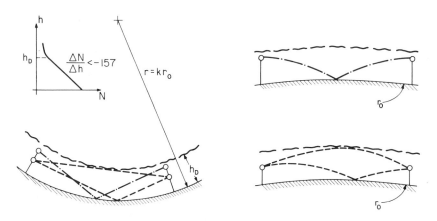

FIGURE 1.15 Examples of surface duct propagation for effective earth radius and true earth radius. (From Ref. 3.)

ference. One effect of a surface duct upon the multipath situation is to provide an increase in effective angle of incidence. This increases the ratio of the diffused to specular amplitudes, and increases the rapidly varying component of the reflected signal. The net result is a total signal whose distribution approaches the Nakagami–Rice distribution (a constant plus a Rayleigh-distributed variable) (Ref. 3).

These surface or ground-based ducts guide or trap the radio waves by the combination of a strong negative refractive index gradient (namely, superrefraction, $\Delta N/\Delta h \leq -157 \, N$ units/km) and a reflecting sea or ground. As such, propagation within the duct can be described in terms of an equivalent linear gradient of refractive index. The corresponding equivalent earth representation is illustrated in Figure 1.15. The field results from phase interference between a direct wave, one to three singly reflected waves, and, for sufficiently strong superrefraction, doubly reflected waves.

Surface duct fading can be reduced by choosing terminal antenna heights to provide adequate Fresnel zone clearance above the ducting layer. This will tend to avoid the situation in Figure 1.14*a*. Likewise, lower antenna heights could achieve the situation in Figure 1.14*b*. In this latter case, diversity reception would also tend to mitigate the problem.

PROBLEMS AND EXERCISES

1. When dealing with radio propagation, why is there a demarcation line at about 10 GHz?

2. Calculate the free space loss for a line-of-sight radiolink 29 miles long (statute miles) operating at 6.135 GHz. Double the frequency and calculate the free space loss. Halve the distance and calculate the free space loss. What is the lesson learned from the last two steps?

3. Calculate the free space loss to a satellite 3200 nautical miles distance and the operating frequency is 14 GHz. What additional loss is there if the satellite is moved to a range of 3450 nautical miles?

4. What causes a radio beam to curve when passing through the atmosphere?

5. What mathematical tool is used to describe the curvature (question 4)?

6. How does the atmospheric dielectric constant vary with altitude?

7. Differentiate between radio line-of-sight and optical line-of-sight.

8. Relate K-factor to true and effective earth radii.

9. What determines K-factor?

10. Convert the N_0 value of 301 N units to the equivalent N_s value at 1.2 km altitude.

11. A line-of-sight radiolink is 12 miles long and operates at 6 GHz. Calculate the 0.6 Fresnel zone clearance for an obstacle 3.6 miles from one end.

12. Describe at least three causes of multipath fading.

13. Describe at least three causes of power fading.

14. What are some of the mitigation techniques that can be used to counter multipath fading (name at least two)?

REFERENCES AND BIBLIOGRAPHY

1. D. C. Livingston, *The Physics of Microwave Propagation*, GT & E Technical Monograph, General Telephone & Electronics Laboratories, Inc., Bayside, NY, May 1967.

2. Military Handbook, "Facility Design for Tropospheric Scatter," MIL-HDBK-417, U.S. Department of Defense, Washington, DC, Nov. 1977.

3. H. T. Dougherty, "A Survey of Microwave Fading Mechanisms, Remedies and Applications," U.S. Department of Commerce, ESSA Technical Report ERL69-WPL 4, Boulder, CO, Mar. 1968.

4. *Engineering Considerations for Microwave Communications Systems*, GTE-Lenkurt, Inc., San Carlos, CA, 1975.

5. Military Handbook, *Design Handbook for Line of Sight Microwave Communication Systems*, MIL-HDBK-416, U.S. Department of Defense, Washington, DC, Nov. 1977.

6. B. R. Bean and E. J. Dutton, *Radio Meteorology*, Dover Publications, New York, 1968.

7. Naval Shore Electronics Criteria, "Line-of-Sight Microwave and Tropospheric Scatter Communication Systems," Navelex 0101, 112, U.S. Department of the Navy, Washington, DC, May 1972.

8. "Radio Wave Propagation: A Handbook of Practical Techniques for Computing Basic Transmission Loss and Field Strength," ECAC-HDBK-82-049 ADA 122-090, U.S. Department of Defense, Electromagnetic Compatibility Analysis Center, Annapolis, MD, Sept. 1982.

9. USAF Technical Order 31Z-10-13, "General Engineering Beyond-Horizon Radio communications," U.S. Air Force, Washington, DC, 1 Oct. 1971.

T W O

LINE-OF-SIGHT RADIOLINKS

2.1 OBJECTIVE AND SCOPE

Line-of-sight (LOS) radiolinks, in the context of this book, provide broadband connectivity for telecommunications using radio equipment with carrier frequencies above 1 GHz. For most applications these radiolinks will be considered a subsystem of a telecommunications network. They will carry one or a mix of the following:

- ☐ Telephone channels
- ☐ Data information
- ☐ Telegraph/telex
- ☐ Facsimile
- ☐ Video
- ☐ Program channels
- ☐ Telemetry (which will be considered a subset of data)

The emitted waveform may be analog (conventionally FM) or digital.

LOS implies a terrestrial connectivity. (A satellite link, by definition, is also LOS from its associated earth terminals.) On such a radiolink the distinguishing feature is line of sight and, in particular, "radio line of sight" (see Section 1.3.2.1). This requires sufficient clearance of intervening terrain on an LOS link such that the emitted beam by the transmitting antenna fully envelops its companion, far-end, receiving antenna. Typically, LOS links, also called hops, are 10–100 km long. This section describes procedures to design LOS links individually, a series of links in tandem and also as a subsystem that is part of an overall telecommunication network. The design of an LOS radiolink is a four-step process:

1. Initial planning and site selection
2. The drawing of a path profile
3. Path analysis
4. Site survey

with iteration among the steps. Commonly resiting is necessary when a path is shown not to be feasible because of terrain, performance, and/or economic factors.

2.2 INITIAL PLANNING AND SITE SELECTION

An LOS microwave route consists of one, several, or many hops. It may carry analog or digital traffic. The design engineer will want to know if the LOS subsystem to be installed is an isolated system on its own such as

☐ A private microwave (radiolink) system
☐ A studio-to-transmitter link
☐ An extension of a CATV headend

or as part of a larger telecommunication network where the link may be part of a backbone route or a "tail" from the backbone.

2.2.1 Requirements and Requirements Analyses

Let us suppose that we are to design a microwave LOS subsystem to provide telecommunication connectivity. The design criteria will be based on the current transmission plan available from the local telecommunication administration. For military systems, the appropriate version of MIL-STD-188 series would be imposed as a system standard. For systems carrying video and related program channel information, if no other standard is available, consult EIA RS-250 latest version and applicable CCIR recommendations.

A transmission plan will state, as a minimum, for analog systems:

☐ Noise accumulation in the voice channel for FDM telephony (see example from CCIR, Table 2.1)
☐ S/N for video and program channel information (CCIR Rec. 567 on 2500-km hypothetical reference circuit, on luminance signal to rms weighted noise ratio: 57 dB for more than 20% of a month and 45 dB for more than 0.1% of a month)

**TABLE 2.1. Noise Accumulation in an FDM Voice Channel Due to Radio
Portion on 2500-km Hypothetical reference Circuit—CCIR Rec. 395-2**

1. That, in circuits established over real links that do not differ appreciably from the
hypothetical reference circuit, the psophometrically weighted noise power at a point of
zero relative level in the telephone channels of frequency-division multiplex radio-relay
systems of length L, where L is between 280 and 2500 km, should not exceed:
    1.1 $3L$ pW 1-min mean power for more than 20% of any month;
    1.2   47,500 pW 1-min mean power for more than ($L/2500$) × 0.1% of any month;
          it is recognized that the performance achieved for very short periods of time is
          very difficult to measure precisely and that in a circuit carried over a real link,
          it may, after installation, differ from the planning objective;
2. that circuits to be established over real links, the composition of which, for planning
reasons, differs substantially from the hypothetical reference circuit, should be planned
in such a way that the psophometrically weighted noise power at a point of zero
relative level in a telephone channel of length L, where L is between 50 and 2500 km,
carried in one or more baseband sections of frequency-division multiplex radio links,
should not exceed:
    2.1   for $50 \leq L \leq 840$ km:
          2.1.1 $3L$ pW + 200 pW 1-min mean power for more than 20% of any month,
          2.1.2   47,500 pW 1-min mean power for more than (280/2500) × 0.1% of any
                month when L is less than 280 km, or more than ($L/2500$) × 0.1% of
                any month when L is greater than 280 km;
    2.2   for $840 < L \leq 1670$ km:
          2.2.1 $3L$ pW + 400 pW 1-min mean power for more than 20% of any month,
          2.2.2   47,500 pW 1-min mean power for more than ($L/2500$) × 0.1% of any
                month;
    2.3   for $1670 < L \leq 2500$ km:
          2.3.1 $3L$ pW + 600 pW 1-min mean power for more than 20% of any month,
          2.3.2   47,500 pW 1-min mean power for more than ($L/2500$) × 0.1% of any
                month;
3. That the following Note should be regarded as part of the Recommendation:
Note 1. Noise in the frequency-division multiplex equipment is excluded. On a 2500 km
hypothetical reference circuit the CCITT allows 2500 pW mean value for this noise in
any hour.

---

[a] The level of uniform-spectrum noise power in a 3.1-kHz band must be reduced by 2.5 dB to
obtain the psophometrically weighted noise power.
Source. CCIR Rec. 395-2.

For digital systems:

    □ Bit error rate (BER), end-to-end for the network from which we can
       derive a BER per link (hop) [see example from CCITT Rec. G.821 Table
       2.2 for Integrated Services Digital Network (ISDN)]

There will also be a requirement for spectrum conservation (i.e., effective
bits/Hz of bandwidth). This information will be an input on which we will
base path analysis.

TABLE 2.2. Error Performance Objectives for International ISDN Connections

T_0 = 1 min		T_0 = 1 sec	
BER in 1min	Percentage of available minutes	BER in 1 sec	Percentage of available seconds
Worse than 1×10^{-6}	Less than 10%	> 0	Less than 8%
Better than 1×10^{-6}	More than 90%	0	More than 92% (% EFS)[a]

Note 1. It is intended that international ISDN connections should meet the requirements in the table for both values of T_0.

Note 2. The limits proposed are based on the best knowledge currently available but are subject to review in the future in the light of further studies. For the time being the limits should be considered as being provisional.
A BER threshold of 1×10^{-5} has been proposed as an alternative to 1×10^{-6} and it is possible that the ultimate value may lie in the range 1×10^{-6}–1×10^{-5}. In considering this threshold, the percentage of available minutes should be kept under review.

Note 3. Total time T_L has not been determined since the period may depend upon the application. A period of the order of any one month is suggested as a reference.

Note 4. The unavailability threshold of 1×10^{-3} may need review, particularly in its effect on some services such as facsimile, where a value of 1×10^{-4} may prove to be advisable.

Source. CCITT Rec. G.821.

[a]EFS = error free seconds

Traffic over the route must be quantified in number of voice channels, video channels, program channels, or gross bit rates for digital systems. Similarly, where applicable, location, routing, and quantification of traffic will be required for drops and inserts along the proposed route. For the case of video and program channels, information bandwidth will be required and, for video, limits of differential phase and gain as well as the manner of handling aural and cue channels.

Once the type of traffic has been stated, an orderwire and telemetry doctrine can be established.

Commonly, the life of a transmission system is 15 years, although many systems remain in operation, often with upgrades, for longer periods. System planning should include future growth out to 15 years with 5-year incremental milestones. Thoughtful provision for future growth during initial installation may well involve a greater first cost but can end up with major savings through the life of the system. These growth considerations will impact

☐ Building size, space requirements, floor loading, prime power, air conditioning
☐ Frequency planning
☐ Installation (wired but not equipped).

Another important factor that may well drive design is compatibility with existing equipment.

2.2.2 Route Layout and Site Selection

Accurate topographic maps are used for route layout and site selection. It is advisable to carry out initial route layout on small-scale topographic (topo) maps such as 1 : 250,000. Drop and insert points should be identified. Potential relay sites are selected that are in apparent LOS of each other. It is incumbent on the design engineer to minimize the number of repeaters along the route. Cost, of course, is a major driver of this requirement. We should not lose sight of some other, equally important reasons. For analog systems, each additional relay inserts noise into the system; for digital systems, each relay adds jitter to the signal and deteriorates error performance.

Drop and insert points, such as telephone exchanges (central offices), are first-choice relay locations, although terrain may not permit this luxury. In such cases, wire, fiberoptic, or radio spurs may be required. Other good candidate locations are existing towers and tall buildings where space may be leased. Economy often limits tower heights to no more than 300 ft. On a smooth earth profile, this limits distance between relay sites to 45 miles assuming $\frac{4}{3}$ earth ($K = \frac{4}{3}$). Figure 2.1 is a height–distance nomogram based on $\frac{4}{3}$ earth. The designer, of course, will take advantage of natural terrain features for relay sites, using prominent elevated terrain.

A final route layout is carried out using large-scale topographic maps with scales of 1 : 25,000 to 1 : 63,000 and contour intervals around 10 ft (3 m). For the United States these maps may be obtained from (Ref. 2):

Director
Defense Mapping Agency
    Topographic Center
Washington, DC 20315

Sales Office
U.S. Geological Survey
Washington, DC 20305

Sales Office
U.S. Geological Survey
Denver, Colorado 80225

Map Information
U.S. Coast and Geodetic Survey
Washington, DC 20350

U.S. Navy Hydrographic Office
Washington, DC 20390

In Canada topographic maps may be obtained from (Ref. 2):

Department of Energy,
    Mines and Resources,
Surveys and Mapping Branch
615 Booth St.
Ottawa, Ontario, Canada

Other map sources include foreign government services such as the British

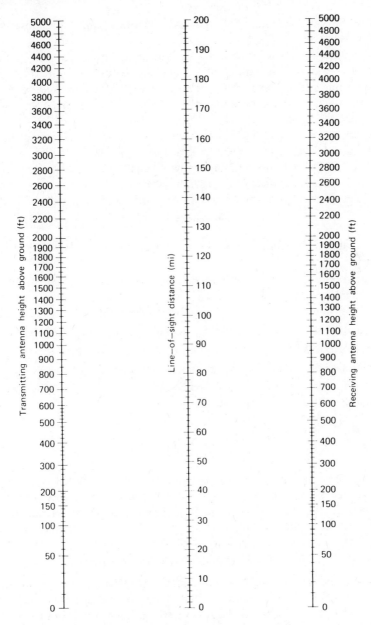

FIGURE 2.1 Nomogram to calculate radio LOS distance. Smooth earth and grazing assumed, $K = \frac{4}{3}$, distance in statute miles, tower heights in feet.

Ordnance Survey, and national agricultural, forest, and soil conservation departments.

In most situations large-scale maps, such as those with scale 1 : 25,000, must be joined together with careful alignment such that an entire hop (link) appears on one sheet. This is done on an open floor space by folding back or clipping map borders and joining them using blank paper backing and double-sided sticky tape.

Once joined, a straight line is drawn on the topo map connecting the two adjacent sites making up a single hop. This line forms the basis (data base) for the path profile to determine required tower heights. The initial site selection was based on LOS clearance of the radio ray beam and drop and insert points. The path profile will confirm the terrain clearance. However, before final site selection is confirmed, the following factors must also be considered:

- □ Availability of land. Is the land available where we have selected the sites?
- □ Site access. Can we build a road to the site that is cost-effective? This may be a major cost driver and may force us to turn to the use of another site where access is easier, but where a taller tower is needed.
- □ Construction restrictions, zoning regulations, nearby airport restrictions.
- □ Level ground for tower and shelter.
- □ Climatic conditions (incidence of snow and ice).
- □ Possibilities of anomalous propagation conditions such as in coastal regions, reflective desert, over-water paths.

2.3 PATH PROFILES

A path profile is a graphical representation of a path between two adjacent radiolink sites in two dimensions. From the profile, tower heights are derived, and, subsequently, these heights can be adjusted (on paper) so that the ray beam reflection point will avoid reflective surfaces. The profile essentially ensures that the proper clearances of path obstructions are achieved.

There are three recognized methods to draw a path profile:

1. *Fully Linear Method.* Common linear graph paper is used where a straight line is drawn from the transmitter site to receiver site giving tangential clearance of equivalent obstacle heights. A straight line is also drawn from the receiver site to the transmitter site. The bending of the radio beam (see Section 1.3.4) is represented by adjustment of each obstacle height by equivalent earth bulge using the equation

$$h = \frac{d_1 d_2}{1.5K} \qquad (2.1)$$

where h is the change in vertical distance in feet from a horizontal reference line, d_1 is the distance in statute miles from one end of the path to obstacle

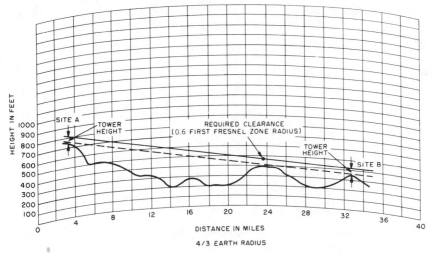

FIGURE 2.2 Illustration of path profile method 2 using $\frac{4}{3}$ earth graph paper.

height in question, and d_2 is the distance from other end of path to the same obstacle. K is the selected K-factor (section 1.3.2.1).

2. $\frac{4}{3}$ *Earth Method.* $\frac{4}{3}$ earth graph paper is required. In this case true values of obstacle height may be used. An example of a profile using $\frac{4}{3}$ graph paper is shown in Figure 2.2. Of course, with this method, the value of K is fixed at $\frac{4}{3}$.

3. *Curvature Method.* Linear graph paper is used. True values of obstacle heights are employed from a reference line or mean sea level (MSL) and a curved line is drawn from transmitter site (arbitrarily one end) to the receiver site and vice versa. The curved line has a curvature KR, where K is the applicable K factor and R is the geometric radius of the earth or 3960 statute miles (6370 km) assuming the earth is a perfect sphere.

Method 1 is recommended because it (1) permits investigation and illustration of the conditions of several values of K to be made on one chart, (2) eliminates the need for special earth curvature graph paper, (3) does not require plotting curved lines, only a straight edge is needed thus facilitating the task of profile plotting. For this method it is convenient to plot on regular 10×10 divisions to the inch (or millimeter paper) graph paper, and B size (11×17-in.) paper is recommended.

The paper requires scaling. For the horizontal scale, usually 2 miles to the inch is satisfactory, permitting a 30-mile path to be plotted on one sheet, B size.

The scale to be used vertically will depend on the type of terrain for the path in question. Where changes in path elevation do not exceed 600–800 ft, a basic elevation scale of 100 ft to the inch will suffice. For a path profile involving hilly country, 200 ft to the inch may be required. Over mountainous country scales of 500 or 1000 ft to the inch should be used. It should be noted

TABLE 2.3. Path Profile Data Base—Path Able Peak to Bakersville[a]

Obstacle	d_1 (mi)	d_2 (mi)	0.6 Fresnel (ft)	EC (ft)[b]	Vegetation	Total Height Extend. (ft)
A	7.5	28.5	43	152	50	245
B	19.4	16.6	53	233	50	336
C	27.0	9.0	46	176	50	272
D	30.0	6.0	39	130	50	219

[a] Frequency band: 6 GHz; K factor 0.92; $D = d_1 + d_2 = 36$ mi; vegetation: tree conditions 40 ft plus 10 ft growth.

[b] EC = earth curvature or earth bulge.

that if the distance scale is doubled, the height scale should be quadrupled to preserve the proper relationship.

Return now to the large scale topo map. The design engineer follows the route line* drawn between the two adjacent sites, which may be called arbitrarily transmitter and receiver sites. Obstacles on the line that will or potentially would interfere with the ray beam are now identified. It is good practice to label each on the map with a simple code name such as letters of the alphabet. For each obstacle annotate on a table the distances d_1 and d_2 and the height determined from the contour line. d_1 is the distance from the transmitter site to the obstacle and d_2 is the distance from the obstacle to the receiver site. Optionally, the site location in latitude and longitude or grid coordinates may also be listed. A typical table is shown in Table 2.3, which applies to an example profile shown in Figure 2.3.

Earth curvature for a particular obstacle point is determined by

$$h = \frac{d_1 d_2}{1.5K} \qquad (2.2a)$$

where h is in feet and d is in statute miles, or

$$h = \frac{d_1 d_2}{12.75K} \qquad (2.2b)$$

where h is in meters and d is in kilometers.

The Fresnel zone clearance may be determined by the equation

$$F_1 = 72.1 \sqrt{\frac{d_1 d_2}{DF_{GHz}}} \qquad (2.3)$$

where F_1 is the radius of the first Fresnel zone; d_1, d_2, and D are in statute

---

*It will be noted that the line drawn is a straight line, which navigators call a rhumb line. Purists in the radio propagation field will correct the author that in fact the beam follows a great circle path. However, for short paths up to 100 km, the rhumb line presents a sufficient approximation. For longer paths, such as in Chapter 3, great circle distances and bearings are used.

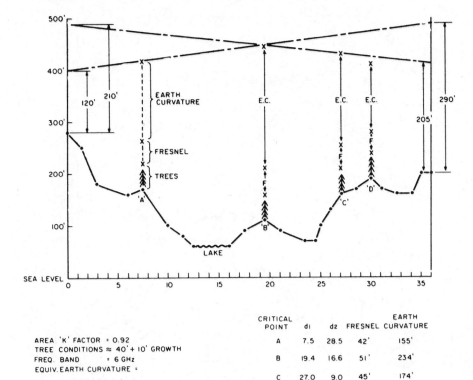

FIGURE 2.3 Example path profile using method 1. (See Table 2.3.)

AREA 'K' FACTOR = 0.92
TREE CONDITIONS ≈ 40' + 10' GROWTH
FREQ. BAND = 6 GHz
EQUIV. EARTH CURVATURE =

$$\frac{0.66\ (d_1 \times d_2)}{'K'}$$

CRITICAL POINT	d₁	d₂	FRESNEL	EARTH CURVATURE
A	7.5	28.5	42'	155'
B	19.4	16.6	51'	234'
C	27.0	9.0	45'	174'
D	30.0	6.0	39'	131'

miles; and F_1 is in feet. For the metric system,

$$F_1 = 17.3 \sqrt{\frac{d_1 d_2}{D F_{\text{GHz}}}} \qquad (2.4)$$

where d_1, d_2, and D are in kilometers and F_1 is in meters.

In Table 2.3, the conventional value of $0.6F_1$ has been used. Section 1.4 shows rationale for other values such as $0.3F_1$.

2.3.1 Determination of Median Value for K Factor

When refractive index gradients $\Delta N/\Delta h$ (defined below) are known, the K factor or effective earth radius factor can be closely approximated from the relationship

$$K \approx \left[1 + \frac{1}{157} \frac{\Delta N}{\Delta h} \right]^{-1} \qquad (2.5)$$

where ΔN may be found in the *World Atlas of Atmospheric Radio Refractivity* (Ref. 4).

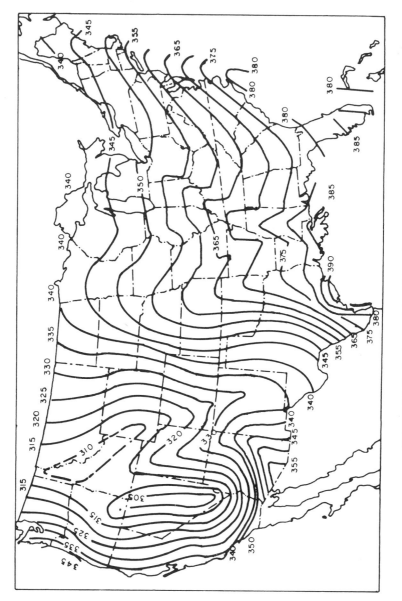

FIGURE 2.4 Sea level refractivity (N_0) maximum for August.

N_s at altitude h_s above MSL can be derived from surface refractivity (N_0) gradients (Figures 1.3 and 1.4) by the following formula:

$$N_s = N_0 \exp(-0.1057 h_s) \qquad (1.17a)$$

where h_s, the altitude above mean sea level, is measured in kilometers. For conditions near the ground the following empirical relationship between N_s and the difference in refractivity ΔN between N_s and N at 1 km above the earth's surface (i.e., $\Delta N/\Delta h$ where $\Delta h = 1$ km):

$$\Delta N \ (1 \ km) = -7.32 \exp(0.005577 N_s) \qquad (1.14)$$

Figure 2.4 give values of N_0 maximums for August for the continental United States.

2.4 REFLECTION POINT

As discussed in Chapter 1, ground reflections are a major cause of multipath fading. These reflections can be reduced or eliminated by the adjustment of tower heights effectively moving the reflection point from an area along the path of greater reflectivity (such as a body of water) to one of lesser reflectivity (such as an area of heavy forest). Of course, for those paths that are entirely over water or over desert, the designer will have to resort to other methods

TABLE 2.4. Approximate Values of R for Various Terrain

Type of Terrain	R^a	Approximate Depth of Even Fresnel Zone Fade (dB)
Heavily wooded, forest land	0 to −0.1	0–2
Partially wooded (trees along roads perpendicular to path, etc.)	−0.1 to −0.4	2–5
Sagebrush, high grassy areas	−0.5 to −0.7	5–10
Cotton with foilage, rough sea water, low grassy areas	−0.7 to −0.8	10–20
Smooth sea water, salt flats, flat earth	−0.9 +	20–40 +

[a] The values of R given in this table are approximate, of course, but they do give an indication of signal degradation to be expected over various terrain should even numbered Fresnel zone reflections occur.

Source. Ref. 3.

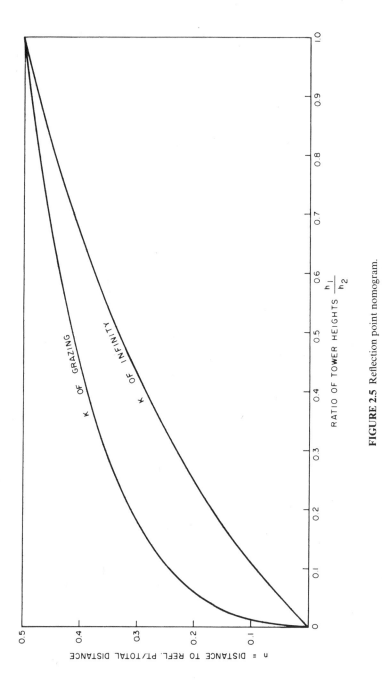

FIGURE 2.5 Reflection point nomogram.

38

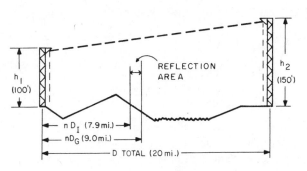

EXAMPLE:
$h_1 = 100'$ $h_2 = 150'$
$D_T = 20$ mi.

Ratio of tower heights:
$$\frac{h_1}{h_2} = \frac{100}{150} = .66$$
Enter .66 on bottom scale of graph E and read distance from shortest tower to point of reflection.
For K = Infinity,
$D_i = .395 \times 20 = 7.9$ mi.
For K = Grazing,
$D_G = .45 \times 20 = 9.0$ mi.

REFLECTION AREA

h_1 (100') h_2 (150')

nD_I (7.9 mi.)
nD_G (9.0 mi.)
D TOTAL (20 mi.)

FIGURE 2.6 Example path.

such as vertically spaced space diversity or frequency diversity to mitigate multipath fading. Table 2.4 provides a guide to the coefficient of reflectivity R for various types of terrain and the fading depths that may be expected for each value when the reflection point is located on a particular type of terrain.

There are several methods to determine the location of the reflection point. The simplest is a graphical method derived from Ref. 5. Here we use Figure 2.5 and an example is given in Figure 2.6.

The term reflection "point" is somewhat confusing. It would indeed be a singular point if we could assume a fixed K-factor. Such an assumption is only valid for a fully homogeneous atmosphere stable across the total path 365 days a year. This is not true because the atmosphere is dynamic, with constantly changing temperature, atmospheric pressure, and humidity across the path with time. For this reason, we assume that the reflection "point" covers a locus on the profile route line whose extremes are determined by extremes of K-factor (i.e., from $K = \infty$ to $K = $ grazing).*

From the path profile, we take the tower heights, h_1 for the transmitter site and h_2 for the receiver site, and determine the ratio h_1/h_2, which is entered on the X-axis of Figure 2.5. On the Y-axis on the figure two values of n are taken, the first for $K = \infty$ and the second for $K = $ grazing. The distances from the nearest site defining the reflection locus are determined from the values nD for each value of n, where D is the total path length.

Example. Assume $h_1 = 100$ ft, $h_2 = 150$ ft, and the path length D is 20 mi.

Ratio of tower heights: $h_1/h_2 = 0.66$

Enter the value 0.66 on the X-axis of Figure 2.6 and read the values for n for $K = \infty$ and for $K = $ grazing. Multiply each value by the path length D. Thus:

$$D_i = 0.39 \times 20 = 7.9 \text{ mi}   \text{(value for } K = \infty)$$
$$D_g = 0.45 \times 20 = 9.0 \text{ mi}   \text{(value fo } K = \text{ grazing)}$$

*$K = $ grazing is its value where the boresight ray grazes the earth for the path of interest.

The reflection "area" is that "area" along the line between 7.9 and 9.0 mi from the shorter tower.

2.5 SITE SURVEY

2.5.1 Introduction

Once the path profile has been completed, the designer should verify the results by a field survey of the sites at each end of the path (hop) and the intervening terrain. Of primary importance is the verification of site locations and conditions that, indeed, the line-of-sight criteria developed on the profile have been met. He or she should be particularly vigilant of structures that have been erected since the preparation of the topographic maps used to construct the profile. Also, maps, especially for emerging nations, may be in error.

2.5.2 Information Listing (Ref. 2)

The following is a general listing of information required from the field survey for a repeater site.

a. *Precise Location of Site.* At least two permanent survey monuments should be placed at each site and the azimuth between them recorded. Their locations should be indicated on a site survey sketch and site photograph. Geographical coordinates of the markers should be determined within $\pm 1''$ and the elevation of each to ± 1.5 m (± 5 ft). All elevation data should be referenced to mean sea level (MSL).

b. *Site Layout Plan.* A sketch of the site should include antenna locations with respect to monument markers as well as shelter location.

c. *Site Description.* Include soil type, vegetation, existing structures, access requirements, leveling or grading requirements, drainage, etc. Use sketch to show distances to property lines, benchmarks, roads, etc. A detailed topographic map of the site should be made. Photographs should be taken to show close-up details as well as general location with respect to surrounding terrain features. For manned sites, an estimate should be made of suitability regarding water supply and sewage disposal. Helpful information in this regard may be obtained from local well drillers, agricultural agents, plumbing contractors, and nearby farmers or ranchers.

d. *Description of Path.* Ideally, the survey team should "walk" the path with a copy of the topo map being utilized. A general description of terrain and vegetation should be annotated while proceeding along the path. Critical points (obstacles) or new critical points (obstacles) are noted, their location (i.e., d_1 and d_2) should be determined within ± 0.2 km (± 0.1 mi), azimuths to 1' of arc, elevations to $\pm 1°$ of path centerline. The surveyor should show by sketch and photograph new construction or other features not correctly indicated on the topo map. From each site, sketches and photographs should be made on path centerline showing azimuth angles to prominent features (references on topo map) and noting elevation angles. A running diary of the survey should be kept showing data obtained and method of verification.

e. *Power Availability*. Give location of nearest commercial transmission line with reference to each site. List name and address of the utility firm. State voltages, phases, line frequency, and main feeder size.

f. *Fuel Supply*. List local sources of propane, diesel fuel, heating oils, and natural gas. Estimate cost of item(s) delivered to site.

g. *Local Materials and Contractors*. List local sources of lumber (if any) and ready-mixed concrete and list names and addresses of local general construction contractors and availability of cranes and earth moving equipment.

h. *Local Zoning Restrictions*. Inquiries should be made as to national regulations or local zoning restrictions that might affect the use of the site or the height of the antenna tower. Give distance from site to nearest airport and determine if site is in a runway approach corridor.

i. *Geologic and Seismic Data*. Determine load-bearing qualities of soil at site, depth to rock and groundwater. Obtain soil samples. Check with local authorities on the frequency and severity of seismic disturbances.

j. *Weather Data*. General climatological data should have been assembled during the initial design studies. This information should be verified/modified by local input information during the site survey stage. Check with local authorities to obtain the following data:

1. Average monthly maximum and minimum temperature.
2. Average monthly precipitation, and extreme short-period totals (day, hour, 5-min intervals, if possible), days/month with rain.
3. Average wind direction and velocity, and direction and velocity of peak gusts.
4. Average and extreme snow accumulation, packing.
5. Flooding data.
6. Occurrence of hurricanes, typhoons, tornadoes.
7. Cloud and fog conditions.
8. Probability of extended periods of very light winds.
9. Icing conditions, freezing rain in particular.
10. Average dewpoint temperature and diurnal variation of relative humidity.

k. *Survey of Electromagnetic Interference (EMI)*. Show sources of "foreign" EMI such as similar system crossing or running parallel to the proposed route line, show locations of "foreign" repeater or terminal sites, EIRPs, antenna patterns, frequencies, bandwidths, spurious emissions (specified). Care must also be exercised to assure that the proposed system does not interfere with existing radio facilities.

2.5.3 Notes on Site Visit

A sketch should be made of the site in a field notebook. The sketch should show the location of the tower and buildings, trees, large boulders, ditches, etc. Measurements are then made with a steel tape and are recorded on the sketch. The sketch should be clearly identified by name, site number, geographical coordinates, and quadrangle map name or number. The sketch should be accompanied by a written description of soil type, vegetation, number, type and size of trees, number of trees to be removed, and approximate location and extent of leveling required. If soil samples are taken, the soil sample points with identifiers should be marked on the sketch.

Photographs should be made of the horizon to the north, east, south, and west from the tower location and along the centerline of the path. The use of a theodolite is recommended to determine azimuths of prominent features on the horizon of each photograph. Indicate the elevation angle to prominent features on the path centerline photograph(s).

A magnetic compass corrected for local declination is suitable for rough site layouts only. Final azimuth references should be determined or confirmed by celestial observation or by accurate radio navigation device.

The latitude and longitude of the tower location on the site should be determined to $\pm 1''$. This can be done by map scaling, traverse survey, triangulation survey, and celestial observations. In most cases sufficient accuracy can be obtained by careful scaling from a $7\frac{1}{2}'$ quadrangle with a device such as the Gerber Variable Scale.

The location should also be described with reference to nearby roads and cities so that contractors and suppliers can be readily directed to the site.

The elevation of the ground at the tower location should be determined to ± 1.5 m (± 5 ft). Differential leveling, or an extension of a known vertical control point (benchmark) by a series of instrument setups, is the most accurate method and is recommended for the final survey or when there is a benchmark close to the site. Trigonometric leveling is used to determine elevations over relatively long distances; it is not ordinarily recommended for site surveys but is useful to determine the elevation of obstacles along the path. Aerial photogrammetry and airborne profile recorders can also be used but tend to be expensive. Barometric leveling or surveying altimetry is the simplest method to determine relative ground elevations and, when carefully done, can provide sufficient accuracy for the final survey.

Access and commercial power entry to the site can be major cost drivers for the system first cost. The sketch map should show the route of the proposed road with reference to the site and existing roads. It is important to show the type of soil, number of trees that will probably have to be removed, degree of slope, rock formations, and approximate length of the road compared to crow line distance to existing road(s). The probability of all-year access must also be determined.

The map sketch should also include the location of the nearest source of commercial prime power, and the proposed route of the prime power exten-

Site Name and Number
Latitude_____Longitude_____(Degrees, Min, Sec)_____
Map reference (most detailed topographic)_____
Nearest town (post office)_____
Access route: (all year?)

Property owner; local contact:
Site sketch_____ Site photograph_____ General description_____
Reference baseline_____ By Polaris_____ Other_____
Antenna No._____   True bearing_____
   Ground elev. MSL_____ Takeoff angle (beam centerline)_____
   Takeoff angles to 45° right and left of centerline_____
    (Significant changes in horizon)
   Critical Points: (include horizon)
    Distance_____ Map elev._____ Survey elev._____
     Tree height_____ Required clearance_____
    Description:
   Horizon sketch_____ Horizon photograph_____

Power availability:
   a. Nearest transmission line_____ b. Voltage_____
   c. Frequency_____ d. Phase_____ e. Operating utility_____
Drinking water source_____ Estimated depth to groundwater_____
Sewage disposal_____ Type and depth of soil on and near site_____
Nearest airport_____ railroad_____ highway_____
   navigable river_____

FIGURE 2.7 Sample checklist for site survey (Ref. 2).

sion should also be included. Prior to the survey an estimate should be made of the prime power requirements of the site including tower lighting and air conditioning (where required). This demand requirement will assist local power company officials in determining power extension cost.

Figure 2.7 is a sample checklist for a site survey.

2.6 PATH ANALYSIS

2.6.1 Objective and Scope

The path analysis or link power budget task provides the designer with the necessary equipment parameters to prepare a block diagram of the terminal or repeater configuration and to specify equipment requirements both quantitatively and qualitatively. We assume that frequency assignments have been made as we did in Section 2.3 or at least assume the frequency band in which assignments will be made by the appropriate regulatory authority.

The discussion that follows in this section is primarily directed toward analog radiolinks. Much of the same approach is also valid for digital radiolink design except that some units of measure will differ. In this section one might say that the bottom line for task output is noise (in dBrnC or pWp) and S/N in the standard voice channel or video channel, whereas in Section 2.11, where digital radiolinks are discussed, we will be dealing with signal energy per bit to noise spectral density ratio (E_b/N_0) and BER on the link. We also assume here that the modulation waveform is conventional FM.

This section will provide us with the tools to derive antenna aperture, receiver front end characteristics, FM deviation, IF/RF bandwidth, transmitter output power, diversity arrangements (if any), and link availability due to propagation. This latter will involve the necessary system overbuild to meet propagation availability requirements in a fading environment. The use of NPR (noise power ratio) as a tool to measure link noise performance will be described. Frequency/bandwidth assignments for analog systems will also be covered.

The analysis described in this chapter is valid for radiolinks in the 1–10 GHz band. Radiolinks operating on frequencies above 10 GHz begin to be impacted by excess attenuation due to rainfall and gaseous absorption. Below 10 GHz the effects of rainfall are marginal, and in many design situations they can be neglected. Chapter 6 describes the design of radiolinks above 10 GHz.

2.6.2 Unfaded Signal Level at the Receiver

It will be helpful to visualize an LOS radiolink by referring to the simplified model shown in Figure 2.8 where a radiolink transmitter and radiolink receiver are separated by a distance D.

Given the path from transmitter to receiver in Figure 2.8 and the transmitter's assigned frequency, we calculate the free space loss (FSL) represented

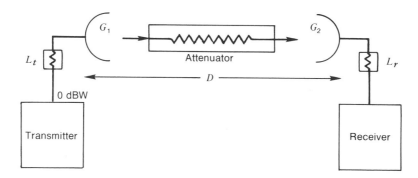

FIGURE 2.8 Simplified model, radiolink path analysis. L_t and L_r are the transmission line losses; G_1 and G_2 are the antenna gains.

by the attenuator. Equations (1.5), (1.7a), and (1.7b) from Section 1.2 apply and are restated here for convenience.

$$\text{FSL}_{dB} = 32.45 + 20 \log D_{km} + 20 \log F_{MHz} \qquad (1.5)$$

where D is measured in kilometers,

$$\text{FSL}_{dB} = 36.58 + 20 \log D_{sm} + 20 \log F_{MHz} \qquad (1.7a)$$

where D is measured in statute miles, and

$$\text{FSL}_{dB} = 37.80 + 20 \log D_{nm} + 20 \log F_{MHz} \qquad (1.7b)$$

where D is measured in nautical miles. If F is stated in gigahertz, add 60 to the value of the constant term.

Example 1. Compute the free space loss for a path 31 statute miles long with a transmit frequency of 6135 MHz.

$$\begin{aligned}
\text{FSL}_{dB} &= 36.58 + 20 \log(31) + 20 \log(6135) \\
&= 36.58 + 29.83 + 75.76 \\
&= 142.17 \text{ dB}
\end{aligned}$$

Example 2. Compute the free space loss for a path 43 km long with a transmit frequency of 4.041 GHz.

$$\begin{aligned}
\text{FSL}_{dB} &= 92.45 + 20 \log(43) + 20 \log(4.041) \\
&= 92.45 + 32.67 + 12.13 \\
&= 137.25 \text{ dB}
\end{aligned}$$

It should be noted that for a fixed path, if the frequency is doubled, approximately 6 dB is added to the free space loss. Conversely, if the transmit frequency is halved, approximately 6 dB is then subtracted from the free space loss. Using similar reasoning, for a fixed frequency path, double the distance between transmitter and receiver and approximately 6 dB is added to the free space loss; halve the distance and the free space loss is 6 dB less.

The total unfaded path loss for an LOS radiolink also includes an attenuation component due to atmospheric gaseous absorption, which is added to the FSL value. Estimates of these component values are 0.3 dB for paths up to 20 miles long; 0.6 dB for paths 20–40 miles long; 1.0 dB for paths 40–60 miles; 1.3 dB for paths 60–80 miles; and 1.6 dB for paths 80–100 miles long. For more accurate estimates consult Figure 3.4.

Turning again to the model in Figure 2.8, the effective isotropic radiated power (EIRP) can be calculated as follows:

$$\text{EIRP}_{dBW} = P_0 + L_t + G_1 \qquad (2.6)$$

where P_0 is the RF power output of the transmitter at the waveguide flange, L_t are the transmission line losses, and G_1 is the gain of the transmit antenna. It should be noted that gains are treated as positive numbers and losses, conventionally, are negative numbers in equation (2.6) and in the following discussion. This permits simple algebraic addition.

Example 3. A microwave radiolink transmitter has an output of 1 W at the waveguide flange, transmission line losses from the flange to the antenna feed are 3 dB, and the gain of the antenna is 31 dB, compute the EIRP in dBW.

$$\text{EIRP} = 0 \text{ dBW} + (-3 \text{ dB}) + 31 \text{ dB}$$

$$= +28 \text{ dBW}$$

Example 4. Compute the EIRP in dBm where the transmitter output power is 200 mW at the waveguide flange, the transmission line losses are 4.7 dB, and the antenna gain is 37.3 dB.

$$\text{EIRP} = 10 \log 200 + (-4.7 \text{ dB}) + 37.3 \text{ dB}$$

$$= +23 \text{ dBm} - 4.7 \text{ dB} + 37.3 \text{ dB}$$

$$= +55.6 \text{ dBm}$$

To calculate the signal power produced by an isotropic antenna at the receiving terminal location in Figure 2.8, the EIRP is algebraically added to the free space loss (FSL) and the gaseous absorption loss L_g. This power level is called the isotropic receive level (IRL). Or, turning to the model in Figure 2.8, the EIRP is the input to the attenuator, and we simply calculate the output of the attenuator or

$$\text{IRL} = \text{EIRP} + \text{FSL}_{\text{dB}} + L_g \tag{2.7}$$

Example 5. Calculate the IRL of a radiolink where the EIRP is $+28$ dBW, the FSL is 137.25 dB, and the atmospheric gaseous absorption loss is 0.6 dB.

$$\text{IRL}_{\text{dBW}} = +28 \text{ dBW} + (-137.25 \text{ dB}) + (-0.6 \text{ dB})$$

$$= -109.85 \text{ dBW}$$

The unfaded receive signal level, RSL, at the receiver input terminal (Figure 2.8) is calculated by algebraically adding the isotropic receive level, the receive antenna gain, G_2, and the transmission line losses at the receiving terminal, L_r, or

$$\text{RSL} = \text{IRL} + G_r + L_r \tag{2.8}$$

where G_r and L_r are the generalized values of the receive antenna gain and the receive system transmission line losses.

To calculate RSL directly, the following formulas apply:

$$\text{RSL} = \text{EIRP} + \text{FSL} + L_g + G_r + L_r \tag{2.9}$$

or

$$\text{RSL} = P_0 + L_t + G_1 + \text{FSL} + L_g + G_2 + L_r \tag{2.10}$$

Example 6. Calculate the RSL in dBW where the transmitter output power to the waveguide flange is 750 mW, the transmission line losses at each end are

3.4 dB, the distance between the transmitter and receiver site is 17 statute miles, and the operating frequency is 7.1 GHz; the antenna gains are 30.5 dB at each end. Assume a gaseous absorption loss of 0.3 dB. First calculate FSL using equation (1.7A).

$$FSL_{dB} = 36.58 + 20\log 17 + 20\log 7100$$

$$= 36.58 + 24.61 + 77.02$$

$$= 138.21 \text{ dB}$$

Next calculate RSL using equation (2.10):

$$RSL = 10\log 0.75 + (-3.4) + 30.5 + (-138.21)$$

$$+ (-0.3) + 30.5 + (-3.4)$$

$$= -1.25 - 3.4 + 30.5 - 138.21 - 0.3 + 30.5 - 3.4$$

$$= -85.56 \text{ dBW}$$

2.6.3 Receiver Thermal Noise Threshold

2.6.3.1 Objective and Basic Calculation

One waypoint objective in the path analysis is to calculate the unfaded carrier-to-noise ratio (C/N). With the RSL determined from Section 2.6.2 and with the receiver thermal noise threshold, we can simply calculate the C/N, where

$$\left(\frac{C}{N}\right)_{dB} = RSL - P_t \tag{2.11}$$

where P_t is the receiver thermal noise threshold. Note that RSL and P_t must be in the same units, conventionally in dBm or dBW.

The equipartition law of Boltzmann and Maxwell states that the available power per unit bandwidth of a thermal noise source is

$$P_n(f) = kT \text{ W/Hz} \tag{2.12}$$

where k is Boltzmann's constant (1.3805×10^{-23} J/K) and T is the absolute temperature of the source in kelvins. At absolute zero the available power in a 1-Hz bandwidth is -228.6 dBW. At room temperature, usually specified as $17°C$ or 290 K,* the available power in a 1-Hz bandwidth is -204 dBW or $-228.6 + 10\log(290)$. Thus

$$P_a = kTBW \text{ W} \tag{2.13}$$

---

*CCIR often uses the value 288 K for room temperature.

where BW is bandwidth expressed in Hz. Expressed in dBW at room temperature

$$P_a = -204 + 10\log(BW) \text{ dBW} \qquad (2.14)$$

Expressed in dBm at room temperature

$$P_a = -174 + 10\log(BW) \text{ dBm} \qquad (2.15)$$

The thermal noise level is frequently referred to as the thermal noise threshold. For a receiver operating at room temperature is a function of the bandwidth of the receiver [in practical systems this is taken as the receiver's IF bandwidth (B_{if})] measured in Hz and the noise figure NF (in dB) of the receiver. Thus, the thermal noise threshold P_t of a receiver can be calculated as follows:

$$P_t = -204 \text{ dBW} + 10\log B_{if} + \text{NF}_{\text{dB}} \qquad (2.16)$$

Example 7. Compute the thermal noise threshold of a receiver with a 12 dB noise figure and an IF bandwidth of 4.2 MHz.

$$P_t = -204 \text{ dBW} + 10\log(4.2 \times 10^6) + 12 \text{ dB}$$

$$= -204 \text{ dBW} + 66.23 \text{ dB} + 12 \text{ dB}$$

$$= -125.77 \text{ dBW} \quad \text{or} \quad -95.77 \text{ dBm}.$$

Example 8. Compute the thermal noise threshold of a receiving system with a 3.1 dB noise figure and an IF bandwidth of 740 kHz.

$$P_t = -204 \text{ dBW} + 10\log(740 \times 10^3) + 3.1 \text{ dB}$$

$$= -204 \text{ dBW} + 58.69 \text{ dB} + 3.1 \text{ dB}$$

$$= -142.2 \text{ dBW} \quad \text{or} \quad -112.2 \text{ dBm}$$

(Note the change in terminology between examples 7 and 8, "receiver" and "receiving system." This will be explained in the next section.)

2.6.3.2 Practical Applications

The majority of LOS radiolink receivers use a mixer for the receiver front end. Such an arrangement is shown (simplified) in Figure 2.9.

To calculate the receiver noise threshold of the mixer, use equation 2.16, where NF is the noise figure of the mixer. Under certain circumstances a path analysis may show a link to be marginal. One means of improvement is to add a low-noise amplifier (LNA), usually a FET-based device, in front of the mixer. This alternative configuration is shown in Figure 2.10. Mixers display

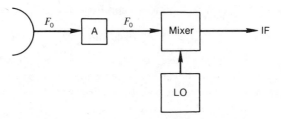

FIGURE 2.9 Simplified functional block diagram of a conventional LOS radiolink (microwave) receiver front end. F_0 = RF operating frequency; A = waveguide devices such as circulator, preselector, etc.; LO = local oscillator; IF = intermediate frequency.

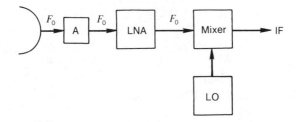

FIGURE 2.10 Alternative receiver front end configuration.

noise figures on the order of 8 to 12 dB; GaAs FET LNAs display noise figures from 1 to 3 dB for systems operating below 10 GHz.

Where front end noise figures are comparatively high, other noise sources, such as antenna noise and ohmic noise, can be neglected. However, if a configuration, like Figure 2.10, is used, where the receiver noise sources can be approximated by the noise figure of the LNA with NF on the order of 1–3 dB, other noise sources may not be neglected. Thus good practice dictates adding 2 dB to the noise figure of the LNA to approximate the noise contributed by other sources. A much more exact method of calculating total noise of a receiving system is given in Chapter 4.

Example 9. A radiolink receiving system has an LNA with a noise figure of 1.3 dB and the IF has a bandwidth of 30 MHz, compute the thermal noise threshold.

$$P_t = -204 \text{ dBW} + 10 \log(30 \times 10^6) + 1.3 \text{ dB} + 2 \text{ dB}$$

$$= -204 + 74.77 + 3.3$$

$$= -125.93 \text{ dBW} \quad \text{or} \quad -95.93 \text{ dBm}$$

2.6.4 Calculation of IF Bandwidth and Peak Frequency Deviation

2.6.4.1 IF Bandwidth

The IF of an FM receiver must accommodate the RF bandwidth, which consists of the total peak deviation spread and a number of generated sidebands. The IF bandwidth can be estimated from Carson's rule

$$B_{if} = 2(\Delta F_p + F_m) \tag{2.17}$$

where ΔF_p is the peak frequency deviation and F_m is the highest modulating frequency given in the middle column of Table 2.6.

2.6.4.2 Frequency Deviation

The value of ΔF_p in equations (2.17) and (2.18) (below) is peak deviation. The peak deviation for a particular link should be taken from equipment manuals for the proposed load (i.e., number of voice channels or TV video). If this information is not available, peak deviation may be calculated using CCIR specified per channel deviation from CCIR Rec. 404-2. See Table 2.5. CCIR recommends a peak deviation of ± 4 MHz for video systems (CCIR Rec. 276-2) without pre-emphasis.

CCIR Rec. 404-2 further states that where pre-emphasis is used, the pre-emphasis characteristic should preferably be such that the effective rms deviation due to the multichannel signal is the same with and without pre-emphasis. (For a discussion of pre-emphasis, see Section 2.6.5.)

To calculate peak deviation when rms per channel deviation Δf (as in Table 2.5) is given, the following formula applies:

$$\Delta F_p = \Delta f(pf)(\text{NLR}_n) \tag{2.18}$$

TABLE 2.5. Frequency Deviation without Pre-emphasis

Maximum Number of Channels	rms Deviation per Channel[a] (kHz)
12	35
24	35
60	50, 100, 200
120	50, 100, 200
300	200
600	200
960	200
1260	140, 200
1800	140
2700	140

[a] For 1 mW, 800 Hz tone at a point of zero reference level.
Source. CCIR Rec. 404-2 (Ref. 1).

where pf is the numerical ratio of peak-to-rms baseband voltage or

$$\text{pf} = \text{antilog}\left(\tfrac{1}{20}PF\right) \qquad (2.19)$$

where PF (peak factor) is the baseband peak-to-rms voltage in dB. Some texts use 12 dB for PF, others use 13 dB. If 12 dB is used for the value of PF, then this represents a value where peaks are not exceeded more than 0.01% of the time, and 13 dB for 0.001% of the time. Normally, NLR_n will be specified at the outset for the number of channels (N) that the system will carry.

CCIR recommends, for FDM telephony baseband, the following

$$\text{NLR}_{\text{dB}} = -1 + 4\log N \qquad (2.20)$$

for FDM configurations from 12 to 240 voice channels and

$$\text{NLR}_{\text{dB}} = -15 + 10\log N \qquad (2.21)^*$$

for FDM configurations where N is greater than 240. For U.S. military systems, the value

$$\text{NLR}_{\text{dB}} = -10 + 10\log N \qquad (2.22)$$

is used.

NLR_n will be given (or calculated) in dB, but in equation (2.18) the numerical equivalent value is used or

$$\text{NLR}_n = \text{antilog}\left(\tfrac{1}{20}\text{NLR}_{\text{dB}}\right) \qquad (2.23)$$

if we assume a PF of 13 dB, then with CCIR loading of FDM telephony channels

$$\Delta F_p = 4.47d\left[\log^{-1}\left(\frac{-1 + 4\log N}{20}\right)\right] \qquad (2.24)$$

for values of N from 12 to 240 voice channels and

$$\Delta F_p = 4.47d\left[\log^{-1}\left(\frac{-15 + 10\log N}{20}\right)\right] \qquad (2.25)$$

for values of N greater than 240. d is the rms test tone deviation from Table 2.5.

Through the use of Figure 2.11, $B_{if} = B_{rf}$ can be derived graphically given the rms per channel deviation from Table 2.5. (The reader will note the variance with Carson's rule for $B_{if} = B_{rf}$.)

Example 10. Compute the peak deviation for a radiolink being designed to transmit 1200 voice channels of FDM telephony.

---

*For North American practice: $\text{NLR}_{\text{dB}} = -16 + 10\log N$ (2.21a).

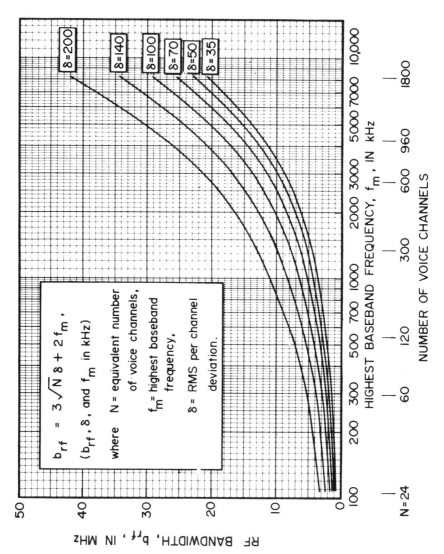

FIGURE 2.11 Dependence of radio frequency bandwidth on baseband width for FDM-FM systems. (From MIL-HDBK-416, Ref. 2.)

52

First calculate the value in dB of the noise load ratio (NLR$_{dB}$) assuming CCIR loading. Use equation (2.21):

$$NLR_{dB} = -15 + 10\log(1200)$$

$$= 15.79 \text{ dB}$$

Calculate the numerical equivalent of this value using equation (2.23):

$$NLR_n = \log^{-1}(15.79/20)$$

$$= 6.16$$

Calculate the peak deviation using equation (2.25):

$$\Delta F_p = 4.47(200)(6.16)$$

where the value for d was derived from Table 2.5,

$$\Delta F_p = 5507 \text{ kHz or } 5.507 \text{ MHz.}$$

2.6.5 Pre-emphasis / De-emphasis

After demodulation in an FM system, thermal noise power (in some texts called "idle noise") is minimum for a given signal at the lowest demodulated baseband frequency and increases at about 6 dB per octave as the baseband frequency increases. This effect is shown in Figure 2.12, which compares thermal noise in an AM system with that in an FM system.

To equalize the noise across the baseband, a pre-emphasis network is introduced ahead of the transmitter modulator to provide increasing attenuation at the lower baseband frequencies. The transmitting baseband gain is then increased so that baseband frequencies above a crossover frequency are increased in level, those below the crossover frequency are lowered in level, and the total baseband energy presented to the modulator is approximately unchanged relative to FM without pre-emphasis.

In the far end receiver of the FM radiolink, a de-emphasis network providing more attenuation with increasing frequency is applied after the demodulator. This network removes both the test-tone slope produced by pre-emphasis and the variable idle noise (thermal noise) slope to provide a more even distribution of signal-to-noise. The two networks must be complementary to ensure a nearly flat frequency versus level response across the baseband.

There are two types of networks that may be encountered in multichannel (FDM) FM systems; the CCIR network and the 6-dB octave network. The CCIR network has much greater present-day application, and henceforth in this text all reference to pre-emphasis will be that meeting CCIR Rec. 275-2 (Ref. 1) for multichannel FDM telephony links and CCIR Rec. 405-1 (Ref. 1) for television transmission. Figure 2.13 gives the pre-emphasis characteristic for telephony, and Table 2.6, which is associated with the figure, shows the characteristic frequencies of pre-emphasis/de-emphasis networks for standard

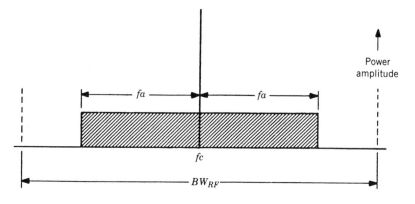

Noise power, AM System

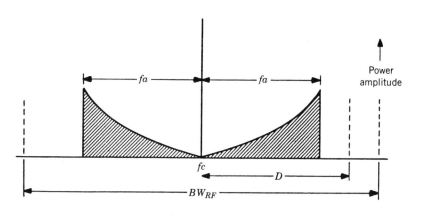

Noise power, FM System

FIGURE 2.12 Illustration of noise power distribution in AM and FM systems.

CCITT FDM multiplex configurations.* Figure 2.14 shows the pre-emphasis characteristic for television (video). Figure 2.15 gives values in decibels for pre-emphasis improvement for FDM telephony links as a function of the number of FDM voice channels transmitted. These values will be used in link analysis to determine signal-to-noise ratios.

---

*The reader should consult CCIR Rec. 380-3 (Ref. 1) and Ref. 5.

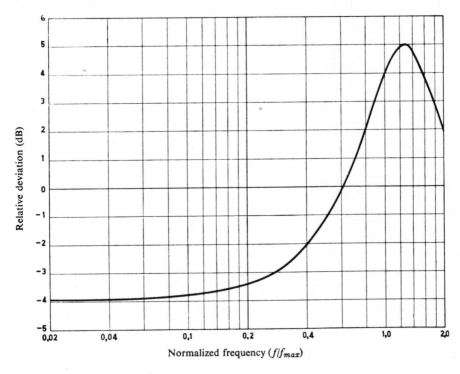

FIGURE 2.13 Pre-emphasis characteristic for telephony. From CCIR Rec. 275-2 (Ref. 1).

TABLE 2.6. Characteristic Frequencies for Pre-emphasis and De-emphasis Networks for Frequency-Division Multiplex Systems

Maximum Number of Telephone Traffic channels[a]	f_{max}^b (kHz)	f_r^c (kHz)
24	108	135
60	300	375
120	552	690
300	1300	1625
600	2660	3325
960	4188	5235
1260	5636	7045
1800	8204	10,255
2700	12,388	15,485

[a] This figure is the nominal maximum traffic capacity of the system and applies also when only a smaller number of telephone channels are in service.

[b] f_{max} is the nominal maximum frequency of the band occupied by telephone channels.

[c] f_r is the nominal resonant frequency of the pre-emphasis or de-emphasis network.

Source. CCIR Rec. 275-2 (Ref. 1).

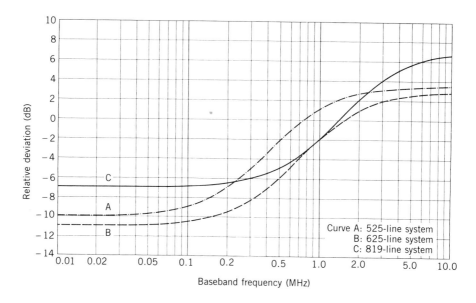

FIGURE 2.14 Pre-emphasis characteristic for television of (*A*) 525-, (*B*) 625-, and (*C*) 819-line systems.

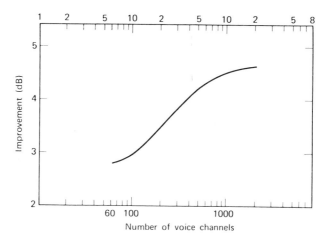

FIGURE 2.15 Pre-emphasis improvement in decibels as a function of the number of FDM telephone channels transmitted. From *Telecommunications Transmission Handbook*, 2nd ed., copyright © 1981 by John Wiley & Sons, Inc. (Ref. 5).

2.6.6. Calculation of Median Carrier-to-Noise Ratio (Unfaded)

The carrier-to-noise ratio (C/N) of an analog radiolink is a function of the receive signal level (RSL) (Section 2.6.2) and the thermal noise threshold (P_t) of the receiver (Section 2.6.3) or

$$\left(\frac{C}{N}\right)_{dB} = RSL - P_t \qquad (2.26)$$

Of course, RSL and P_t must be expressed in the same units, either dBm or dBW.

Example 1. Compute the unfaded C/N in decibels of a radiolink transmitting 960 voice channels of FDM telephony. The design of the link is based on CCIR recommendations and the link is 35 km long. The transmitter output to the waveguide flange is 750 mW on a frequency of 6.1 GHz, and the receiver noise figure is 9 dB. Assume antenna gains on each end of 30 dB and the transmission line losses at each end to be 2.1 dB. Let L_g (gaseous atmospheric loss) equal 0.3 dB.

Calculation of RSL: use equation (2.10),

$$RSL_{dBW} = +10\log(0.750) + (-2.1\,dB) + 30\,dB + FSL + (-0.3\,dB)$$

$$+30\,dB + (-2.1\,dB)$$

Compute FSL: use equation (1.5),

$$FSL_{dB} = 32.45\,dB + 20\log(35) + 20\log(6100)$$

$$= 32.45 + 30.88 + 75.71$$

$$= 139.04\,dB$$

$$RSL_{dBW} = -1.25\,dBW - 2.1\,dB + 30\,dB - 139.04\,dB$$

$$-0.3\,dB + 30\,dB - 2.1\,dB$$

$$= -84.79\,dBW$$

Note that in an equation, losses are conventionally expressed with a minus sign and gains are expressed with a plus sign.

Calculation of P_t in dBW: use equation (2.16),

$$P_t \doteq -204\,dBW + 9\,dB + 10\log B_{if}$$

Compute B_{if} using Carson's rule, equation (2.17). The maximum modulating baseband frequency is 4188 kHz for 960 voice channels from Table 2.6,

$$B_{if} = 2(\Delta F_p + 4188\,kHz)$$

Calculate ΔF_p: use equation (2.25). First calculate NLR.

$$\text{NLR}_{dB} = -15 + 10 \log N$$

where $N = 960$,

$$\text{NLR}_{dB} = -15 + 29.82$$

$$= 14.82 \text{ dB}$$

Convert NLR to its numerical equivalent:

$$\text{NLR}_n = \log^{-1}(14.82/20)$$

$$= 5.51$$

$$\Delta F_p = 4.47(200)(5.51) \quad \text{(note: the value 200 kHz/channel}$$

$$\text{is derived from Table 2.5)}$$

$$= 4926 \text{ kHz}$$

$$B_{if} = 2(4926 + 4188) \quad (\text{kHz})$$

$$= 18.228 \text{ MHz}$$

$$P_t = -204 + 9 + 10 \log(18.228 \times 10^6)$$

$$= -204 + 9 + 72.61$$

$$= -122.39 \text{ dBW}$$

$$\left(\frac{C}{N}\right)_{dB} = \text{RSL} - P_t$$

$$= -84.79 \text{ dBW} - (-122.39 \text{ dBW})$$

$$= 37.6 \text{ dB}$$

2.6.7 Calculation of Antenna Gain

To achieve a required C/N for a radiolink, a primary tool at the designer's disposal is the sizing of the link antennas. For nearly all applications described in this text, the antenna will be based on the parabolic reflector.

The gain efficiencies of most commercially available parabolic antennas are on the order of 55–65%. It is generally good practice on the first cut link analysis to assume the 55% value. In this case the antenna gain can be calculated from the reflector diameter and the operating frequency by the following formula(s):

$$G_{dB} = 20 \log B + 20 \log F + 7.5 \qquad (2.27a)$$

where B is the parabolic reflector diameter in feet and F is the operating frequency in gigahertz. In the metric system, the following formula applies:

$$G_{dB} = 20 \log B + 20 \log F + 17.8 \qquad (2.27b)$$

Example 12. A radiolink requires an antenna with a 32-dB gain. The link operates at 6 GHz. What diameter parabolic antenna is required?

$$32 \text{ dB} = 20 \log B + 20 \log 6.0 + 7.5 \text{ dB}$$

$$20 \log B = 24.5 - 15.6$$

$$\log B = 8.9/20$$

$$B = 2.8 \text{ ft}$$

A more detailed discussion of antennas and the calculation of transmission line losses may be found in Chapter 7.

2.7 ESTIMATION OF FADE MARGIN AND MITIGATION OF FADING EFFECTS

2.7.1 Rationale

Up to this point we have been dealing with the calculation of unfaded signal levels at the far end receiver. A fixed FSL was assumed. On most short links, in the order of 3 mi (5 km) or less, only FSL need be considered.* As the link length (hop distance) increases, fading becomes a major consideration.

Fading is defined (Ref. 1) as "the variation with time of the intensity or relative phase, or both, of any of the frequency components of a received radio signal due to changes in the characteristics of the propagation path with time." During a fade the RSL decreases. This results in a degradation of C/N, thus a reduction in signal-to-noise ratio of the demodulated signal and, finally, an increase in noise in the derived voice channel. The various causes of fading were described qualitatively in Section 1.6. This section gives several methods of quantifying fading so that the link design engineer can overbuild the link to maintain the C/N above a stated level for a specified percentage of time.

Time availability (of a link or several links in tandem) is the period of time during which a specified noise performance is equaled or exceeded. Some texts

---

*It is recommended that a minimum of 5-dB margin be applied for equipment aging and miscellaneous other losses.

use the term propagation reliability to describe time availability. For radio-links, the time availability is commonly specified in the range from 0.99 to 0.99999 or 99% to 99.999% of the time.

There are essentially three methods available to the link design engineer to mitigate the effects of fading listed below in declining order of desirability:

1. Overbuild the link by the use of:
   a. Larger antennas.
   b. Improved receiver noise performance.
   c. Higher transmitter power output.
2. Use of diversity.
3. Resite or shorten distance between sites.

2.7.2 Calculating Fade Margin

2.7.2.1 Rayleigh Fading Assumption

Some texts (Ref. 6) assume fading is (1) entirely due to multipath phenome-non, and, as a result, (2) the worst case fading can be described by a Rayleigh distribution. A Rayleigh fading distribution is plotted in Figure 2.19.

For Rayleigh fading, the link margins given in Table 2.7 are required versus time availability. Extrapolation from Table 2.7 can be made for any time availability. A link requiring 99.95% time availability would require a 33-dB margin. This tells us that if the minimum unfaded C/N for the link were specified as 20 dB, the link would require 20 dB + 33 dB or a C/N = 53 dB to meet the objective 99.95% time availability. Allowing the validity of the pure Rayleigh assumption, the unavailability of the link is $1 - 0.9995$ or 0.0005. A year has 8760 hr or 8760×60 min. Thus the total time in a year when the C/N would be less than 20 dB would be $0.0005 \times 8760 \times 60$ or 262.8 min.

2.7.2.2 Path Classification Method 1

This method of quantifying fade margin is based on Ref. 7 and applies only to those paths that are over land and with unobstructed LOS conditions. The

TABLE 2.7. Fade Margins for Rayleigh Fading

Time Avail (%)	Fade Margin (dB)
90	8
99	18
99.9	28
99.99	38
99.999	48

method is empirical and was developed from CCIR reports as well as test data from the numerous LOS radiolinks installed by Siemens.

Siemens has classified LOS radiolink paths into three types—A, B, and C—depending on the characteristics of the paths. The curves presented in Figures 2.16A through 2.16C provide required fade margins for fading depths exceeded during 1% of the time in any month with severe fading as a function of frequency and path length. Conversion to lower probabilities of exceeding fading depth can be made using Figure 2.17.

Type A paths have comparatively favorable fading characteristics where the formation of tropospheric layers is a rare occurrence and where calm weather is a relatively rare occurrence. Type A paths are over hilly country, but not over wide river valleys and inland waters; and in high mountainous country with paths high above valleys. Type A paths are also characterized as being between a plain or a valley and mountains, where the angle of elevation relative to a horizontal plane at the lower site exceeds about 0.5°.

Type B paths with average fading characteristics are typically over flat or slightly undulating country where tropospheric layers may occasionally occur. They are also over hilly country, but not over river valleys or inland waters. Type B paths are also characterized as being in coastal regions with moderate temperatures, but not over the sea or also over those steeply rising paths in hot and tropical regions.

Type C paths have adverse fading and are characterized over humid areas where ground fog is apt to occur, particularly those paths that are low over flat country, such as wide river valleys and moors. They are also typically near the coast in hot regions and generally are those paths in tropical regions without an appreciable angle of elevation.

Rather than use Figures 2.16 and 2.17, the following formulas apply for rough estimations of the probability of fading exceeding the fading depth A:

$$\text{Type A:} \quad W = 16 \times 10^{-7} f d^2 \times 10^{-A/10} \qquad (2.28a)$$

$$\text{Type B:} \quad W = 8 \times 10^{-7} f d^{2.5} \times 10^{-A/10} \qquad (2.28b)$$

$$\text{Type C:} \quad W = 2 \times 10^{-7} f d^3 \times 10^{-A/12} \qquad (2.28c)$$

where W = probability that a fade depth A is exceeded during 1 year
f = radio carrier frequency in gigahertz
d = path length in kilometers
A = fading depth in decibels

The probability of exceeding a specified fading depth during an unfavorable month is

$$W_{\text{month}} = (12/M) \times W_{\text{year}} \qquad (2.29)$$

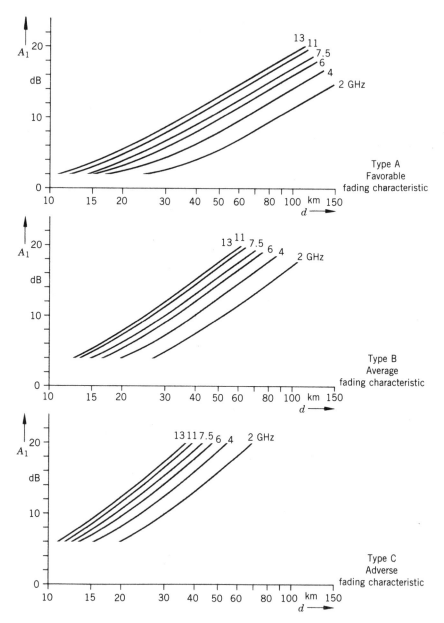

FIGURE 2.16 Fade margins A_1 as a function of path length d with unobstructed LOS (without precipitation attenuation). Typical values for 1% of the time of a month with severe fading. From *Planning and Engineering of Radio Relay Links*, Siemens-Heyden. Copyright 1976 Siemens Aktiengesellschaft, FRG (Ref. 7).

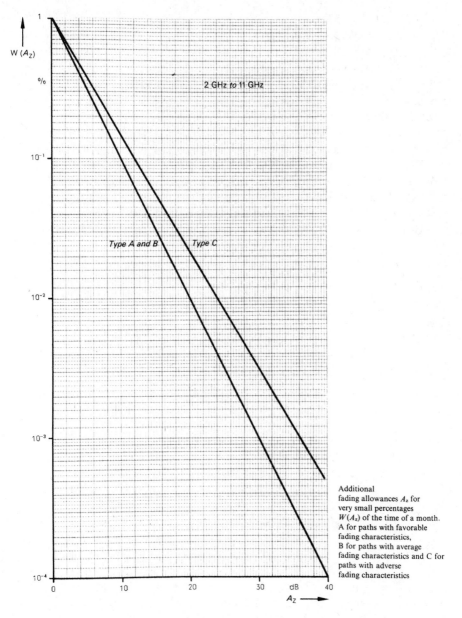

FIGURE 2.17 Additional fade margin to Figure 2.16, A_z for very small percentages $W(A_z)$ of the time of a month for paths type A, B, and C. From *Planning and Engineering of Radio Relay Links*, Siemens-Heyden, Copyright 1976 by Siemens Aktiengesellschaft, FRG (Ref. 7).

where M is the number of months in any year with intense multipath fading (for Europe, $M \approx 3$).

The approximations for formulas (2.28) and (2.29) are valid with the following assumptions: $W \leq 10^{-2}$; $f =$ from 2 to 15 GHz; $d =$ from 20 to 80 km; and $A \geq 15$ dB. Over-sea paths may be characterized (with caution) as Type C paths.

Example 13. A particular path has been specified to have a time availability of 99.95%; the operating frequency is 4.0 GHz and the path is 40 km long. What fade margin is required?

Use equation (2.28b). It should be noted that "the probability of exceeding a specified fading depth" or W is really the path unavailability. This equals $1 - (0.9995)$ or $W = 0.0005$. Then

$$0.0005 = 8 \times 10^{-7} \times 4 \times 40^{2.5} \times 10^{-A/10}$$

Solve for A and

$$10^{-A/10} = 0.0005/(8 \times 10^{-7} \times 4 \times 40^{2.5})$$

$$10^{-A/10} = 0.01544$$

$$-A/10 = \log_{10}(0.01544)$$

$$-A/10 = -1.8114$$

$$A = 18.11 \text{ dB}$$

which fairly well conforms to Figure 2.17.

2.7.2.3 Path Classification Method 2

This method is based on an empirical formula developed by Barnett (Ref. 8). It is similar in many respects to Method 1 in that it classifies paths by terrain and climate, and it is used to estimate the percentage of time within a year P_{mf} that fades *exceed* a specified depth below free space (M_f) for a given path and frequency. The formula applies only to paths in the United States and does not specifically consider beam penetration angle through the atmosphere or beam clearance of terrain:

$$P_{mf}(\%) = 6.0 \times 10^{-5} abfd^3 \times 10^{-M_f/10} \qquad (2.30)$$

where

$$a = \begin{cases} 4 & \text{for very smooth terrain including over-water} \\ 1 & \text{for average terrain with some roughness} \\ \frac{1}{4} & \text{for mountainous, very rough, or very dry terrain} \end{cases}$$

$$b = \begin{cases} \frac{1}{2} & \text{Gulf coast or similar hot, humid areas} \\ \frac{1}{4} & \text{normal interior temperate or northern climate} \\ \frac{1}{8} & \text{mountainous or very dry climate} \end{cases}$$

f = frequency in gigahertz

d = path length in kilometers

M_f = fading depth exceed below free space level, in decibels

Example 14. What will the path unavailability be for a path 50 km long over flat terrain in a relatively humid region, operating at 6 GHz with a 40 dB fade margin?

$$P_{mf}(\%) = 6.0 \times 10^{-5}(4)\left(\tfrac{1}{2}\right)(6.0)(50^3) \times 10^{-40/10}$$

$$= 0.009\%$$

The path availability then is $(1.00000 - 0.00009)$ or 99.991%.

2.7.3 Notes on Path Fading Range Estimates

Three methods have been described in Section 2.7.2 for first-order rough estimations of fade margin. Other methods are available using Hoyt, Nakagami–Rice distributions, and others (Ref. 2). Even with these tools at our disposal, experience has shown that fade margins and other design improvements based on fading estimates are insufficient. The best tool is long-term testing, over a full year. For most projects, such testing is not economical. It is not unknown to install a link that subsequently does not meet specifications because fade depth and fading rate are greater than predicted. Such links require upgrading, often at the expense of the contractor.

Satisfactory methods of estimating fading probability on specific paths from climatological statistics are not available at present. Statistical studies of refractivity (Section 1.3) can, however, provide information on relative gradient probabilities for different areas, and indicate the seasonal changes to be expected (Ref. 9). The low-level refractivity gradients that are of most importance to LOS radiolink propagation are very sensitive to variations in local weather conditions. Therefore the information obtained on site surveys can be very useful during this stage of link design. Operating experience on many microwave links (Ref. 2) shows that

- □ Fading is more likely on paths across flat ground than on paths over rough terrain.
- □ There is less fading on paths across dry ground than on paths across river valleys, wet or swampy terrain, or irrigated fields.
- □ Calm weather favors atmospheric stratifications that may result in deep fading. These conditions occur more often in broad, protected river valleys than over open country.
- □ Fading is likely near the center of large, slow-moving anticyclones (high-pressure areas). These are more likely to occur in summer and fall, in the northern hemisphere, than in winter and spring.
- □ Expect fading to occur more frequently and with greater severity in summer than in winter for mid-latitudes (temperate zones).
- □ Paths with takeoff angles greater than about 0.5° are less susceptible to fading, and the effect of refractivity gradients is negligible where takeoff angles exceed 1.5°.

2.7.4 Diversity as a Means to Mitigate Fading

2.7.4.1 General

The first and most economic step to achieve required fade margins is to overbuild the link by using larger aperture antennas, improved receiver noise performance (i.e., use of an LNA in front of the mixer), and/or a higher transmitter output power. Under certain circumstances, the link design engineer will find that it is uneconomic to overbuild the link further. The next step, then, is to examine diversity as another method to mitigate fading, and, at the same time, add a "diversity improvement factor," which can be translated directly to some 3 dB or more of system gain.

Diversity is based on providing separate paths to transmit redundant information. These paths may be in the domain of space, frequency, or time. In essence, the idea is that a fade occurring at time T_a on one of the redundant paths does not occur on the other. The ability to mitigate fading effects by diversity reception is a function of the correlation coefficient of the signals on the redundant paths. The higher the decorrelation, the more effective the diversity. Generally, when the correlation coefficient is ≤ 0.6, full diversity improvement can be entirely achieved (Ref. 9).

The most commonly used methods of diversity are frequency and space diversity to minimize the effects of multipath fading. (In Chapter 5 another form of "space" diversity is introduced called "spatial diversity" or path diversity.) Troposcatter and diffraction links (Chapter 3) often use a combination of frequency and space diversity, achieving still greater protection against the effects of fading.

Frequency diversity uses two different frequencies to transmit the same information. With space diversity, the same frequency is used, but two receive antennas separated vertically on the same tower receive the information over two different physical paths separated in space. These two types of diversity are shown conceptually in Figures 2.18a and 2.18b.

Compared to space diversity, frequency diversity is somewhat less expensive to implement and has some operational and maintenance advantages. Its principal drawback is regulatory. As an example, the U.S. Federal Communications Commission (FCC) (Ref. 2) rules prohibit frequency diversity for common carriers unless sufficient evidence can be shown that frequency diversity is the only way to obtain the required system reliability. This ruling was established to preserve centimetric frequencies for working radio channels owing to frequency congestion in the centimetric bands and the demand for working frequencies. Therefore, in the United States, the first alternative should be space diversity.

2.7.4.2 Frequency Diversity

Frequency diversity offers two advantages. Not only does it provide a full order of diversity and resulting diversity gain, but it also provides a fully redundant path, improving equipment reliability. (See Section 2.13.)

To achieve maximum fading decorrelation, the separation in frequency of the two transmit frequencies must be in the order of 3–5%. However, because of congestion and lack of frequency assignments in highly developed countries, separations of 2% are more common, and some systems operate satisfactorily with separations under 1%. Figure 2.19 shows approximate worst-case multipath fading (Rayleigh) and diversity improvement for frequency diversity systems with different frequency separations. The figure shows that a 14–19 dB improvement may be expected for a link with a time availability requirement of 99.99% over the same link with no diversity, assuming Rayleigh fading.

2.7.4.3 Space Diversity

Because of the difficulty of obtaining the second diversity frequency to transmit redundant information, vertical space diversity may be the easier of the two alternatives, and in some cases it may be the only diversity alternative open to the designer. In fact, experience is now showing that space diversity has considerably lower correlation coefficients, with consequently much larger diversity improvements than was earlier believed.

It should be noted that a space diversity arrangement can also provide full equipment redundancy when automatically switched hot standby transmitters are used (see Section 7.2.5). However, this arrangement does not provide a separate end-to-end operational path as does the frequency diversity arrangement.

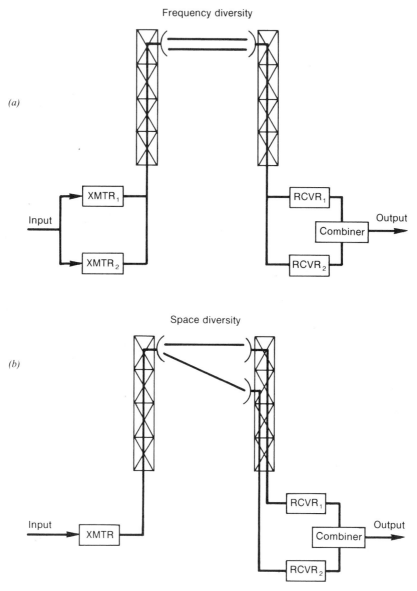

FIGURE 2.18 Simplified functional block diagrams distinguishing frequency and space diversity operation on LOS radiolink.

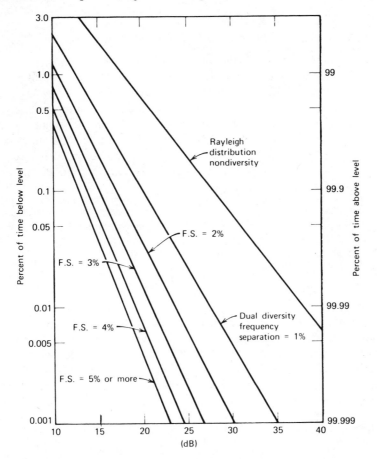

FIGURE 2.19 Frequency diversity improvement for various frequency spacings (in percent) compared to an equivalent nondiversity path. Rayleigh fading is assumed. From *Telecommunication Transmission Handbook*, 2nd ed., Copyright by John Wiley & Sons, Inc., 1981. (Ref. 5).

Of principal concern in the design of space diversity on a particular path is the amount of vertical separation. Reference 2 suggests a rule-of-thumb separation distance of 200 wavelengths or more. One wavelength at 6 GHz is 5 cm. The required separation, then, would be 5×200 cm or 10 m (about 33 ft). Reference 8 suggests 60 ft at 2 GHz, 45 ft at 4 GHz, 30 ft at 6 GHz, and 15–20 ft at 12 GHz. It will be appreciated that both antennas must meet the path profile clearance criteria, and the result will be taller towers.

Reference 8 provides a formula (modified from Vigants) to calculate the space diversity improvement factor I_{sd}, where

$$I_{sd} = \frac{7.0 \times 10^{-5} f s^2 \times 10^{\overline{F}/10}}{D} \tag{2.31}$$

where f = frequency in gigahertz

   s = vertical antenna spacing in feet between antenna centers

   D = path length in statute miles

   $\bar{F}$ = fade margin in decibels associated with the second antenna. The barred F factor is introduced to cover the situation where the fade margins are different on the upper and lower paths of the vertically spaced antennas. In such a case F will be taken as the larger of the two fade margins and will be used to calculate the unavailability (U_{nd}) in the computation for the nondiversity path. $\bar{F}$ in equation (2.31) will be taken as the smaller fade margin of the two, if different.

Reference 8 recommends that one first calculates the nondiversity path unavailability P_{mf} [from equation (2.30)], then calculate the space diversity improvement factor I_{sd}. The diversity outage (unavailability) or fade probability (U_{div}) is given by

$$U_{div} = \frac{P_{mf}}{I_{sd}} \qquad (2.32)$$

Example 15. Consider a 30-mile (48.3 km) path with average terrain that includes some roughness and where the climate is inland temperate. The operating frequency of the radiolink is 6.7 GHz and the fade margin incorporated is 40 dB. Calculate the unavailability and path availability for the nondiversity case and for the space diversity case with 40-ft vertical spacing.

Calculate P_{mf} using equation (2.30);

$$P_{mf}\,(\%) = 6.0 \times 10^{-5}(1)\left(\tfrac{1}{4}\right)(6.7)(48.3)^{3}(10)^{-40/10}$$

$$= 6.0 \times 10^{-5} \times 1.675 \times 112678.6 \times 10^{-4}$$

$$= 0.0011\% \text{ or } 0.000011$$

This corresponds to a path availability of $1 - 0.00011$ or 99.9989%. Calculate I_{sd} from equation (2.31):

$$I_{sd} = \frac{7.0 \times 10^{-5}(6.7)(40)^{2} \times 10^{4}}{30}$$

$$\approx 250$$

Substitute this value into equation (2.32):

$$U_{div} = 0.000011/250$$

$$= 0.000000044$$

The path availability for space diversity is then 1-0.000000044 = 0.999999956 or 99.9999956%.

Figure 2.20 is a useful nomogram to determine the space diversity improvement factor denoted I_{sd}. The nomogram is taken from CCIR Rep. 376-3.

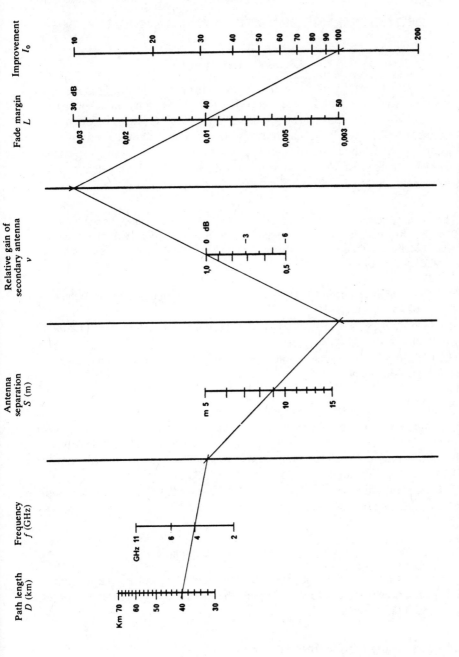

FIGURE 2.20 Nomogram to determine the space diversity improvement factor I_{sd} or the reduction in outage time due to multipath fading. This nomogram assumes perfect comparative switching. From CCIR Rep. 376-3. (Ref. 1).

71

2.8 THE ANALYSIS OF NOISE ON AN FM RADIOLINK

2.8.1 Introduction

In the design of a radiolink or system of radiolinks, a noise requirement will be specified. In the case of a system carrying FDM telephony, noise power in the derived voice channel will be specified. For a system carrying video, the requirement will probably be stated as a weighted signal-to-noise ratio.

Because C/N at the receiver input varies with time, noise power is specified statistically. Specifications are typically based on Table 2.1 and derive from the following exerpt from CCIR Rec. 393-3 (Ref. 1):

The noise power at a point of zero relative level in any telephone channel on a 2500 km hypothetical reference circuit for frequency division multiplex radio relay systems should not exceed the values given below, which have been chosen to take account of fading:

☐ 7500 pW0p, psophometrically weighted one minute mean power for more than 20% of any month;

☐ 47,500 pW0p, psophometrically weighted mean power for more than 0.1% of any month;

☐ 1,000,000 pW0, unweighted (with an integrating time of 5 ms) for more than 0.01% of any month.

If we were to connect a psophometer at the end of a 2500 km (reference) circuit made up of homogeneous radio relay sections, in a derived voice channel, we should read values no greater than these plus 2500 pW0p mean noise power due to the FDM equipment contribution, or, for example, not to exceed 10,000 pW0p total noise for more than 20% of any month, etc.

The problem addressed in this section is to ensure that the radiolink design can meet this or other similar criteria.

2.8.2 Sources of Noise in a Radiolink

Figure 2.21 shows three basic noise sources assuming voice channel input (insert) on one end of the link and voice channel output (drop) at the other end of the link. These are:

1. Load-invariant noise (thermal noise).
2. Load-dependent noise (intermodulation noise).
3. Interference noise.

The modem in the Figure refers to the FDM modulator and demodulator.

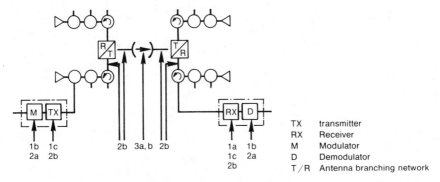

					TX	transmitter
					RX	Receiver
1b	1c		2b	3a, b	2b	1a 1b	M	Modulator
2a	2b					1c 2a	D	Demodulator
						2b	T/R	Antenna branching network

1   Load-invariant noise (basic noise):
1a   Receiver thermal noise (cause of loss-dependent noise in the signal channel)
1b   Basic noise of modem equipment
1c   Basic noise of RF equipment
2   Load-dependent noise
2a   Intermodulation noise of modem equipment
2b   Intermodulation noise of RF equipment and antenna systems
3   Noise from other RF channels of the same system
3a   Noise due to cochannel interference
3b   Noise due to adjacent-channel interference

FIGURE 2.21 Noise contributors on a radiolink. From *Planning and Engineering of Radio Relay Links*, Siemens/Heyden. Reprinted with permission. (Ref. 7).

2.8.3 FM Improvement Threshold

FM is wasteful of bandwidth when compared to AM-SSB, for instance. However, this "waste" of bandwidth is compensated for by an improvement in thermal noise power when the input signal level (RSL) reaches FM improvement threshold (i.e., when C/N ≃ 10 dB). In other words, we are giving up bandwidth for a thermal noise improvement.

If we were to draw a curve of signal-to-noise power ratio at the output of the FM demodulator versus the carrier-to-noise ratio at the input (Figure 2.22), we will note three important points of reference:

1. Thermal noise threshold (see Section 2.6.3).
2. FM improvement threshold (10 dB above thermal noise threshold P_{fm}).
3. Saturation (where compression starts to take place).

The first reference point, thermal noise threshold (P_t) is only a way-point in link calculations. The second point, FM improvement threshold, where

$$P_{fm} = P_t + 10 \text{ dB} \qquad (2.33)$$

will be used as a refernce on which link calculations will be based. In equation

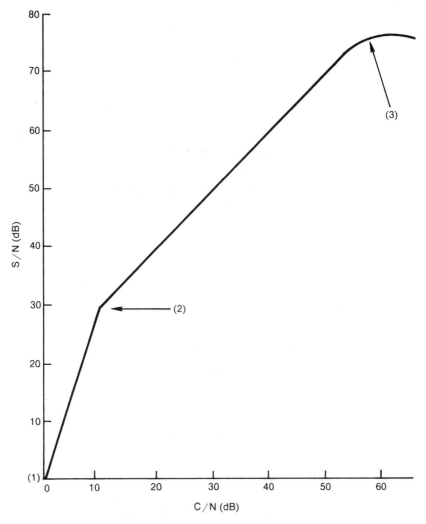

FIGURE 2.22 A typical plot of signal-to-noise power ratio at the output of an FM demodulator versus the carrier-to-noise ratio at the input of the demodulator. Point 1 is the thermal noise threshold, point 2 is the FM improvement threshold, and point 3 is saturation.

(2.33) substitute the value of P_t taken from equation (2.16) or

$$P_{fm} = -204 \text{ dBW} + 10 \log B_{if} + NF_{dB} + 10 \text{ dB} \qquad (2.34)$$

Example 16. A radiolink receiver has a noise figure of 8 dB and an IF bandwidth of 12 MHz, what is the FM improvement threshold in dBW? Use equation (2.34).

$$P_{fm}(\text{dBW}) = -204 \text{ dBW} + 10 \log(12 \times 10^6) + 8 \text{ dB} + 10 \text{ dB}$$

$$= -115.21 \text{ dBW}$$

2.8.4 Noise in a Derived Voice Channel

2.8.4.1 Introduction

On an FM radiolink, noise is examined during conditions of fading and for the unfaded or median noise condition. For the low RSL or faded condition, Ref. 8 states that "the noise in a derived (FDM) voice channel at FM threshold, falls approximately at, or slightly higher than the level considered to be the maximum tolerable noise for a telephone channel in the public network. By present standards, this maximum is considered to be 55 dBrnc0 (316,200 pWp0). In industrial systems, a value of 59 dBrnc0 (631,000 pWp0) is commonly used as the maximum acceptable noise level." FM threshold, therefore, is the usual point where fade margin is added (Section 2.7.2) to achieve the unfaded RSL. In many systems, however, the CCIR guideline of 3 pWp/km may not be achieved at the high-signal-level condition, and the unfaded RSL may have to be increased still further.

At low RSL, the primary contributor to noise is thermal noise. At high-signal-level conditions (unfaded RSL), there are three contributors:

☐ Thermal noise (discussed below).
☐ Radio equipment IM noise (Section 2.9.5.4.2).
☐ IM noise due to antenna feeder distortion (Section 2.9.6).

Each are calculated for a particular link for unfaded RSL and each value is converted to noise power in mW, pW, or pWp and summed. The sum is then compared to the noise apportionment for the link from Table 2.1 and Section 2.1.

2.8.4.2 Calculation of Thermal Noise

The conventional formula (Ref. 12) to calculate signal-to-noise power ratio on FM radiolink (test tone to flat weighted noise ratio for thermal noise) is

$$\frac{S}{N} = \frac{RSL}{(2ktbF)[\Delta f//f_c]^2} \qquad (2.35)$$

where b = channel bandwidth (3.1 kHz)
Δf = channel test tone peak deviation (as adjusted by pre-emphasis)
f_c = channel (center) frequency in baseband in kilohertz
F = the noise factor of the receiver (i.e., the numeric equivalent of the noise figure in decibels)
$kt = 4 \times 10^{-18}$ mW/Hz
RSL = receive signal level in milliwatts
S/N = test tone-to-noise ratio (numeric equivalent of S/N in decibels)

In the more useful decibel form, equation (2.35) is

$$\left(\frac{S}{N}\right)_{dB} = RSL_{dBm} + 136.1 - NF_{dB} + 20 \log\left(\frac{\Delta f}{f_c}\right) \quad \text{(flat)} \quad (2.36)$$

In a similar fashion (Ref. 8) the noise power in the derived voice channel can be calculated as follows:

$$P_{dBrnc0} = -RSL_{dBm} - 48.1 + NF_{dB} - 20 \log(\Delta f//f_c) \quad (2.37)$$

and

$$P_{pWp0} = \log_{10}^{-1}\left[\frac{-RSL_{dBm} - 48.6 + NF_{dB} - 20 \log(\Delta f//f_c)}{10}\right] \quad (2.38)$$

Note: Values for Δf, the *channel* test tone *peak* deviation (Table 2.5): 200 kHz rms deviation for 60 through 960 VF channel loading and 140 kHz rms deviation for transmitters loaded with more than 1200 voice channels. Per channel peak deviation is 282.8 and 200 kHz, respectively.

Example 17. A 50-hop is to be designed to CCIR noise recommendations based on Table 2.1. What level RSL is required for 47,500 pWp0? (It should be noted that this level must be met 99.99% of the time.) The link is designed for 300 FDM channels and the highest channel center frequency is 1248 kHz. The noise figure of the receiver is 10 dB.

Use equation (2.37) and set 47,500 pWp0 equal to the value of the right-hand side of that equation.

$$47,500 \text{ pWp0} = \log_{10}^{-1}\left[\frac{-RSL - 48.6 + 10 \text{ dB} - 20 \log(\Delta f//f_c)}{10}\right]$$

Calculate the value of

$$20 \log(\Delta f//f_c) = 20 \log(282.8/1248)$$

$$= -12.9 \text{ dB}$$

then

$$10 \log(47,500) = -RSL - 48.6 + 10 + 12.9$$

$$RSL = -72.47 \text{ dBm}$$

Table 2.10 presents equivalent values of dBrnc, pWp, and signal-to-noise power ratio for the standard voice channel.

2.8.4.3 Notes on Noise in the Voice Channel

Flat noise must be distinguished from weighted noise. Theoretically, flat noise power is evenly distributed across a band of interest such as the voice channel. As a theoretical example, suppose the noise power in a voice channel at 1000 Hz was measured at −31 dBm. We would then expect to measure this value at all points in the channel. In practice, this is impossible because of band-limiting characteristics of the medium such as FDM channel bank filters.

A weighted channel does not display a uniform response. A weighted channel displays certain special frequency and transient response characteristics. Two such "weightings" reflect response of the typical human ear. These are C-message and psophometric weightings.

In North America, C-message weighting is used, and its reference frequency is 1000 Hz. In Europe and many other parts of the world psophometric weighting is used where the reference frequency is 800 Hz. The characteristics of each weighting type are slightly different as shown in the response curves in Figure 2.23 Weighting networks are used in test equipment to simulate these weighting characteristics, either C-message or psophometric.

As was previously pointed out, the practical voice channel response is not flat, but has rolloff characteristics that are a function of the media's bandpass

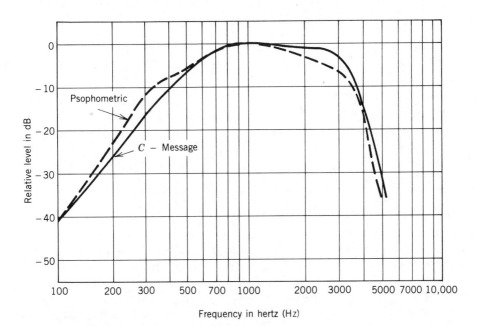

FIGURE 2.23 Comparison between psophometric and C-message weighting. From *Reference Manual for Telecommunication Engineering*, John Wiley & Sons, Inc., 1985. Reprinted with permission (Ref. 22).

characteristics. To simulate this quasiflat response for the voice channel, a 3-kHz flat network is used to measure power density (under "flat" conditions) of Gaussian noise. The network has a nominal low-pass response, which is down 3 dB at 3 kHz and rolls off at 12 dB per octave. (Ref. 15).

The noise measurement unit used nearly universally in North America is the dBrnc0. The "c" implies C-message weighting and the "0" means that the measurement is referenced to the 0 test level point (0 TLP).

In Europe and in much of the CCITT/CCIR documentation, the weighted noise measurement unit is the pWp, picowatts psophometrically weighted (pWp0 or pW0p when referenced to the 0 TLP). Alternative measurement units are dBp and dBmp. The final "p" implies psophometric weighting.

Some guidelines follow:

☐ A voice channel with psophometric weighting has a 2.5 dB noise improvement over a flat channel.

☐ dBrnc (dB above reference noise with C-message weighting): The reference frequency/level is a 1000 Hz tone at -90 dBm (Ref. 13).

☐ pWp (picowatts of noise power with psophometric weighting): The noise power reference frequency/level is an 800 Hz tone where 1 pWp = -90 dBm (pWp = pW × 0.56) (Ref. 13).

☐ dBmp (psophometrically weighted noise power measured in dBm), where, with an 800 Hz tone 0 dBmp = 0 dBm. For flat noise in the band 300–3400 Hz, dBmp = dBm $-$ 2.5 dB (Ref. 13).

☐ 0 dBrnc = -88.5 dBm. Commonly, the -88.5 dBm value is rounded off to -88 dBm, thus 0 dBrnc = -88.0 dBm (Ref. 13).

☐ dBrnc = dBmp + 90 dB: It should be noted that C-message weighting and psophometric weighting in fact vary by 1 dB (-1.5 dB and -2.5 dB) and this equivalency has an inherent error of 1 dB. However, the equivalency is commonly accepted in the industry (Ref. 13).

☐ dBrnc0 = $10 \log$ pWp0 + 0.8 dB = dBmp + 90.8 dB = 88.3 (dB) $-$ $(S/N)_{\text{dB (flat)}}$ (Ref. 8).

2.8.5 Noise Power Ratio (NPR)

2.8.5.1 Introduction

Up to this point we have only dealt with thermal noise in a radiolink. In an operational analog radiolink a second type of noise can be equally important. This is intermodulation noise (IM noise).

IM noise is caused by nonlinearity when information signals in one or more channels give rise to harmonics or intermodulation products that appear as unintelligible noise in other channels. In an FDM/FM radiolink, nonlinear noise in a particular channel varies as the multiplex signal level and the position of the channel in the multiplex baseband spectrum. For a fixed multiplex signal level and for a specific FDM channel, nonlinear noise is constant.

For low receive signal levels at the far end FM receiver, such as during conditions of deep fades, thermal noise limits performance. During the converse condition, when there are high signal levels, IM noise may become the limiting performance factor.

In an FM radio system nonlinear (IM) noise may be attributed to three principal factors: (1) transmitter nonlinearity, (2) multipath effects of the medium, and (3) receiver nonlinearity. Amplitude and phase nonlinearity are equally important in contribution to total noise and each should be carefully considered.

2.8.5.2 Methods of Measurements

There are three ways to measure IM noise:

1. One-tone test where harmonic distortion is measured.
2. Two-tone test where specific IM products can be identified with a spectrum analyzer and levels accurately measured with a frequency selective voltmeter.
3. White noise loading test, where, among other parameters, NPR can be measured.

Method 3 will be subsequently described and its application discussed.

2.8.5.3 White Noise Method of Measuring IM Noise in FDM/FM Radiolinks

A common method of measuring total noise on an FM radiolink under maximum (traffic) loading conditions consists of applying a "white noise" signal at the baseband input port of the FM transmitter. A white noise generator produces a noise spectrum approximating that produced by the FDM equipment. The output noise level of the generator is adjusted to a desired multiplex composite baseband level (composite noise power). Then a notched filter is switched in to clear a narrow slot in the spectrum of the noise signal and a noise analyzer is connected to the output of the system. The analyzer can be used to measure the ratio of the composite noise power to the noise power in the cleared slot. The noise power is equivalent to the total noise (e.g., thermal plus IM noise) present in the slot bandwidth. Conventionally, the slot bandwidth is made equal to that of a single FDM voice channel. A typical white noise test setup is shown in Figure 2.24.

The most common unit of noise measurement in white noise testing is NPR, which is defined as follows (Ref. 13):

> NPR is the decibel ratio of the noise level in a measuring channel with the baseband fully loaded ... to ... the level in that channel with all of the baseband noise loaded except the measuring channel.

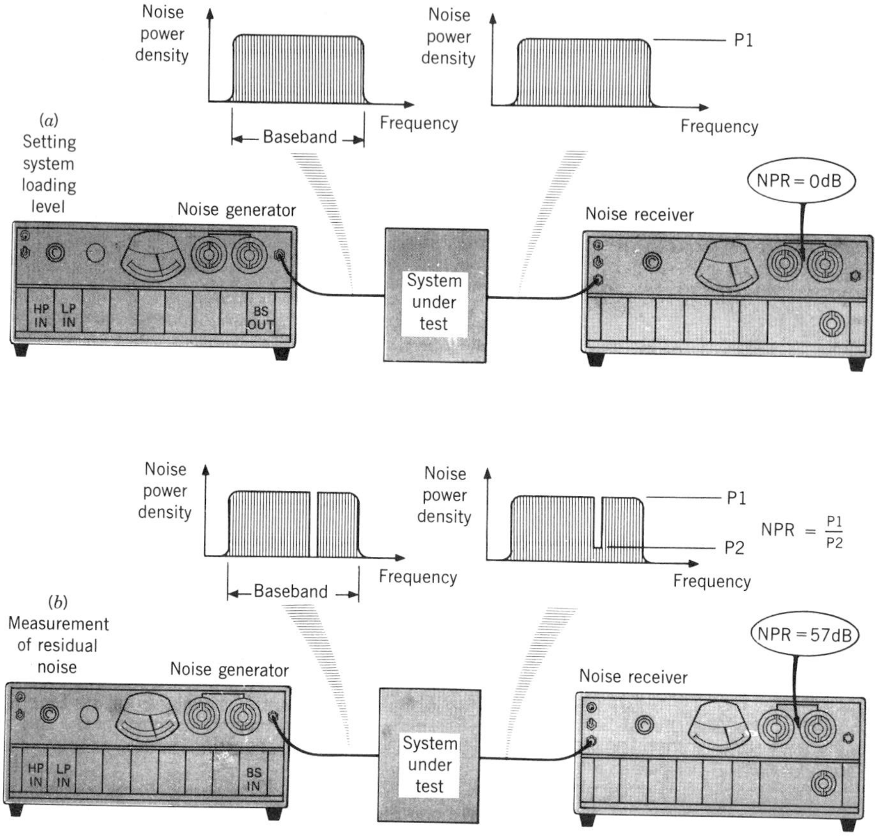

FIGURE 2.24 White noise testing. Courtesy of Marconi Instruments, St. Albans, Hertfordshire, England (Ref. 13).

The notched (slot) filters used in white noise testing have been standardized for radiolinks by CCIR in Rec. 399-3 (Ref. 1) for 10 common FDM baseband configurations. Available measuring channel frequencies and high and low baseband cutoff frequencies are shown in Table 2.8. Table 2.9 gives the stop band (slot) filter rolloff characteristics. In an NPR test usually three different slots are tested separately: high frequency, mid-band, and low frequency.

When an NPR measurement is made at high RF signal levels, such as when the measurement is made in a back-to-back configuration, the dominant noise component is equipment IM noise. This parameter can be used as an approximation of the equipment IM noise contribution. This value together with stated equivalent noise loading should also be available from manufacturer's published specifications on the equipment to be used. Modern, new radiolink terminal equipment (i.e., that equipment that accepts an information baseband for modulation and demodulates an RF signal to baseband) should display an

TABLE 2.8. CCIR Measurement Frequencies for White Noise Testing

System Capacity (Channels)	Limits of Band Occupied by Telephone Channels (kHz)	Effective Cut-off Frequencies of Band-Limiting Filters (kHz) High pass	Low Pass	Frequencies of Available Measuring Channels (kHz)
60	60–300	60 ± 1	300 ± 2	70 270
120	60–552	60 ± 1	552 ± 4	70 270 534
300	{ 60–1300 / 64–1296 }	60 ± 1	1296 ± 8	70 270 534 1248
600	{ 60–2540 / 54–2660 }	60 ± 1	2600 ± 20	70 270 534 1248 2438
960	{ 60–4028 / 64–4024 }	60 ± 1	4100 ± 30	70 270 534 1248 2438 3886
900	316–4188	316 ± 5	4100 ± 30	534 1248 2438 3886
1260	{ 60–5636 / 60–5564 }	60 ± 1	5600 ± 50	70 270 534 1248 2438 3886 5340
1200	316–5564	316 ± 5	5600 ± 50	534 1248 2438 3886 5340
1800	{ 312–8120 / 312–8204 / 316–8204 }	316 ± 5	8160 ± 75	534 7600 1248 2438 3886 5340
2700	{ 312–12,336 / 316–12,388 / 312–12,388 }	316 ± 5	12,360 ± 100	534 7600 1248 2438 11 700 3886 5340

Source. CCIR Rec. 399-3 (Ref. 1).

TABLE 2.9. CCIR Stop-Band Filter Rolloff Characteristics

Center Frequency f_c (kHz)	Bandwidth (kHz), in relation to f_c, over which the Discrimination Should be at Least 70 dB	55 dB	30 dB	3 dB	Bandwidth (kHz), in Relation to f_c, outside which the Discrimination Should not Exceed 3 dB	0.5 dB
70	{ ±1.5	±2.2 / ±1.7	±3.5 / ±2.0	—	± 12 / ± 5	± 18 / ± 10
270	±1.5	±2.3	±2.9	—	± 8	± 24
534	±1.5	±3.5	±7.0	—	± 15	± 48
1248	±1.5	±4.0	±11.0	—	± 35	±110
2438	±1.5	±4.5	±19	—	± 60	±220
3886	{ ±1.5 / ±1.5	±15.0 / ±1.8	±30.0 / ±3.5	— / ±8.0	±110 / ± 12	±350 / ±100
5340	±1.5	±2.2	±4.0	±8.5	± 14	±150
7600	±1.5	±2.4	±4.6	±9.5	± 16	±200
11,700	±1.5	±3.0	±7.0	±11.0	± 20	±300

Source. CCIR Rec. 399-3 (Ref. 1).

NPR of at least 55 dB when tested back-to-back. It should be noted, however, that diversity combining can improve NPR. This is because in equal gain and maximal ratio combiners the signal powers are added coherently, whereas the IM noise contribution, which is similar (for this discussion) to other noise, is added randomly. Reference 2 allows a 3 dB improvement in NPR when diversity combining is used on a link.

Up to this point NPR has been treated for terminal radio equipment or baseband repeaters. If the designer is concerned with heterodyne repeaters (Section 7.2.3.), the white noise test procedure as previously described cannot be carried out per se, and the designer should rely on manufacturer's specifications. Alternatively, about 4 dB improvement (Ref. 2) in NPR over baseband radio equipment may be assumed, or for a new IF repeater of modern design, an NPR of at least 59 dB should be achieved.

2.8.5.4 Application

2.8.5.4.1 THERMAL NOISE MEASUREMENT

A distinction has been made between thermal (idle) noise, which is not a function of traffic loading level and IM noise, which is caused by system nonlinearity. It can be seen that in Figure 2.25, with the removal of the white noise loading, the remaining noise being recorded in the noise receiver is thermal noise (assuming that there are no spurious signals being generated).

Thermal or idle noise in a test channel is defined by BINR (baseband intrinsic noise ratio), which is the decibel ratio of noise in a test channel with the baseband fully loaded and stop band or slot filters disconnected to the noise in the test channel with all noise loading removed.

The difference between NPR and BINR, therefore, indicates the amount of noise present in the system due to IM noise and crosstalk.

2.8.5.4.2 DERIVING SIGNAL-TO-NOISE RATIO AND IM NOISE POWER

Specifying the noise in a test channel by NPR provides a relative indication of IM noise and crosstalk. An alternative is to express the noise in decibels relative to a specified absolute signal level in a test channel. In this case we can define the signal-to-noise power ratio (S/N) as the decibel ratio of the level of the standard test tone to the noise in a standard channel bandwidth (3100 Hz) within the test channel or

$$\left(\frac{S}{N}\right)_{dB} = NPR + BWR - NLR \qquad (2.39)$$

where NLR (Noise Load Ratio) is defined in formulas (2.20)–(2.22) or

$$\text{NLR} = 10\log\left(\frac{\text{equivalent noise test load power}}{\text{voice channel test tone power}}\right) \qquad (2.40)$$

$$\text{BWR} = 10\log\left(\frac{\text{occupied baseband of white noise test signal}}{\text{voice channel bandwidth}}\right) \qquad (2.41)$$

where BWR is Bandwidth Ratio.

Example 18. A particular FM radiolink transmitter and receiver back-to-back display an NPR of 55 dB. They have been designed and adjusted for 960 VF channel operation and will use CCIR loading. What is the S/N in a voice channel?

Consult Table 2.8. The baseband occupies 60–4028 kHz.

$$\text{BWR} = 10\log[(4028 - 60)/3.1] \quad (\text{kHz})$$

$$= 10\log[3968/3.1]$$

$$= 31.07 \text{ dB}$$

$$\text{NLR} = -15 + 10\log(960)$$

$$= 14.82 \text{ dB}$$

$$\text{S/N} = 55 \text{ dB} + 31.07 \text{ dB} - 14.82 \text{ dB}$$

$$= 71.25 \text{ dB}$$

To calculate flat noise in the test channel, the following expression applies:

$$P_{\text{tcf}} = \log^{-1}\left(\frac{90 - (\text{S/N})_{\text{dB}}}{10}\right) \quad (\text{pW0}) \qquad (2.42)$$

and to calculate psophometrically weighted noise:

$$P_{\text{tcp}} = 0.56\log^{-1}\left(\frac{90 - (\text{S/N})_{\text{dB}}}{10}\right) \quad (\text{pWp0}) \qquad (2.43)$$

Example 19. If S/N in voice channel is 71.25 dB, what is the noise level in that channel in pWp0?

Use equation (2.43).

$$P_{tcp} = 0.56 \log^{-1}\left(\frac{90 - 71.25}{10}\right)$$

$$= 0.56 \times 74.99$$

$$= 41.99 \text{ pWp0}$$

The results of the calculations in this subsection provide the input for the radio equipment IM noise contribution for total noise in the voice channel under unfaded conditions.

Table 2.10 gives approximate equivalents for S/N in the standard voice channel and its respective values of noise power in dBrnc0 and pWp0.

2.8.5.5 Balance between IM Noise and Thermal Noise

After a path analysis is completed, the designer may find that the specified worst channel noise requirements have not been met. One alternative would be to move the reference operating point as described in Section 2.8.3. Another alternative would be to adjust the frequency deviation to ensure its optimum operating point.

We know that as the input level to an FM transmitter is increased, the deviation is increased. As the deviation is increased, the FM improvement threshold becomes more apparent. In effect, we are trading off bandwidth for thermal noise improvement. However, with increasing input levels, the IM noise of the system increases. There is some point of optimum input to a wideband FM transmitter where the thermal noise in the system and the IM noise level have been balanced so the total noise has been optimized. This concept is illustrated in Figure 2.25.

2.8.5.6 Standardized FM Transmitter Loading for FDM Telephony

Several standards have evolved, mostly based on the work done by Holbrook and Dixon of Bell Telephone Laboratories, for wideband FM transmitter loading of FDM baseband signals. These standard formulas assume that the FDM voice channels will carry telephony speech traffic with a 25% activity factor. From the formula(s) we derive the total system loading in dBm0, which would be used for white noise testing and for link analysis. The formulas are

$$P_{tl} = -15 + 10 \log N \quad \text{(CCIR)} \tag{2.44}$$

$$P_{tl} = -16 + 10 \log N \quad \text{(North American practice)} \tag{2.45}$$

where P_{tl} is the test load power level in dBm0 and N is the number of FDM

TABLE 2.10. Approximate Equivalents for S/N and Common Noise Level Values[a]

dBrnc0	pWp0	dBm0p	S/N (flat)	NPR	dBrnc0	pWp0	dBm0p	S/N (flat)	NPR
0	1.0	−90	88	71.6	30	1,000	−60	58	41.6
1	1.3	−89	87	70.6	31	1,259	−59	57	40.6
2	1.6	−88	86	69.6	32	1,585	−58	56	39.6
3	2.0	−87	85	68.6	33	1,995	−57	55	38.6
4	2.5	−86	84	67.6	34	2,520	−56	54	37.6
5	3.2	−85	83	66.6	35	3,162	−55	53	36.6
6	4.0	−84	82	65.6	36	3,981	−54	52	35.6
7	5.0	−83	81	64.6	37	5,012	−53	51	34.6
8	6.3	−82	80	63.6	38	6,310	−52	50	33.6
9	7.9	−81	79	62.6	39	7,943	−51	49	32.6
10	10.0	−80	78	61.6	40	10,000	−50	48	31.6
11	12.6	−79	77	60.6	41	12,590	−49	47	30.6
12	15.8	−78	76	59.6	42	15,850	−48	46	29.6
13	20.0	−77	75	58.6	43	19,950	−47	45	28.6
14	25.2	−76	74	57.6	44	25,200	−46	44	27.6
15	31.6	−75	73	56.6	45	31,620	−45	43	26.6
16	39.8	−74	72	55.6	46	39,810	−44	42	25.6
17	50.1	−73	71	54.6	47	50,120	−43	41	24.6
18	63.1	−72	70	53.6	48	63,100	−42	40	23.6
19	79.4	−71	69	52.6	49	79,430	−41	39	22.6
20	100	−70	68	51.6	50	100,000	−40	38	21.6
21	126	−69	67	50.6	51	125,900	−39	37	20.6
22	158	−68	66	49.6	52	158,500	−38	36	19.6
23	200	−67	65	48.6	53	199,500	−37	35	18.6
24	252	−66	54	47.6	54	252,000	−36	34	17.6
25	316	−65	63	46.6	55	316,200	−35	33	16.6
26	398	−64	62	45.6	56	398,100	−34	32	15.6
27	501	−63	61	44.6	57	501,200	−33	31	14.6
28	631	−62	60	43.6	58	631,000	−32	30	13.6
29	794	−61	59	42.6	59	794,300	−31	29	12.6

[a] This table is based on the following commonly used relationships, which include some rounding off for convenience.

Correlations between columns 3 and 4 are valid for all types of noise. All other correlations are valid for white noise only. $\text{dBrnc0} = 10 \log_{10} \text{pWp0} = \text{dBm0p} + 90 = 88 - \text{S/N (flat)} = 7.16 - \text{NPR}$.

Source. Extracted from EIA RS-252-A (Ref. 12).

channels. Equations (2.44) and (2.45) are valid only when N is equal to or greater than 240 channels.

For systems with 12–240 FDM channels, the following formula applies:

$$P_{\text{tl}} = -1 + 4 \log N \quad \text{(CCIR)} \qquad (2.46)$$

It should be stressed that the loading level values derived are for systems

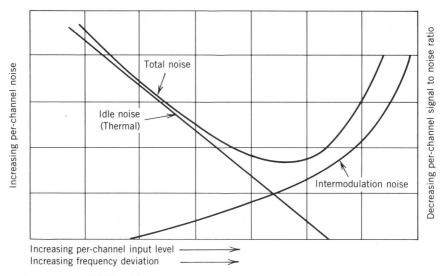

FIGURE 2.25 Noise performance versus frequency deviation.

carrying speech telephony with the 25% activity factor. There is allowance for pilot tones and a small number of data channels. These data channels would have a test tone level of -13 dBm0, constant amplitude, and 100% activity factor.

The U.S. Department of Defense (MIL-STD-188-100, Ref. 23) (also see Ref. 13) uses a loading formula that permits unrestricted data loading:

$$P_{tl} = -10 + 10 \log N \quad \text{(dBm0)} \tag{2.47}$$

If a system is to be designed for CCIR or North American standard loading and will carry a significant number of data/telegraph channels, it is good practice to calculate each category separately and sum the power levels. Assume -13 dBm0 loading for each data and composite telegraph channel. Now suppose a system were to carry 900 FDM VF channels of which 800 are dedicated to conventional telephony and 100 for data/telegraph, then for the telephony portion

$$P_{tl} = -15 + 10 \log 800 \quad \text{(dBm0)}$$

$$= +14 \text{ dBm0 or 25 mW}$$

and for the data/telegraph portion

$$P_{tl} = -13 + 10 \log 100 \quad \text{(dBm0)}$$

$$= +7 \text{ dBm0 or 5 mW}$$

Add the milliwatt values (i.e., $25 + 5 = 30$), then

$$P_{tl} = 10 \log 30$$

$$= +14.77 \text{ dBm0 composite loading.}$$

2.8.5.7 Standardized FM Transmitter Loading for TV

The FM transmitter is loaded with an input power level necessary to achieve a peak deviation of ± 4 MHz (8 MHz total) referred to the nominal peak-to-peak amplitude of the video-frequency signal. (Refer to CCIR Rec. 276-2.)

2.8.6 Antenna Feeder Distortion

Antenna feeder distortion or echo distortion is caused by mismatches in the transmission line connecting the radio equipment to the antenna. These mismatches cause echos or reflections of the incident wave. Similar distortion can be caused by long IF runs, however, in most cases, this can be neglected.

Echo distortion actually results from a second signal arriving at the receiver but delayed in time by some given amount. It should be noted that multipath propagation may also cause the same effect. In this case, though, the delay time is random and continuously varying, thus making analysis difficult, if not impossible.

The level of the echo signal is an inverse function of the return loss at each end of the transmission line and its terminating device (i.e., the antenna at one end and the communication equipment at the other). An echo signal so generated will be constant, since the variables that established it are constant. Thus the distortion created by the echo will be constant but contingent on modulation. In other words, if the carrier were unmodulated, there would be no distortion due to echo. When the carrier is modulated, echo appears. Figure 2.26 illustrates echo paths on a transmission line.

Estimation of echo distortion, expressed as signal-to-distortion ratio, is at best an approximation. One of the most straightforward methods, as described in Ref. 9, is presented below. Inputs required for its calculation are the type and length of transmission line, usually waveguide for the applications covered in this text, the return losses (or VSWR) of the communications equipment, and the antenna return loss (or VSWR). These values should be provided in the equipment specifications, but typically are

Antenna VSWR: $1.05 : 1$ to $1.2 : 1$

Waveguide VSWR: $1.03 : 1$ to $1.15 : 1$

Equipment return loss: 26–32 dB

Calculations for echo distortion are performed separately for each end of the link. Procedures for its calculation are as follows.

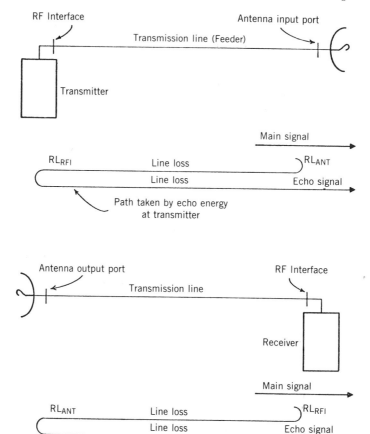

FIGURE 2.26 Illustration of principal signal echo paths.

From Figure 2.27 using the value for the corresponding length of waveguide run in feet and FDM channel configuration, determine the echo distortion gross contribution to noise. The value must now be reduced by two times the waveguide losses. (The echo signal travels up to the antenna and back down again as in Figure 2.27.) This noise value (in dBrnc0 or dBmp0) is further reduced by the equipment return loss and the composite antenna waveguide return loss.

To calculate return loss (dB) from VSWR, proceed as follows:

$$RL_{dB} = 20 \log\left(\frac{1}{\rho}\right) \qquad (2.48)$$

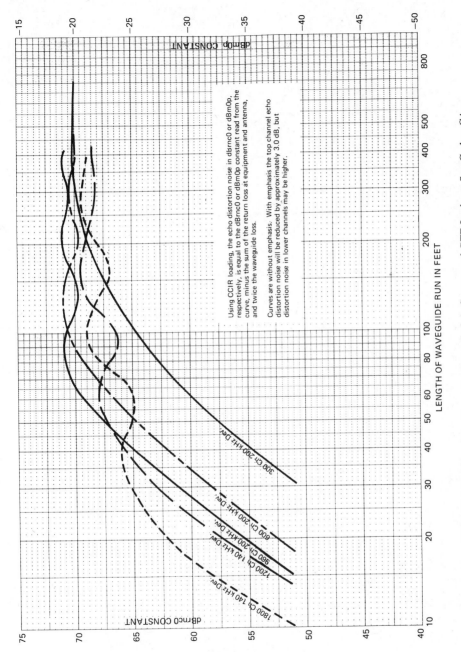

Using CCIR loading, the echo distortion noise in dBrnc0 or dBm0p, respectively, is equal to the dBrnc0 or dBm0p constant read from the curve, minus the sum of the return loss at equipment and antenna, and twice the waveguide loss.

Curves are without emphasis. With emphasis the top channel echo distortion noise will be reduced by approximately 3.0 dB, but distortion noise in lower channels may be higher.

FIGURE 2.27 Echo distortion noise. From Ref. 8. Courtesy of GTE-Lenkurt, San Carlos, CA.

89

where ρ is the reflection coefficient, and

$$\rho = \frac{\text{VSWR} - 1}{\text{VSWR} + 1} \qquad (2.49)$$

Example 20. The signal-to-distortion ratio (S/D) expressed in dBrnc0 for a 1200-channel system operating at 6 GHz with 60 ft of rectangular waveguide is 66 dBrnc0; waveguide loss is 1.05 dB; equipment return loss is 30 dB; and the waveguide/antenna composite return loss is 26 dB (equivalent to a VSWR of 1.1 : 1). Calculate the noise contribution in dBrnc0 for the worst VF channel (top channel) for this end of the link.

P_e is the value for the echo distortion noise contribution:

$$P_e = 66 - 2.1 \text{ dB} - 30 \text{ dB} - 26 \text{ dB}$$

$$= 7.9 \text{ dBrnc0}$$

Calculate the value in dBm0p. The value from Figure 2.28 is -24.5 dBm0p:

$$P_e = -24.5 - 2.1 \text{ dB} - 30 \text{ dB} - 26 \text{ dB}$$

$$= -82.6 \text{ dBm0p}$$

Calculate the equivalent value in pWp:

$$P_e = \log^{-1}\left(\frac{-82.6}{10}\right) \times 10^9$$

$$= 5.62 \text{ pWp}$$

If exactly the same parameters were valid for the other end of the link, we could double the value above (i.e., 5.62×2 pWp) and 11.24 pWp would then be added arithmetically to the value calculated for the worst VF channel noise in Section 2.8.4 or 2.8.5.

2.8.7 Total Noise in the Voice Channel

To calculate the total noise in the voice channel, convert the thermal noise at the unfaded RSL (Section 2.8.4) and intermodulation noise (Sections 2.8.5 and 2.8.6) to absolute values of noise power such as mW, pW, or pWp and add, reconverting the sum to decibels if desired.

2.8.8 Signal-to-Noise Ratio for TV Video

Radiolinks carrying exclusively video or video and a program channel are generally thermal noise limited. Here we are dealing with a broadband S/N, which is defined as the ratio of the peak-to-peak signal to the rms thermal noise in the video baseband. The value for S/N is a function of the RSL, the receiver noise figure, the video bandwidth and peak deviation, emphasis (when used), and the weighting function.

The following relations give a value of S/N in decibels. In all cases, assume a peak deviation of ± 4 MHz. For North American systems, the video baseband width is 4.3 MHz and for other systems the baseband bandwidth varies from 4 to 10 MHz.

For North American video systems, the following relations may be used (Ref. 8):

$$\frac{S}{N} = RSL - NF_{dB} + 118 \quad \text{(unweighted, unemphasized)} \tag{2.50}$$

$$\frac{S}{N} = RSL - NF_{dB} + 126.5 \quad \text{(EIA emphasis, EIA color weighting)} \tag{2.51}$$

The following was taken from Ref. 8, which states that these relations are valid for monochrome TV transmission only and cover television systems likely to be found in countries outside of North America:

$$\frac{S}{N} = RSL - NF_{dB} + A \tag{2.52}$$

where A is made up of two terms (constants). The first constant represents the unemphasized, unweighted S/N value. The second constant term is the combined effect of emphasis and weighting:

$$A = 115.7 + 17.3 \quad \text{(525 lines, 4 MHz—Japan)} \tag{2.52a}$$

$$A = 112.8 + 16.2 \quad \text{(625 lines, 5 MHz)} \tag{2.52b}$$

$$A = 110.5 + 18.1 \quad \text{(625 lines, 6 MHz)} \tag{2.52c}$$

$$A = 112.8 + 13.5 \quad \text{(819 lines, 5 MHz)} \tag{2.52d}$$

$$A = 103.8 + 16.1 \quad \text{(819 lines, 10 MHz)} \tag{2.52e}$$

RSL is the receive signal level in dBm and NF is the receiver noise figure in dB.

2.9 PATH ANALYSIS WORKSHEET AND EXAMPLE

2.9.1 Introduction

A path analysis worksheet is a useful tool to use to carry out a path analysis or "link budget." The bottom line, so to speak, of the worksheet for an analog radiolink is the noise in the worst FDM channel, usually the highest or top voice channel in the baseband.

The worksheet sets out in tabular form the required calculations starting with transmitter power output, the various losses and gains from the transmitter outward through the medium, receive antenna system to the input port of the far end receiver, thence the receiver characteristics, pre/de-emphasis and diversity improvements, and then carries on down to the noise in the worst channel. An equivalent worksheet could be used for TV transmission to derive weighted S/N. These values are then compared to the specified values (Section 2.2.1) and the link parameters are then reviewed and, if necessary, adjusted accordingly.

2.9.2 Sample Worksheet

A sample worksheet is provided in Table 2.11. It is filled in for a hypothetical path where it is assumed that a path profile has been completed beforehand and site surveys have been carried out. The far right-hand column of the table gives reference section numbers in the text and/or numbered notes. These explanatory notes follow the table.

Specific equipment implementation for such a configuration is described in Chapter 7.

2.10 FREQUENCY ASSIGNMENT, COMPATIBILITY AND FREQUENCY PLAN

2.10.1 Introduction

A major task in the planning and implementation of a radiolink involves radio frequency assignment, granting of a radio license, EMI (electromagnetic interference) studies for compatibility, and development of a frequency plan. Frequency assignments and licenses to operate radio transmitters are granted by a national regulatory authority such as the Federal Communications Commission in the United States.

Nearly all countries in the world are members of the International Telecommunications Union (ITU) and thus are signatories to the "Radio Regulations" issued by the ITU. The assignment of frequency bands is governed by the Radio Regulations. Because many radiolinks are transnational or are integral parts of extensions of an international network, national regulatory authorities take guidance from CCIR Recommendations.

Table 2.12 from CCIR OP 14-3, Annex I, gives basic guidelines on available frequency bands up to 13 GHz and cross-references applicable CCIR Recommendations.

2.10.2 Frequency Planning—Channel Arrangement

A radiolink system may be a single-thread, low-capacity system or a multiple-thread, high-capacity system with spurs, or, initially, a single-thread system

TABLE 2.11. Path Analysis Worksheet

System Identifier *Amber*   Link *Charlie* to *Delta*

Signal type (TV, FDM, composite) *FDM*   FDM chan. *1200*

Loading *+15.79* (dBm0) Baseband config. *60-5564*   (Table 2-8)

Frequency *6100* MHz   (2.11)

Spec worst chan median noise − S/N *400* (dBrnc0/pWp0)   (2.2)

1. SITE (A) *Charlie* TO SITE (B) *Delta*
2. Lat/Long (A) *43° 43′ N/90° 50′ W* Lat/Long (B) *44° 11′ N/90° 20′ W*
3. Path length *41.29* (km/sm)   (2.3)
4. Site elevation (A) *185* (m/ft) (B) *250* (m/ft)   (2.3)
5. Tower Height (A) *110* (m/ft) (B) *95* (m/ft)   (2.3)
6. Azimuth from true north (A) *37° 21′* (B) *217° 42′*   (2.3)
7. Transmitter power output at flange (A) *10* W; *0* dBW   (2.6.2)
8. Transmission line losses (A):

8A.	W/G type, *EW-64* W/G length *110* W/G loss	*1.67*	dB	(Sect 7)
8B.	Flex guide loss	*0.1*	dB	(Sect 7)
8C.	Transition/connector losses	*0.4*	dB	(Sect 7)
8D.	Directional cplr loss	*0.2*	dB	(Sect 7)
8E.	Circulator or hybrid losses	*0.5*	dB	(Sect 7)
8F.	Radome loss	*0.6*	dB	(Sect 7)
8G.	Other losses	*0.0*	dB	(Sect 7)
8H.	Total trans. line losses	*3.47*	dB	(2.6.2)

9. Antenna (A) diameter *8* (m/ft) gain   *41.25* dB   (2.6.7)
10. EIRP   *+37.38* dBW   (2.6.2)
11. Free space loss (FSL)   *144.6* dB   (1.2; 2.6.2)
12. Atmospheric absorption   *0.7* dB   (2.62)
13. Unfaded isotropic rec. level (B)   *−107.52* dBW   (2.6.2)
14. Antenna (B) diameter *8* (m/ft) gain   *41.25* dB   (2.6.7)
15. Transmission line losses (B):

15A.	W/G type *EW-64* W/G length *95′* W/G loss	*1.45*	dB	(Sect 7)
15B.	Flexguide loss	*0.13*	dB	(Sect 7)
15C.	Transition/connector losses	*0.4*	dB	(Sect 7)
15D.	Directional coupler loss	*0.0*	dB	(Sect 7)
15E.	Circulator or hybrid losses	*0.5*	dB	(Sect 7)
15F.	Radome loss	*0.6*	dB	(Sect 7)
15G.	Other losses	*0.0*	dB	(Sect 7)
15H.	Total transmission line losses	*3.08*	dB	(2.6.2)

16. Unfaded RSL   *−69.35* dBW   (2.6.2)
17. Receiver noise threshold calculation (B):   (2.6.3)

17A.	Receiver noise figure	*8*	dB	(2.6.3)
17B.	RMS per channel deviation	*140*	kHz	(2.6.4) (Table 2.5)
17C.	Peak channel deviation	*200*	kHz	(2.9.4)
17D.	Carrier peak deviation	*3855*	MHz	(2.6.4.2)
17E.	Highest baseband frequency	*5.564*	MHz	(Table 2.86)
17F.	B_{if}	*18.836*	MHz	(2.6.4.1)
17G.	Receiver thermal noise threshold	*−123.25*	dBW	(2.6.3)

93

TABLE 2.11. (*Continued*)

18.	Receiver FM improvement threshold	-113.25	dBW	(2.8.3)
19.	Reference threshold	-103.71	dBW	(2.8.4)
	19A. diversity improvement/fading	0	dB	(2.7.4)
20A.	Fade margin w/o diversity	34.3	dB	(2.7.2)
20B.	Fade margin with diversity	0	dB	(2.7.4)
21.	C/N unfaded	53.9	dB	(2.6.6)
22.	Link unavail to ref level 0.01 %, link avail.	99.99	%	(2.7.2)
23.	Calculation of S/N in worst channel			
	24A. Pre-emphasis improvement	4.5	dB	(2.6.5)
	24B. Diversity unfaded improvement	0	dB	(2.8.3.4)
	24C. Link calculated S/N	64.36	dB	(2.8.4)
	24D. NPR	55	dB	(2.8.5)
	24E. IM noise contribution	37.43	pWp0	(2.9.5)
	24F. Sum of echo noise contrib.	8.3	pWp0	(2.8.6)
	24G. Thermal noise contrib.	206.1	pWp0	(2.8.4)
	24H. Total noise worst channel	251.83	pWp0	(2.8.4)
	24I. Specified noise worst channel	400	pWp0	(2.2)
	24J. Margin	2.0	dB	

EXPLANATORY NOTES TO TABLE 2.11
(Referenced to title and/or line number)

Loading. This is the level of the input test signal to the FM transmitter. See Section 2.8.5.6. For TV loading, see Section 2.8.5.7. In this case for FDM loading with 1200 VF channels, formula 2.43 was used.

$$P_{tl} = -15 + 10 \log 1200$$

$$= +15.79 \text{ dBm0}$$

Spec worst channel median noise. This is the noise allotment for the link that is specified at the outset. The link is 41.29 miles long or 66.4 km. From Section 2.2, 200 pWp + (3 pWp) (66.4) = 400 pWp. The reader is cautioned regarding the initial 200 pWp. It can only be used once for short and medium haul sections up to 840 km long made up of one or more hops or links. Sections between 840 and 1670 km, use 400 pWp + 3 pWp/km and from 1670 to 2500 km, use 600 pWp + 3pWp/km.

3, 6. *Path length and azimuth from each site.* For ordinary LOS radiolinks, rhumb line distance and azimuths suffice. Nevertheless, we used great circle distance and azimuths because, in the United States, among other locations, the FCC license application requires great circle azimuths and distances. It will be appreciated that final antenna alignment will be done by a rigger at the direction of the site installation engineer where the antenna on each end will be adjusted for peak RSL before being bolted down.

7. *Transmitter output power in dBW (or dBm).* This must be specified at a point in the transmitter subsystem where it can be conveniently measured.

8, 15. Transmission lines and related devices are described in Section 7.2.2. In this case, EW-64 elliptical waveguide was used for the principal waveguide run with about 0.015 dB/ft. Other losses are typical for the example. *8H* and *15H* are the sums of the transmission line losses in decibels.

10. EIRP [equation (2.6)]: The work must be done consistently with either dBm or dBW. In this case sum items 7, 8H, and 9:

$$\text{EIRP}_{\text{dBW}} = 0 \text{ dBW} - 3.47 \text{ dB} + 41.25 \text{ dB} = +37.78 \text{ dBW}$$

11. Free space loss [equation (1.7)];

$$\text{FSL}_{\text{dB}} = 36.58 + 20 \log(6100 \text{ MHz}) + 20 \log(41.29)$$

$$= 144.6 \text{ dB}$$

13. Isotropic receive level [equation (2.7)]. Algebraically add items 10, 11, and 12:

$$\text{IRL}_{\text{dBW}} = +37.78 \text{ dBW} - 144.6 - 0.7 = -107.52 \text{ dBW}$$

16. Unfaded RSL [equation (2.8)]. Algebraically add items 13, 14, and 15H:

$$\text{RSL}_{\text{dBW}} = -107.52 + 41.25 \text{ dB} - 3.08 \text{ dB} = -69.35 \text{ dBW}$$

17. Receiver noise threshold calculation. First calculate in 17D the peak carrier deviation using rms deviation of 140 kHz/channel from Table 2.5. Use equation (2.25).

$$\Delta F_p = 4.47(140) \left[\log^{-1} \left(\frac{-15 + 10 \log 1200}{20} \right) \right]$$

$$= 4.47(140)(6.160)$$

$$= 3855 \text{ kHz}$$

Calculate B_{if} using equation (2.17):

$$B_{if} = 2(3855 + 5564)$$

$$= 18.838 \text{ MHz}$$

where 5564 kHz is the highest modulating frequency. calculate receiver noise threshold using equation (2.16), NF = 8 dB:

$$P_t = -204 \text{ dBW} + 10 \log(18.838 \times 10^6) + 8 \text{ dB}$$

$$= -204 + 72.75 + 8$$

$$= -123.25 \text{ dBW}$$

EXPLANATORY NOTES TO TABLE 2.11. (*Continued*)

18. The FM improvement threshold is the value in 17G + 10 dB or −113.25 dBW.

19. Reference Threshold. It is upon this threshold which we add fade margin (item 20). CCIR Rec. 393-1 states that "the additional objective on the 2500 km reference circuit should not exceed 1,000,000 pW0 unweighted...for more than 0.01% of any month." 1,000,000 pW0 corresponds to 562,000 pWp0 (see Section 2.8.4.1). We can calculate the RSL for this value of noise in the voice channel by using equation (2.38) and add the value for emphasis for a 1200-channel system from Figure 2.15. This is 4.5 dB; thus

$$562,000 = \log^{-1}\left[\frac{-\text{RSL} - 48.6 + 8 - 4.5 + 28.89}{10}\right]$$

$$57.5 = -\text{RSL} - 48.6 + 8 - 4.5 + 28.89$$

$$\text{RSL} = -73.71 \text{ dBm} \quad \text{or} \quad -103.71 \text{ dBW}$$

20. Calculate fade margin (in this case without diversity). Turn to Section 2.7.2.3 and use equation (2.30). Path length 41.29 miles or 66.4 km = d. d^3 = 292,755. a = 1, b = $\frac{1}{4}$, f = 6.1, and P_{mf} = 0.00995%. The fade margin is 34.3 dB and then the unfaded RSL (item 16) should be −69.41 dBW. However, it was calculated (using 8-ft aperture antennas) as −69.35 dBW. This latter value will be shown to be the appropriate value for the link in item 23. Availability = 1 − 0.0000995 or 99.99%.

21. C / N unfaded. Use equation (2.26):

$$\frac{C}{N} = -69.35 \text{ dBW} - (-123.25 \text{ dBW})$$

$$= 53.9 \text{ dB}$$

or item 16 minus item 17G.

22. The link unavailability from item 20 was established as 0.01%. Link availability is then 1 − 0.0001 or 99.99%. It should be remembered that link time availability/unavailability is system related and is discussed further in Section 2.12.

24A. Pre-emphasis improvement for 1200-channel FDM operation. From figure 2.15. The value is 4.5 dB, and this value will be used below.

24C. Link calculated S/N. This is S/N of this one link alone, not part of the system. Use equation (2.36). This equation does not reflect pre-emphasis improvement:

$$\frac{S}{N} = -39.35 \text{ dBm} + 136.1 - 8 + 20\log(\Delta f/f_c)$$

$$20\log(\Delta f/f_c) = 20\log(200/5564)$$

$$= -28.89$$

$$\frac{S}{N} = -39.35 \text{ dBm} + 136.1 - 8 - 28.89$$

$$= 59.86 \text{ dB}$$

and with emphasis of 4.5 dB,

$$\frac{S}{N} = 64.36 \text{ dB}$$

RSL from equation 2.36 is the unfaded RSL from item 16.

24D. NPR is given by the equipment manufacturer (or measured). Given as 55 dB.

24E. *IM noise contribution.* Use equation (2.39) where NLR is 15.79 dB.

$$\frac{S}{N} = 55 + 10\log(5564/3.1) - 15.79$$

$$= 55 + 32.54 - 15.79$$

$$= 71.75 \text{ dB}$$

Now use equation (2.43) and substitute 71.75 dB for S/N:

$$P_{\text{tcp}} = 0.56\log^{-1}\frac{90 - 71.75}{10}$$

$$= 37.43 \text{ pWp}$$

24F. *Sum of echo noise contributions.* Use Figure 2.28 to obtain values for the waveguide lengths given, 1200 VF channels with 140 kHz/channel rms deviation. For transmit, the value is -22 (dBm0p) and for receive the value is -23. Take transmission line losses from item 8H and 15H. Use methodology of Section 2.8.6.

$$P_{\text{et}} = -22 - 30 - 26 - 5.74 \text{ dB}$$

$$= -83.74 \text{ dBm0p} \quad \text{or} \quad 4.23 \text{ pWp}$$

For the receive side the value is 4.07 pWp. Sum the two values and the total is 8.3 pWp.

Note: Values for equipment return loss are given by manufacturer as 30 dB and for composite waveguide/antenna return loss as 26 dB.

24G. Thermal noise contribution. Use equation (2.38). value of RSL is unfaded value. Use pre-emphasis improvement value from 24A or 4.5 dB and modify equation (2.36):

$$P_{\text{pWp0}} = \log^{-1}\left[\frac{39.35 - 48.6 + 8 - 4.5 + 28.89}{10}\right]$$

$$= \log^{-1}(23.14/10)$$

$$= 206.1 \text{ pWp0}$$

24H. *Total noise worst channel.* Sum the values in pWp0 of 24E, 24F, and 24G or total noise $= 37.43 + 8.3 + 206.1 = 251.83$ pWp. Compare this against the value allotted for the link or 400 pWp0. The margin is derived by calculating the RSL when given the noise value of 400 pWp minus the value from 24H or 251.83 pWp0. The margin is 2.0 dB.

TABLE 2.12. CCIR Recommendations for Preferred Radio-Frequency Channel Arrangements for Radio-Relay Systems, Used for International Connections[a,b]

CCIR Recommendation	Maximum capacity in Analogue Operation of Each Radio Carrier (Telephone channels or the Equivalent)	Capacity of Each Digital Channel[h]	Preferred center Frequency f_0 (MHz)[c]	Width of Radio-Frequency Band Occupied (MHz)
283-3	60/120/300/960[g]	Low, medium	1,808	200
			2,000	200
			2,203	200
			2,586	200
			1,903	400
382-2	600/1800		2,101	400
			4,003.5[d]	400[d]
383-1	600/1800		6,175	500
384-2	1260/2700		6,770	680
385-1	60/120	300	7,575	300
386-1	300/960[e]		8,350[e]	300[e]
387-3	600/1800	Low, medium	11,200	1000
497-1	960	Medium	12,996[f]	500

[a] The Recommendations referred to above apply to LOS and near LOS systems. For transhorizon systems, it has not yet been possible to formulate preferred radio-frequency channel arrangements, but the attention of the Administrative Radio Conference is drawn to Recommendation 388 and to Report 286.

[b] Attention should also be drawn to Recommendation 389-2, Study programme 4A-1/9 and to Report 284-1.

[c] Other center frequencies may be used by agreement between the Administrations concerned.

[d] In some countries, mostly in a large part of Region 2 and in certain other areas, a reference frequency $f_r = 3700$ MHz is used at the lower edge of a band 500 MHz wide (see Annex to Recommendation 382-2).

[e] In some countries a maximum capacity of 1800 telephone channels or the equivalent on each radio-frequency carrier may be used with a preferred center frequency of 8000 MHz. The width of the radio-frequency band occupied is 500 MHz (see Recommendation 386-1, § 7 and Annex).

[f] Reference frequency.

[g] The 960-channel capacity can only be used with the centre frequency 2586 MHz.

[h] The definition of the terms low- and medium-capacity digital systems is given in Report 378-3.

Source. CCIR OP 14-3, Annex I (Ref. 1).

with future growth requirements. The system must also coexist with other nearby systems, which will be discussed further on.

One objective when drawing up the frequency plan is to minimize inter-channel interference or what is also called cosystem interference. Also, the national regulatory authority will require that the system be spectrum conservative.

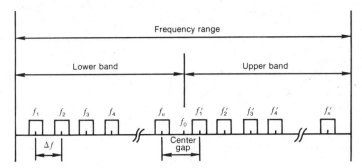

FIGURE 2.28 A generalized CCIR frequency arrangement.

Figure 2.28 shows a generalized CCIR frequency arrangement for a multiple thread (i.e., multiple RF channel) link. It will be appreciated that in most applications (some video links may prove the exception, such as an STL) we are working with frequency pairs, a "go" and a "return" channel. A frequency pair is designated, arbitrarily, with f_1 for a "go" channel, and its companion "return" channel is designated f_1', f_2 and f_2', etc.

CCIR practice divides a band (Table 2.12 and Figure 2.29) into two halves separated by a center guard band, which is larger than or equal to the separation or spacing between the center frequencies of two adjacent channels. Thus, from a single site in one direction (i.e., site B to site C) one half of the band is for transmit ("go") channels and one half for receive ("return") channels.

Where two or more RF channels are to be provided over a route, frequencies should first be assigned from the odd-numbered group of channels or from the even-numbered group of channels, but not from both, since this would require the use of two antennas at each end of each section. As an example, when all the channels from the odd-numbered group have been assigned, further expansion from the even-numbered group would be provided by means of a second antenna with polarization orthogonal to that of the first antenna.

2.10.3 Some Typical CCIR Channel Arrangements

2.10.3.1 1800 FDM Channel Systems or Equivalent in the 6-GHz Band

Figure 2.29 shows the RF channel arrangement for radiolinks operating in the 6-GHz band as recommended in CCIR Rec. 383-1.

In the figure, let f_0 be the center frequency of the frequency band occupied; let f_n be the center frequency of one RF channel in the lower half of the band, and let f_n' be the center frequency of one RF channel in the

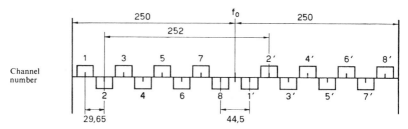

FIGURE 2.29 RF channel arrangement for radiolinks operating in the 6-GHz band. All frequencies are in MHz. From CCIR Rec. 383-1 (Ref. 1).

upper half of the band. Then the frequencies of the individual channels are expressed by the following relationships:

□ Lower half of the band: $f_n = f_0 - 259.45 + 29.65n$ (MHz)

□ Upper half of the band: $f_n' = f_0 - 7.41 + 29.65n$ (MHz)

where $n = 1, 2, 3, 4, 5, 6, 7,$ or 8.

The go and return channels on a given section should preferably use the polarizations as shown below:

	Go						Return			
$H(V)$	1	3	5	7			2′	4′	6′	8′
$V(H)$		2	4	6	8	1′	3′	5′	7′	

When a dual polarization antenna is used accommodating both transmit and receive functions and not more than four RF channels are on a single antenna, it is preferred that the channel frequencies be selected by either making $n = 1$, 3, 5, and 7 in both halves of the band or making $n = 2, 4, 6,$ and 8 in both halves of the band; or, alternatively, $n = 1, 3, 5,$ and 7 in the lower half of the band and $n = 2, 4, 6,$ and 8 in the upper half of the band. If a second similar antenna is used for four more channels, the channel frequencies may be selected by making $n = 2, 4, 6,$ and 8 in the lower half of the band and $n = 1$, 3, 5, and 7 in the upper half of the band; but if only three more channels are required, the channel frequencies may be selected by making $n = 2, 4,$ and 6 in the lower half of the band and $n = 3, 5,$ and 7 in the upper half of the band to avoid difficulty in separating frequencies 8 and 1′.

When additional RF channels, interleaved between those of the main pattern, are required, the values of the center frequencies of these RF channels should be 14.825 MHz below those of the corresponding main channel frequencies. However, in systems for 1800 channels or the equivalent, it may not be practical because of the bandwidth of the modulated carrier to use interleaved frequencies.

When up to 16 go and 16 return RF channels are required, each with a capacity of up to 600 VF channels on the same route simultaneously, different polarizations should be used alternately for adjacent RF channels in the same half of the band.

2.10.3.2 A Configuration for the 11-GHz Band for Analog Television and FDM Telephony Transmission for 600–1800 VF Channels and Digital Systems of Equivalent Bandwidth

CCIR Rec. 387-3 covers a configuration for the 11-GHz band where a 1-GHz transmission bandwidth is available permitting up to 12 go and 12 return analog channels, or up to 22 go and 22 return digital channels. Figures 2.30 and 2.31 show the CCIR recommended configurations for this band for analog transmission.

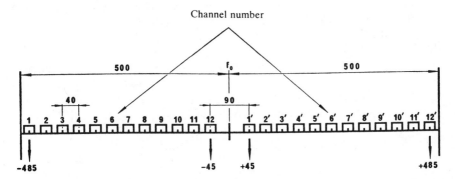

FIGURE 2.30 Radio-frequency channel arrangement for radio-relay systems operating in the 11-GHz band (main pattern). (All frequencies are in MHz.)

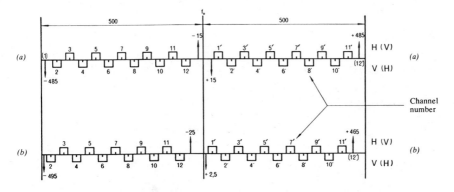

FIGURE 2.31 Radio-frequency channel arrangement for main and auxiliary radio-relay systems operating in the 11-GHz band. (All frequencies are in MHz.) (a) Main pattern. (b) Interleaved pattern. From CCIR Rec. 387-3. (Ref. 1).

For radiolink systems with a maximum capacity of 1800 FDM telephony channels or the equivalent operating in the 11-GHz band, the preferred RF channel arrangement for analog transmission is derived as follows. Let f_0 be the frequency in megahertz of the center of the operating band; let f_n be the center frequency of one RF channel in the lower half of the band and f_n' be the center frequency of one RF channel in the upper half of the band. The frequencies in the MHz of the individual channels then are expressed by the following relationship:

□ Lower half of the band: $f_n = f_0 - 525 + 40n$ (MHz)
□ Upper half of the band: $f_n' = f_0 + 5 + 40n$ (MHz)

where $n = 1, 2, 3, 4, 5, 6, 7, 8, 9, 10, 11$, or 12. Figure 2.31 applies.

When additional analog RF channels, interleaved between those of the main pattern, are required, the values of the center frequencies of these RF channels should be 20 MHz below those of the corresponding main channel frequencies (Figure 2.31). It should be noted that channel 1 of the interleaved pattern in the lower half of the band is beyond the lower extremity of a 1000-MHz band and may therefore not be available for use.

When analog RF channels are also required for auxiliary radiolink systems, the preferred frequencies for 11 go and 11 return channels including two pairs of auxiliary channels in both the main and interleaved patterns should be derived by making $n = 2, 3, 4, \ldots, 12$ in the lower half of the band and $n = 1, 2, 3, \ldots, 11$ in the upper half of the band. The radio frequencies in megahertz for the auxiliary systems should be selected as shown below:

	Main Pattern	Interleaved Pattern
Lower half of the band	$f_0 - 485$	$f_0 - 495$
	$f_0 - 15$	$f_0 - 25$
Upper half of the band	$f_0 + 15$	$f_0 + 2.5$
	$f_0 + 485$	$f_0 + 465$

Figure 2.31 applies; it also shows a possible polarization arrangement.

If only three go and three return channels are accommodated on a common transmit–receive antenna, it is preferable that the channel frequencies (MHz) be selected by making

$n = 1, 5, 9$ or
$n = 2, 6, 10$ or
$n = 3, 7, 11$ or
$n = 4, 8, 12$

(all combinations in both halves of the band). Otherwise, for channel arrange-

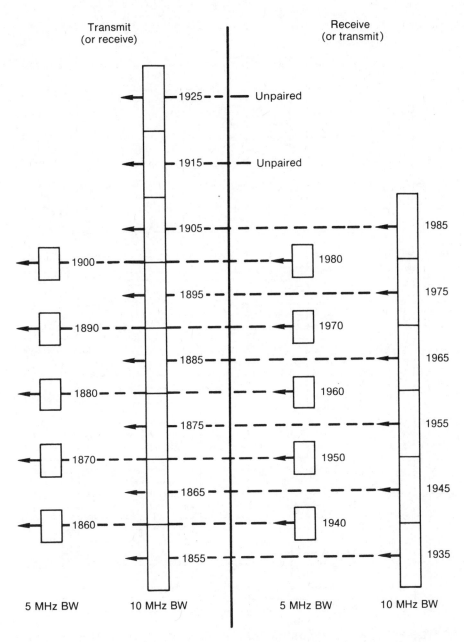

Transmit
(or receive)

Receive
(or transmit)

1925 — — Unpaired

1915 — — Unpaired

1905 — — 1985

1900 — — 1980

1895 — — 1975

1890 — — 1970

1885 — — 1965

1880 — — 1960

1875 — — 1955

1870 — — 1950

1865 — — 1945

1860 — — 1940

1855 — — 1935

5 MHz BW   10 MHz BW   5 MHz BW   10 MHz BW

FIGURE 2.32 FCC frequency plan, band 1850–1990 MHz. From Ref. 14.

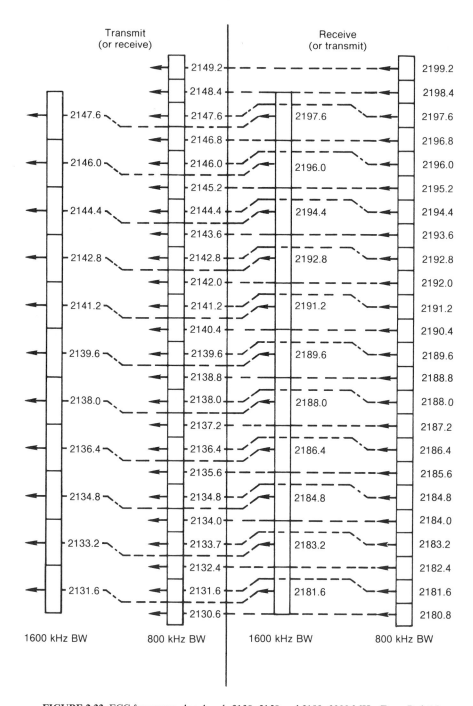

FIGURE 2.33 FCC frequency plan, bands 2130–2150 and 2180–2200 MHz. From Ref. 14.

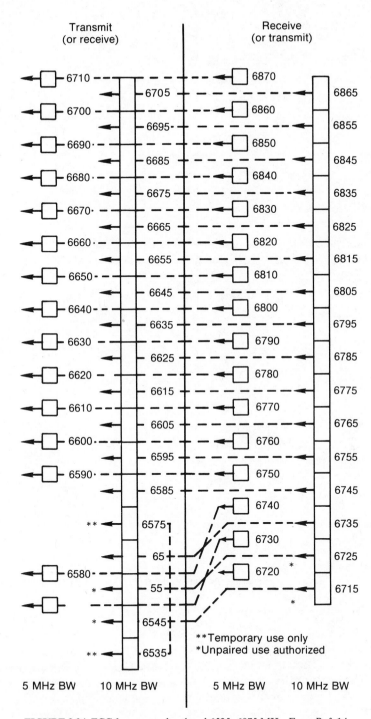

FIGURE 2.34 FCC frequency plan, band 6525–6875 MHz. From Ref. 14.

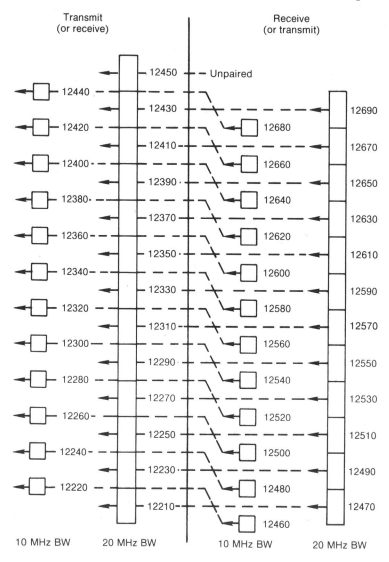

FIGURE 2.35 FCC frequency plan, band 12,200–12,700 MHz. From Ref. 14.

ments where more than three go and three return channels are required, all to channels should be in one half of the band and all return channels should be in the other half of the band.

It is further recommended that for adjacent analog RF channels in the same half of the band, different polarizations should be used alternately.

The preferred center frequency in the 11-GHz band is 11,200 MHz.

TABLE 2.13. FCC Standard Parameters (Applicable to Figures 2.32 through 2.35)

Frequency band	1.85–1.990 GHz		2.13–2.15 and 2.18–2.2 GHz		6.525–6.875 GHz			12.2–12.7 GHz
Carrier frequency spacing	10 MHz with 10 MHz interstitial (Figure 2.32)		0.8 MHz with 1.6 MHz Cochannel (Figure 2.33)		10 MHz with 10 MHz interstitial (Figure 2.34)			20 MHz with 20 MHz interstitial (Figure 2.35)
Bandwidths	5 MHz	10 MHz	0.8 MHz	1.6 MHz	5 MHz	10 MHz	10 MHz	20 MHz
Traffic capacity FDM	120	300 480 600	24 48	72 96	120	300 480 600	300	300 600 1200
Other	Digital	Digital	Digital	Digital	Digital	Digital	Digital	Video/digital
FDM per channel rms deviation (kHz)	200	200 200 140	35 25	60 47	200	200 200 140	200	200 200 140
Freq stab[a] operational low stability	±0.002% ±0.02%		±0.0003%		±0.002% ±0.02%			±0.002% ±0.02%
Freq tolerance (FCC requirements)	0.002%		0.001%		0.005%			0.005%

[a] Operational stability used is based on most probable performance of modern transmitters over normal environmental range. Low stability represents potentials from grandfathered systems.

Source. Reference 14.

2.10.4 Several FCC Frequency Plans

Figures 2.32 through 2.35 (on pages 103–106) give current FCC (U.S. Federal Communications Commission) frequency plans for the frequency bands 1850–1990 MHz, 2130–2150 and 2180–2200 MHz, 6525–6875 MHz, and 12,200–12,700 MHz, respectively. Table l2.13 provides standard parameters applicable to the figures. (From Ref. 14.)

2.11 DIGITAL RADIO SYSTEMS

2.11.1 Introduction

The implementation of digital LOS radiolinks is accelerating primarily due to the transition of the telephone network to an all-digital network. Many available texts (Refs. 5 and 15) demonstrate the rationale for the transition in that thermal and IM noise accumulation can be disregarded over the network, and noise becomes an isolated problem between points of regeneration. This is an overwhelming advantage over analog transmission, where the primary concern of the transmission engineer is noise accumulation.

Other arguments presented for going all-digital are based on its compatibility with digital information transmission requirements such as telephone signaling, data transmission, digitized voice, programming information, and facsimile.

Several very important factors representing the other side of the coin must also be highlighted. The digital radiolink engineer must not only be cognizant of these factors but also must understand them well. The digital network is based on a PCM waveform, which, when compared to its analog FDM counterpart, is wasteful of bandwidth. A nominal 4-kHz voice channel on an FDM baseband system occupies about 4-kHz of bandwidth (Ref. 17, Chapter 3). On an FDM/FM radiolink, by rough estimation, we can say it occupies about 16 kHz.

In conventional PCM baseband system, allowing 1 bit per Hz of bandwidth, a 4-kHz voice channel roughly requires 64 kHz (64 kbps) of bandwidth. This is derived using the Nyquist sampling rate of 8000/sec (4000 Hz × 2) and each sample is assigned an 8-bit code word, thus 8000 × 8 bits per second or 64 kbps.

RF bandwidth is at a premium and, as a result, it is incumbent on the digital radiolink engineer to select a waveform that conserves bandwidth, achieving, essentially, more bits per hertz. He or she will also find that many national regulatory agencies require a minimum number of digital voice channels per unit of bandwidth.

This section first introduces some sample regulatory requirements and then discusses modulation techniques that are bandwidth conservative so that these national requirements can be met. It then describes methods of link analysis to achieve specified digital network performance. The discussion will rely heavily on the previous sections demonstrating that much of the approach used on analog radiolink design is also applicable to digital radiolink design. The unit of digital radiolink performance is BER rather than S/N and noise accumulation, which were the measures for analog radiolink design.

2.11.1.1 Energy per Bit per Noise Density Ratio, E_b / N_0

The efficiency of a digital communication system in the presence of wideband noise with a single-sided noise spectral density N_0 is commonly measured by the received information bit energy to noise density ratio (E_b/N_0) required to achieve a specified error rate. This ratio can be expressed in terms of received signal level power C (RSL) by

$$\frac{E_b}{N_0} = \frac{C}{N_0(R_{bps})} \tag{2.53}$$

E_b/N_0 is normally expressed in decibels. If we let C, the modulated carrier level, be RSL (receive signal level), then

$$\frac{E_b}{N_0}(\text{dB}) = \text{RSL}_{\text{dBW}} - 10\log(\text{bit rate}) - N_0 \qquad (2.54)$$

For a perfect uncooled receiver (e.g., a receiver operating at room temperature, 290 K),

$$N_0 = -204 \text{ dBW} + NF_{\text{dB}} \qquad (2.55)$$

where NF_{dB} is the receiver noise figure. We will appreciate, of course, that N_0 is the noise in 1 Hz of bandwidth.

Now we can restate equation (2.54), substituting the value of N_0 from (2.55) or

$$\frac{E_b}{N_0} = \text{RSL}_{\text{dBW}} - 10\log(\text{bit rate}) - (-204 \text{ dBW} + NF_{\text{dB}})$$

or

$$\frac{E_b}{N_0} = \text{RSL}_{\text{dBW}} + 204 \text{ dBW} - 10\log(\text{bit rate}) - NF_{\text{dB}} \qquad (2.56)$$

2.11.2 Regulatory Issues

The U.S. regulatory agency, the FCC, has long recognized that conventional digital modulation schemes (such as FSK, BPSK/QPSK, etc.) were not bandwidth conservative and ruled in Part 21.122 of "Rules and Regulations" (Ref. 16) that:

Microwave transmitters employing digital modulation techniques and operating below 15 GHz shall, with appropriate multiplex equipment, comply with the following additional requirements:

1. The bit rate in bits per second shall be equal to or greater than the bandwidth specified by the emission designator in hertz (e.g., to be acceptable, equipment transmitting at a 20-Mb/s rate must not require a bandwidth greater than 20 MHz). Except that the bandwidth used to calculate the minimum rate shall not include any authorized guardband.
2. Equipment to be used for voice transmission shall be capable of satisfactory operation within the authorized bandwidth to encode at least the

following number of voice channels:

Frequency Band (MHz)	Allowable Bandwidthwidth (MHz)	Minimum Capacity of Encoded Voice Channels
2110–2130	3.5	96
2160–2180	3.5	96
3700–4200	20	1152
5925–6425	20	1152
10,700–11,700	40	1152

The FCC has the following rule on emission limitation (Part 21.106) (Ref. 16):

When using transmissions employing digital modulation techniques

(2) (i) For operating frequencies below 15 GHz, in any 4-kHz band, the center frequency of which is removed from the assigned frequency by more than 50% up to and including 250% of the authorized bandwidth: As specified by the following equation but in no event less than 50 dB:

$$A = 35 + 0.8(P - 50) + 10\log_{10}B \qquad (2.57)$$

where A = attenuation (dB) below the mean output power level

P = percent removed from carrier frequency

B = authorized bandwidth (MHz)

(attenuation greater than 80 dB is not required).

(ii) For operating frequencies above 15 GHz, in any 1 MHz band, the center frequency which is removed from the assigned frequency by more than 50% up to and including 250% of the authorized bandwidth: As specified by the following equation but in no event less than 11 dB:

$$A = 11 + 0.4(P - 50) + 10\log_{10}B \qquad (2.58)$$

(attenuation greater than 56 dB is not required).

(iii) In any 4-kHz band, the center frequency of which is removed from the assigned frequency by more than 250% of the authorized bandwidth: at least $43 + 10\log_{10}$ (mean power output in watts) dB or 80 dB, whichever is the least attenuation . . . .

The reference FCC Part 21.106 describes the "FCC mask" often referenced in the literature as the FCC Docket 19311 mask. This spectral mask is shown in Figure 2.36.

Whereas, conventionally, bandwidths most often describe 3 dB points, the 19311 mask describes 50 dB points at about plus or minus half the authorized bandwidth. As it can be seen clearly, a packing ratio considerably greater than

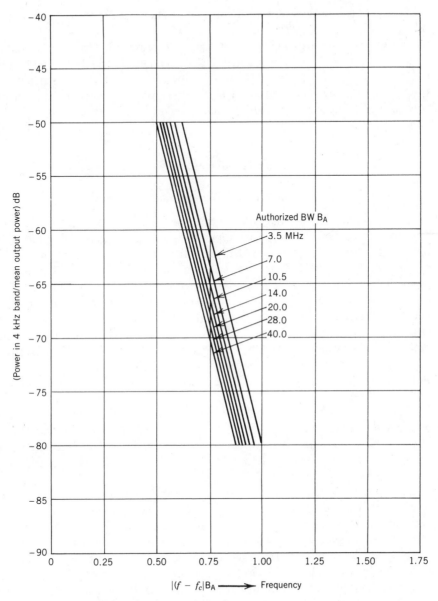

FIGURE 2.36 FCC Docket 19311 Spectrum Masks. From Ref. 17. (f_c = center frequency; f = frequency of interest.)

1 bit/Hz must be achieved to comply with the FCC spectral mask. This subject is discussed in Section 2.11.3.

CCIR also provides guidance on spectral occupancy for digital radio systems. Reference should be made to CCIR Recs. 497-1 and 387-3 and CCIR Reps. 607-1, 608-1, and 609-1 (Ref. 1). However, CCIR is less specific on spectral occupancy versus bit rate (packing ratio) and numbers of equivalent voice channels. It deals more with frequency allocation for specific operational bands.

2.11.3 Modulation Techniques and Spectral Efficiency

2.11.3.1 Introduction

There are three generic modulation techniques available: AM, FM, and PM. Terminology of the industry often appends the letters SK to the first letter of the modulation type, such as ASK meaning amplitude shift keying, FSK meaning frequency shift keying (classified as digital FM), and PSK meaning phase shift keying. Any of the three basic modulation techniques may be two level or multilevel. For the two-level case, one state represents the binary "1" and the other state a binary "0." For multilevel or M-ary systems there are more than two levels or states, usually a multiple of 2, with a few exceptions, such as partial response systems, duo-binary being an example. Four-level or 4-ary systems are in common use, such as QPSK. In this case, each level or state represents two information bits or coded symbols. For eight-level systems such as 8-ary FSK or 8-ary PSK, three bits are transmitted for each transition or change of state and for a 16-level system (16-ary) four bits are transmitted for each change of state or transition. Of course, in M-ary systems some form of coding or combining is required prior to modulation and decoding after demodulation to recover the original bit stream. Conceptually, a typical QPSK modulator is shown in Figure 2.37.

Let us consider that each of the three basic modulation techniques may be represented by a modulated sinusoid. At the far end receiver some sort of detection must be carried out. Coherent detection requires a sinusoidal reference signal extremely closely matched in both frequency and phase to the received carrier. This phase frequency reference may be obtained from a transmitted pilot tone or from the modulated signal itself. Noncoherent detection, being based on waveform characteristics independent of phase (e.g., energy or frequency), does not require a phase reference. FSK commonly uses noncoherent detection.

After detection in the receiver there is usually some device that carries out a decision process, although in less sophisticated systems this process may be carried out in the detector itself. Some decision circuits make decisions on a baud-by-baud basis. Others obtain some advantage by examining the signal over several baud intervals prior to making each "baud" decision. The

4-PHASE-SHIFT KEYING (PSK)

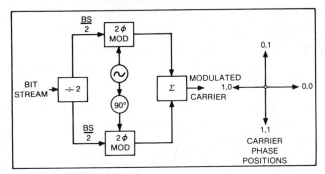

- HAS BEST NOISE-EFFICIENCY PERFORMANCE
- 4-LOGIC LEVEL SYSTEM

FIGURE 2.37 Conceptual block diagram of a QPSK modulator.

observation interval is the portion of the received waveform examined by the decision device.

2.11.3.2 Selection of a Spectrally Conservative Modulation Technique

There are three aspects to be considered in selecting a particular modulation technique:

1. To meet the spectral efficiency requirements (i.e., a certain number of megabits for a given bandwidth, as described in Section 2.11.2).
2. To meet realizable performance requirements (see Sections 2.2.1 and 2.11.5).
3. As constrained by a complexity factor having economic impact on the system.

Let us now follow an elementary mathematical analysis that will clarify and assist the design engineer in selecting the type of modulation that will meet the needs of a particular link or system. [Taken from CCIR Rep. 378-3 (Ref. 1).] Assume that the only source of errors is thermal noise in the receiver. The normalized carrier-to-noise ratio W may be expressed as

$$W = 10 \log_{10}\left(\frac{W_{in}}{W_n f_n}\right) \tag{2.59}$$

where W_{in} = received maximum steady-state signal power (i.e., the highest
value of the mean power during any bit period) (valid for FSK,
PSK, binary ASK)

W_n = noise power density at the receiver input

f_n = bandwidth numerically equal to the bit rate B of the binary
signal before the modulation process

The bit rate B is the rate of the incoming serial bit stream, taking into
account bits added for server channel(s), redundancy added for error control
(FEC), and so forth.

The normalized carrier-to-noise ratio can be related to the more familiar
carrier-to-noise ratio C/N by the following expression:

$$W = 10 \log_{10}\left(\frac{C}{N} \times \frac{B_{eq}}{B}\right) \qquad (2.60)$$

where B_{eq} = equivalent noise bandwidth of receiver.

The necessary bandwidth for a given class of emission is defined by CCIR
as the width of the frequency band that is just sufficient to ensure the
transmission of information at the rate and quality (error rate) required under
specified conditions. In digital radio systems one desirable condition to be met
is that the previously defined power requirement W, which is variable with
receiver bandwidth, should be at its minimum value. The necessary bandwidth
in megahertz is then by given by:

$$B_n = \begin{cases} FBR & \text{for DSB modulation} \\ 0.5FBR & \text{for SSB modulation} \\ 0.6FBR & \text{for VSB modulation} \end{cases} \qquad (2.61)$$

where F = a design factor depending on the implementation approach. For
efficient modulation methods F is generally between 1 and 2;
values of F below 1 are possible, but only at the expense of
increasing intersymbol interference.

B = bit rate of the binary signal (Mbps) before the modulation process.

R = symbol rate (Mbaud/B).

Table 2.14 compares various modulation types assuming no intersymbol or
other types of interference. For each modulation type it gives values of W and
B_n and includes comments on complexity. CCIR Rep. 378-3, from which this
table has been taken, states that the value of F may differ for different variants
in the table. The introduction of the factor F is an attempt to make an
allowance for the compromise that must be effected in any practical system.

TABLE 2.14. CCIR Comparison of Common Modulation Methods

System	Variant	W (dB)[b]	Necessary Bandwidth	Remarks[d]
Amplitude modulation	Full-carrier binary double-sideband with envelope detection	17	FB	Simple, wasteful of bandwidth, high signal power
	Double-sideband, suppressed-carrier two binary channels in quadrature with coherent detection	10.5	$FB/2$	Fairly complex, tolerant to distortion
	Double-sideband, suppressed-carrier, two binary channels in quadrature with differentially coherent detection	12.8	$FB/2$	Fairly simple, fairly sensitive to distortion
	Single-sideband binary, suppressed-carrier[a]	10.5	05. FB	Complex, loss of low baseband frequencies
	Vestigial-sideband binary, suppressed carrier, with coherent detection[a]	11.3	0.6 FB	Fairly complex
	Vestigial-sideband binary, reduced carrier, with coherent detection[a]	11.8	0.6 FB	Fairly simple
	Vestigial-sideband binary, suppressed carrier, 50% amplitude modulation with envelope detection[a]	17.8	0.6 FB	Simple, subject to pulse distortion, high signal power
Phase modulation, with coherent detection[c]	Two-level	10.5	FB	Fairly simple, tolerant to distortion, wasteful of bandwidth
	Four-level	10.5	$FB/2$	Fairly simple, tolerant to distortion
	Eight-level	13.8	$FB/3$	Complex, economic of bandwidth, sensitive to distortion
Phase modulation, with differentially coherent detection[c]	Two-level	11.2	FB	Simple, fairly tolerant to distortion, wasteful of bandwidth
	Four-level	12.8	$FB/2$	Fairly simple, fairly sensitive to distortion
	Eight-level	16.8	$FB/3$	Complex, high signal power, economic of bandwidth, sensitive to distortion
Frequency modulation, with discriminator detection[f]	Two-level	13.4	FB	Simple, wasteful of bandwidth
	Three-level (duo-binary)	15.9	FB^c	Fairly simple
	Four-level	20.1	$FB/2$	Fairly simple, high signal power
	Eight-level	25.5	$Fb/3$	Complex, high signal power, economic of bandwidth
Other modulation methods with coherent detection	Two three-level class 1 partial response channels in quadrature	13.5	$Fb/2^c$	Fairly simple, economic of bandwidth
	16-level quadrature amplitude modulation[a]	17	$FB/4$	Fairly simple, economic of bandwidth, sensitive to distortion

[a] The maximum steady-state signal power depends on the shape of the modulating pulses. These figures are therefore based on average power.

[b] $P_e = 10^{-6}$.

[c] All digital phase modulation may be obtained directly by phase modulation or indirectly by methods of amplitude or frequency modulation.

[d] Reconsideration of the validity of the remarks is desirable.

[e] The design factor F in this case can be close to the value of 1. This effective reduction in necessary bandwidth is achieved at the expense of a greater number of transmitted levels for a given number of input levels, or equivalently, at a greater value of W for a given error rate.

[f] The adaptation of analogue frequency radio relay systems for the transmission of digital signals seems feasible at the present time for gross bit rates in the medium capacity range. For frequency shift keying (FSK) the bandwidth given by the relation $FD + BR$ is used by one Administration and includes the peak frequency deviation D.

Source. CCIR Rep. 378-3 (Ref. 1); Courtesy of ITU-CCIR.

This same CCIR Report classifies digital capacity into three categories as follows:

☐ Small capacity with gross bit rates up to and including 10 Mbps.

☐ Medium capacity with gross bit rates ranging from 10 to 100 Mbps.

☐ Large capacity with gross bit rates greater than 100 Mbps.

As far as the three digital capacity categories are concerned, the modulation methods that have reasonable characteristics from the aspects previously considered in Table 2.14 are:

1. AM, full carrier, DSB, with envelope detection, is wasteful of bandwidth and power and should not be used.

2. FM, two-level, with discriminator detection, is also wasteful of bandwidth and may be considered under certain situations for small capacity systems when permitted by national regulation.

3. PM, two-level, with coherent or differentially coherent detection. These methods of modulation are fairly simple to implement but comparatively wasteful of bandwidth and are suitable for small capacity systems where national regulations permit.

4. PM, four-level (QPSK), with coherent or differentially coherent detection. These are suitable methods for all information bit rates where national regulation is not too restrictive of bandwidth utilization. For high-capacity digital radiolinks the more desirable demodulation method may be coherent detection because it is more tolerant of interference. 4-ary FSK and QPSK are in common use in many locations on digital radiolinks.

5. FM, three-level (duo-binary), four-level or eight-level with discriminator detectors, is preferred for the simple adaptation to digital transmission of FM (analog) radiolink systems using FDM. In such cases a large carrier-to-noise ratio will be available and the method of modulation need not be changed.

6. PM, eight-level (8-ary-PSK), with coherent detection, is particularly suitable for medium capacity systems operating below 12 GHz. The method is attractive when systems are to be mixed with existing analog channel arrangements because it offers good spectrum efficiency even when using a single polarization. Although it is theoretically more sensitive to distortion than four-level CPSK, comparable performance can be achieved by optimal equipment filtering and equalization.

7. Partial response coding techniques may be attractive to reduce bandwidth occupancy by reducing the value of the design factor F with some increase in equipment complexity. This reduction in bandwidth is achieved at the expense of a greater number of transmitted levels for a given number of input levels, or equivalently, at a greater value of W for a given error rate.

2.11.3.3 Spectrally Efficient Modulation Waveforms

Table 2.15 compares some digital modulation waveforms with regard to their theoretical and practical spectral efficiency (i.e., bandwidth utilization) and performance. Only the higher logic level options (i.e., ≥ 8) in Table 2.15 are reasonable candidates to meet the FCC requirements for spectral occupancy by digital radiolink transmitters. Reference 17 states that "M-QAM is concluded to be the most desirable modulation technique of those considered." We will also briefly cover higher logic levels (≥ 8-ary).

An ideal digital radio system using quadrature modulation is shown in Figure 2.38. The system shown is ideal in the sense that the filters used are bandlimited to the Nyquist band $\pm \frac{1}{2}T$. The transmitted RF spectrum resulting from the use of such a filter is thus ideally bandlimited to a bandwidth of $1/T$, where T is the transmitted symbol rate (i.e., the rate of choosing a *pair* of impulses to input the baseband filters is $1/T$). Furthermore, it is assumed that each pair of impulses carries B bits of information such that there are 2^B distinct pairs of impulse weights, one pair of which is chosen for transmission each symbol time. The bit rate is thus B/T in such a system. The packing ratio or bits per hertz of RF bandwidth capacity C_I is thus

$$C_I = B \tag{2.62}$$

TABLE 2.15. Comparison of Some Digital Modulation Waveforms

Waveform	Theoretical (b/Hz)	Practical (b/Hz)	BER $= 1 \times 10^{-4}$	
			E_b/N_0(theoretical) (dB)	E_b/N_0 (practical) (dB)
OOK (coherent detector)	1	0.8	11.4 dB	12.5 dB
QAM	2	1.7	8.4 dB	9.5 dB
FSK	1	0.8	12.5 dB	11.8[a] dB
BPSK (coherent detector)	1	0.8	8.4 dB	9.4 dB
QPSK	2	1.9	8.4 dB	9.9 dB
8-ary PSK	3	2.6	11.8 dB	12.8 dB
16-ary PSK	4	2.9	16.2 dB	17.2 dB
16-ary APK (4-QAM)	4	3.1	13.1 dB	13.4 dB
32-ary APK (8-QAM)	6	4.5	17.8 dB	18.4 dB
64-ary APK (16-QAM)	8		22.4 dB	

[a] Discriminator detection.

Source. First seven items from Ref. 24; last three items from Ref. 17.

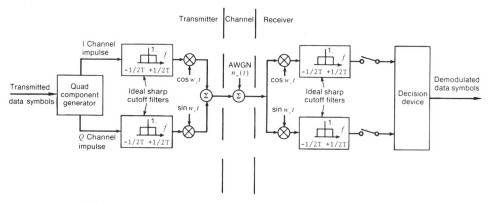

FIGURE 2.38 Ideal digital radio system using quadrature modulation.

where C_I = ideal system capacity in bits/sec/Hz

         B = number of bits carried per RF symbol

(*Note*: In many texts and in other parts of this book the transmitted symbol rate is equivalent to the baud rate.)

As shown in Figure 2.38 the two baseband signals resulting from the pair of impulses are used to modulate quadrature carriers. At the far end receiver, the RF signal is downconverted back to baseband by the use of coherent quadrature mixer references. Because of the ideal Nyquist character of the channel, samples of the baseband signals will be the transmitted impulse weights corrupted by the channel noise, but with zero intersymbol interference. The sampled quadrature components at the receiver are fed to a decision device that optimally decides which quadrature pair was transmitted given the received pair of samples.

Space diagrams of two sets of 2^B quadrature pair systems are shown in Figure 2.39—8-ary PSK and 4-QAM. Note that the 4-QAM has 16 logic states.

Figure 2.39 shows the eight pairs of impulse weights employed in 8-ary PSK. The pairs are chosen equispaced around a circle in the quadrature two-dimensional space. At the symbol sample times, the RF amplitude is a constant and the phase is one of the eight phases. At the far end receiver, the decision device decides the transmitted pair is that nearest in phase to the received pairs' phase and outputs the corresponding three bits of information. It is to be noted that the RF envelope is constant only at the symbol sampling times; at other times between samples, the envelope fluctuates in a complex manner. Figure 2.39 shows the 8-ary PSK case, but, in general, there can be any number of points equispaced around the circle.

Figure 2.40 shows the performance for ideal M-ary PSK for various values of M from $M = 4$ to $M = 128$ and corresponding values of C_I. (From Ref. 17.)

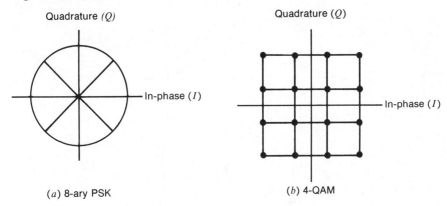

(a) 8-ary PSK

(b) 4-QAM

FIGURE 2.39 Signal space diagrams.

Figure 2.39b shows another example of an I and Q (in-phase and quadrature) implementation from Figure 2.38. In this case the 16 pairs of quadrature components generated at the transmitter fall on a 4×4 square grid. These points are generated when 4-ary PAM (phase-amplitude modulation) impulses are conveyed on each of the two quadrature channels. This type of modulation is called *quadrature amplitude modulation* (QAM), with a prefix number designating the number of PAM levels on *each* quadrature component. Thus, the modulation of Figure 2.39b is designated 4-QAM, and the generalized modulation of this type is called M-QAM. The ideal performance of M-QAM is shown in Figure 2.41 for values of $M = 2$, 4, 8, and 16. Since there are M^2 signal points for $M = $ QAM, the number of bits b carried per RF symbol is

$$B = 2 \log_2 M \tag{2.63}$$

Equations (2.62) and (2.63) are of an ideally bandlimited M-QAM waveform as

$$C_I = 2 \log_2 M \text{ bits sec}^{-1} \text{ Hz}^{-1} \tag{2.64}$$

Values for C_I are shown in Figure 2.42.

When one compares Figures 2.41 and 2.42, the superiority of M-QAM over M-PSK waveforms is noted for the same theoretical capacity (packing ratio), and as the capacity increases, the superiority increases. For example, 64-ary PSK at 10^{-5} error rate requires an E_b/N_0 of about 28.5 dB (6 bits/Hz packing ratio) and for the same ideal packing ratio for M-QAM or 8-QAM, an E_b/N_0 of about 18.6 dB is required for the same error rate.

The reader should also note that for either the M-PSK or M-QAM scheme, as logic levels continue to increase, certain implementation impracticality is approached. For example, if we were to use 16-QAM to obtain an ideal 8

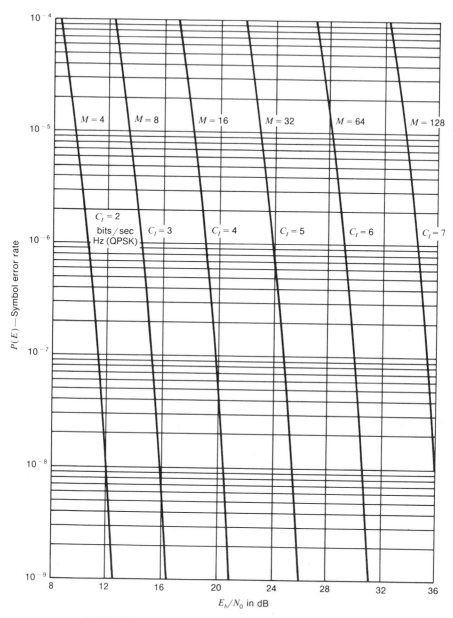

FIGURE 2.40 Ideal *M*-ARY PSK performance. From Ref. 17.

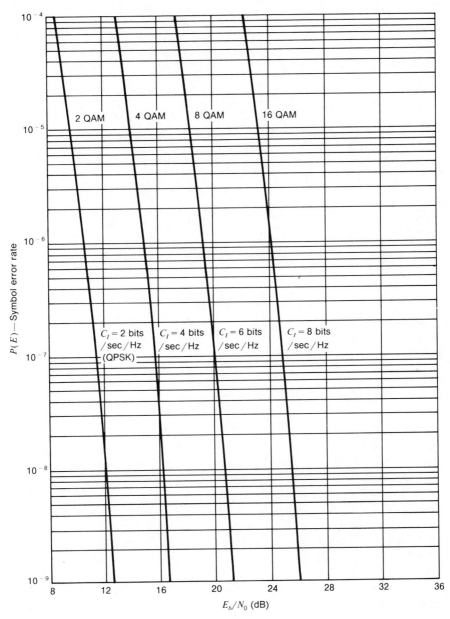

FIGURE 2.41 Ideal *M*-QAM performance. From Ref. 17.

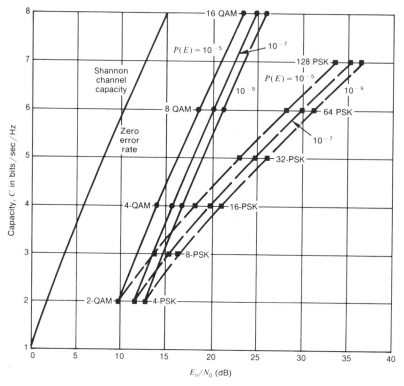

FIGURE 2.42 Channel capacity comparisons M-QAM and M-PSK ideal systems.

bits/Hz packing ratio, it would require radio transmitters and receivers capable of maintaining good resolution among the 256 signal points on a 16×16 grid. Reference 17 states that this requirement is too stringent on all the system-distorting elements and would be beyond the present state-of-the-art (1979). However, there is at least one equipment manufacturer in the United States that now has a 16-QAM product fielded.

Figure 2.42 shows plots comparing M-QAM and M-PSK ideal systems. The figure gives values of C_I in bits/sec/Hz versus E_b/N_0 for error rates of 10^{-5}, 10^{-7}, and 10^{-9}. At the left in the figure is a plot of Shannon's channel-capacity curve. Shannon's curve represents a theoretical bound on the absolute maximum capacity at zero error rate for a given E_b/N_0 for an infinitely complex digital modulation/demodulation transmission system, but it represents unattainable performance in a practical system. It should be noted in the figure that the ideal M-QAM schemes track parallel to Shannon's capacity bound, being about 8 dB away at 10^{-5} error rate.

Having now considered ideal spectrally efficient digital radio systems, we now turn to the practical. Higher-logic-level digital radio systems that are bandwidth limited (such as by FCC Docket 19311) have equipment- and system-design constraints forcing the ideal into the practical domain. Some of the equipment-design constraints that are encountered in real systems are:

1. The effect of nonideal sharp cutoff spectral shaping filters.
2. The effect of additional linear distortion created by realistic filters.
3. The effect of signaling with a peak-power-limited amplifier.
4. The relative efficiency between preamplifier filtering and postamplifier filtering.
5. The effect of practical nonlinearities encountered with real amplifiers on system performance.
6. Technique(s) required for adapting realistic amplifier nonlinearities in order to render the amplifier linear for the purpose of supporting highly bandwidth-efficient digital data radio transmission.
7. The effect on performance of practical imperfections in an implementation of a bandwidth-efficient modem including a baseband equalizer for counteracting linear distortion caused by realistic filter characteristics.

Many system problems arise at the receiver, which must resolve which signal point was transmitted. Outside disturbances tend to deteriorate the system by masking or otherwise confusing the correct signal point degrading error performance.

One such outside disturbance is additive Gaussian noise. The bandwidth-efficient modem consequently requires a higher signal-to-noise ratio for a given symbol rate as the number of bits per second per hertz is increased.

Another type of outside disturbance is created by cochannel or adjacent-channel interference. Here the bandwidth-efficient modem is more vulnerable to interference since less interference power is required to push the transmitted signal point to an adjacent point, thus resulting in a hit causing errors at the receiver.

For the system engineer, one of the most perplexing outside disturbances is caused by the transmission medium itself. This is multipath distortion causing signal dispersion. This problem is dealt with, in part, in the next subsection.

2.11.4 Mitigation Techniques for Multipath Fading

In analog radiolink systems, multipath fading results in an increase in thermal noise as the RSL drops. In digital radio systems, however, there is a degradation in BER during periods of fading that is usually caused by intersymbol interference due to multipath. Even rather shallow fades can cause relatively destructive amounts of intersymbol interference. This interference results from frequency-dependent amplitude and group delay changes. The degradation depends on the magnitude of in-band amplitude and delay distortion. This, in turn, is a function of fade depth and time delay between the direct and reflected signals.

Five of the most common methods to mitigate the effects of multipath (Ref. 20) on digital radiolinks are:

☐ System configuration (i.e., adjusting antenna height to avoid ground reflection; implementation of space and/or frequency diversity).

☐ Use of IF combiners in diversity configurations.

☐ Use of baseband switching combiners in diversity configuration.

☐ Adaptive IF equalizers.

☐ Adaptive transversal equalizers.

System-configuration techniques have been described previously in this section.

An optimal IF combiner for digital radio receiving subsystems can be designed to adjust adaptively to path conditions. One such combining technique, the maximum power IF combiner, vectorially adds the two diversity paths to give maximum power output from the two input signals. This is done by conditioning the signal on one path with an endless phase shifter, which rotates the phase on this path to within a few degrees of the signal on the other diversity path prior to combining. The output of this type of combiner can display in-band distortion that is worse than the distortion on either diversity path alone, but functions well to keep the signal at an acceptable level during deep fades on one of the diversity paths.

A minimum-distortion IF combiner operation is similar in most respects to the maximum power IF combiner but uses a different algorithm to control the endless phase shifter. The output spectrum of the combiner is monitored for flatness such that the phase of one diversity path is rotated and, when combined with the second diversity path, produces a comparatively flat spectral output. The algorithm also suppresses the polarity inversion on the group delay, which is present during nonminimum phase conditions. One disadvantage of this combiner is that it can cancel two like signals such that the signal level can be degraded below threshold.

Reference 18 suggests a dual algorithm combiner that functions primarily as a maximum power combiner and automatically converts to a minimum-distortion combiner when signal conditions warrant. Using space diversity followed by a dual-algorithm combiner can give improvement factors better than 150.

Adaptive IF equalizers attempt to compensate directly at IF for multipath passband distortion. Digital radio transmitters emit a transmit spectra of relatively fixed shape. Thus, various points on the spectrum can be monitored, and when distortion is present, corrective action can be taken to restore spectral fidelity. The three most common types of IF adaptive equalizers are shape-only equalizers, slope and fixed notch equalizers, and tracking notch equalizers.

Another equalizer is the adaptive transversal equalizer, which is efficient at canceling intersymbol interference due to signal dispersion caused by multipath. The signal energy dispersion can be such that energy from a digital transition or pulse arrives both before and after the main bang of the pulse. The equalizer uses a cascade of baud delay sections that are analog elements to which the symbol or baud sequence is inputted. The "present" baud or symbol is defined as the output of the Nth section. Sufficient sections are required to encompass those symbols or bauds that are producing the distortion. These transversal equalizers provide both feed forward and feedback information. There are both linear and nonlinear versions. The nonlinear version is sometimes called a decision feedback equalizer. Reference 18 reports that both the IF and transversal equalizers show better than three times improvement in error rate performance over systems without such equalizers.

2.11.5 Performance Requirements on Digital Radiolinks

2.11.5.1 CCIR Service Quality Performance Objectives

The measure of quality of service for a digital transmission system is BER. CCIR provides interim guidelines for BER in CCIR Rep. 378-3 (Ref. 1). In this CCIR report error rate is broken down into two categories:

☐ Low error rate value 80% of the time.
☐ High error rate value not to exceed between 0.1% and 0.01% of the time (*note*: a firm value has yet to be recommended).

For a 2500-km hypothetical reference circuit, the low-error-rate value* is 1×10^{-7}; the high error rate value should lie in the range of 1×10^{-3} to 1×10^{-6}. CCIR states that the high-error-rate value requires further study. However, for error rates higher than 1×10^{-3} for systems carrying telephony, dropout can occur due to loss of supervisory signaling.

2.11.5.2 Performance Factor for Voice (PFV)

Guiffrida of Bell Telephone Laboratories has developed a figure of merit for digital radio systems carrying voice traffic. It is defined as follows:

$$\text{PFV} = \frac{\text{time BER worse than } 10^{-3}}{\text{time principal signal path faded more than 30 dB}} \quad (2.65)$$

*Does not include error due to multiplex equipment.

2.11.6 Link Calculations

The procedure for link analysis closely follows that for an analog radiolink (Section 2.6) or simply:

☐ Calculate EIRP
☐ Algebraically add FSL and other losses due to the medium (P_L) such as gaseous absorption loss
☐ Add receiving antenna gain (G_r)
☐ Algebraically add line losses (L_{Lr})

from which we derive the RSL in dBW or dBm:

$$\text{RSL}_{\text{dBW}} = \text{EIRP}_{\text{dBW}} + \text{FSL} + P_L + G_r + L_{Lr} \qquad (2.66)$$

RSL, as defined in this text, is the signal level at any given time at the input to the first active stage of a receiver chain whether an LNA or a mixer.

Whereas on an analog radiolink, the measure of quality is S/N and noise in the derived voice channel, the measure of quality on a digital link is BER. To derive a value for BER, we must first calculate E_b/N_0. E_b/N_0 expresses received signal energy per bit per hertz of thermal noise. (See Section 2.11.1.1.)

One approach is to break E_b/N_0 down into E_b and N_0:

$$E_b = \frac{\text{RSL}}{\text{bit rate}}$$

or logarithmically

$$E_b = \text{RSL}_{\text{dBW}} - 10\log(\text{BR}) \qquad (2.67)$$

where BR = bit rate

$$N_0 = kT \text{ or } -228.6 \text{ dBW} + 10\log T_{sys} \qquad (2.68)$$

where T_{sys} is the effective noise temperature of the receiving system, or for a system operating at room temperature

$$N_0 = -204 \text{ dBW} + \text{NF}_{\text{dB}} \qquad (2.55)$$

where NF_{dB} is the receiver noise figure.

In Chapter 4 and in many other similar works the ratio C/N_0 is used. Here it should be noted that C and RSL are synonomous. Thus we can express

$$\frac{C}{N_0} = \text{RSL} - (-204 \text{ dBW}) - \text{NF}_{\text{dB}} \qquad (2.69a)$$

or

$$\frac{C}{N_0} = \text{RSL} - (228.6 \text{ dBW}) - 10\log T_{sys} \qquad (2.69b)$$

and

$$\frac{E_b}{N_0} = \frac{C}{N_0} - 10\log(\text{BR}) \tag{2.70}$$

Furthermore,

$$\frac{E_b}{N_0} = \text{RSL}_{\text{dBW}} - (-204 \text{ dBW}) - \text{NF}_{\text{dB}} - 10\log(\text{BR}) \tag{2.71a}$$

or

$$\frac{E_b}{N_0} = \text{RSL}_{\text{dBW}} - (-228.6 \text{ dBW}) - 10\log T_{sys} - 10\log(\text{BR}) \tag{2.71b}$$

During the process of carrying out a link analysis, we establish at the outset the desired E_b/N_0 derived from curves such as those in Figures 2.40 and 2.41. Of course, the curves in the two figures are for *ideal* systems. For a real system we need *practical* curves, which can be obtained from the modem or equipment manufacturer. The difference between ideal values for E_b/N_0 and practical values is on the order of 0.5 to over 5 dB. This difference is referred to as modulation implementation loss or system degradation. Table 2.16 (from Ref. 19) gives a budget for contributors to modulation implementation loss.

TABLE 2.16. Typical Degradation Budget for 90-Mbps Digital Radiolink

Cause[a]	Degradation (dB)
A. Modem, AWGN back-to-back	
1. Phase and amplitude errors of modulator	0.1
2. ISI caused by filters	1.0
3. Carrier recovery phase noise	0.1
4. Differential encoding/decoding	0.3
5. Jitter (imperfect sampling instants)	0.1
6. Excess noise bandwidth of receiver	
(demodulator)	0.5
7. Other hardware impairments	
(temperature variations, aging, etc)	0.4
Modem total	2.5 dB
B. RF Channel imperfections	
1. AM/PM conversion of the quasilinear	
output stage	1.5
2. Band limitation and group delay	0.3
3. Adjacent RF channel interference	1.0
4. Feeder echo distortion	0.2
Channel total	3.0 dB
Total degradation	5.5 dB

[a]AWGN = additive white Gaussian noise; ISI = intersymbol interference.

Source. Reference 19, courtesy of Prentice-Hall.

Example 21. Assume a 15-mile path with the operating frequency at 6 GHz on a link designed to transmit 90 Mbps (1344 VF channels of North American PCM). 16-QAM modulation was selected. The modulation implementation loss is 4.7 dB and the desired BER per hop is 1×10^{-9}. Turning to Figure 2.41, we find that the ideal E_b/N_0 for a BER $= 1 \times 10^{-9}$ is 21.2 dB. Assume no fading and zero margin, receiver noise figure of 5 dB, waveguide losses at each end of 1.5 dB, and atmospheric absorption of 0.2 dB. Find reasonable transmitter output power and antenna apertures to meet these conditions.

Calculate N_0. Use equation (2.55):

$$N_0 = -204 \text{ dBW} + 5 \text{ dB}$$

$$= -199 \text{ dBW}$$

Adding the ideal E_b/N_0 (21.2 dB) to the modulation implementation loss (4.7 dB) gives the practical E_b/N_0 required to achieve a BER of 1×10^{-9}. The energy per bit E_b, must then be 21.2 + 4.7 dB or 25.9 dB higher in level than N_0. We express this as

$$E_b = -199 \text{ dBW} + 25.9 \text{ dB}$$

$$= -173.1 \text{ dBW}$$

The receive signal level (RSL) is $10 \log(\text{BR})$ greater than -173.1 dBW:

$$\text{RSL}_{\text{dBW}} = -173.1 \text{ dBW} + 10 \log(90 \times 10^6)$$

$$= -173.1 \text{ dBW} + 79.54 \text{ dB}$$

$$= -93.56 \text{ dBW}.$$

This is the minimum required RSL to meet specifications with 0-dB margin.

We now calculate a "trial" RSL to see how much gain or loss we must add to the system to meet the required value. Using equation (2.66):

$$\text{RSL}_{\text{dBW}} = \text{EIRP}_{\text{dBW}} + \text{FSL} + P_L + G_r + L_{Lr}$$

where, as we know, $L_{Lr} = 1.5$ dB. Now assume the output power to the transmitter is 0 dBW (e.g., 1 W), a reasonable value, and calculate FSL using Equation (1.7a) for the 15-mile hop at 6 GHz:

$$\text{FSL} = 36.58 + 20 \log 15 + 20 \log 6000$$

$$= 135.66 \text{ dB}$$

Then $\text{RSL}_{\text{dBW}} = \text{EIRP}_{\text{dBW}} + (-135.66 \text{ dB}) + (-0.2 \text{ dB}) + G_r + (-1.5 \text{ dB})$. Simplifying the above expression gives

$$\text{RSL}_{\text{dBW}} = \text{EIRP} - 137.36 \text{ dB} + G_r$$

G_r is the gain of the receive antenna; let G_t be the gain of the transmit antenna. Let the aperture of each antenna be 2 ft, then from equation (2.27a)

$$G_t = G_r = 20 \log B + 20 \log F + 7.5$$

where B is the diameter (aperture) of the parabolic dish in feet and F is the frequency in gigahertz (assuming 55% antenna efficiency). Thus

$$G_t = G_r = 29 \text{ dB}.$$

Calculate the EIRP from equation (2.6):

$$\text{EIRP}_{dBW} = P_0 + L_{Lt} + G_t$$

where L_{Lt} = waveguide loss from transmitter to antenna = 1.5 dB. Thus

$$\text{EIRP}_{dBW} = 0 + (-1.5 \text{ dB}) + 29 \text{ dB}$$

$$= +27.5 \text{ dBW}$$

The "trial run"

$$\text{RSL} = +27.5 \text{ dBW} - 137.36 \text{ dB} - 1.5 \text{ dB} + 29 \text{ dB}$$

$$= -82.36 \text{ dBW}$$

However, the specified RSL = -93.56 dBW, and we find that we have $-82.36 - (-93.56 \text{ dBW})$ or 11.2 dB margin. If we desire a 0-dB margin as specified (which on real systems we would not), we have 11.2 dB either by reducing antenna aperture of the transmit and receive antennas, reducing transmitter output power, or a combination of both.

By reducing antenna diameter at both ends by 25% (e.g., using a 1.5-ft aperture rather than 2-ft), we would reduce the antenna gain by about 3 dB or a total of 6 dB for both antennas. The remaining 5.2 dB could be used to reduce transmitter power output by that amount, and the new output would be -5.2 dBW (Remember that we set it at 0 dBW in the preceding calculations.) Convert -5.2 dBW to watts by

$$P_0(\text{watts}) = \log^{-1}\left(\frac{\text{power}_{dBW}}{10} \right)$$

$$= 302 \text{ mW}$$

2.12 RADIOLINK AVAILABILITY

2.12.1 Introduction

Availability, usually expressed in percentage or as a decimal, defines the time a system or terminal is meeting its operational requirements. Equipment availability is expressed by the familiar equation

$$A = \frac{\text{MTBF}}{\text{MTBF} + \text{MTTR}} \tag{2.72}$$

where MTBF is the mean time between failures and MTTR is the mean time to repair. MTBF and MTTR are usually expressed in hours.

As one can see, this equation only treats equipment failure and its repair time; it does not reflect outage due to fading. By restating the equation, we can cover the general case:

$$A = \frac{\text{uptime}}{\text{uptime} + \text{downtime}} \tag{2.73}$$

Example 22. If a system has an uptime of 10,000 hr and 10 hr of downtime, then

$$A = \frac{10,000}{10,000 + 10}$$

$$= 0.999 \quad \text{or} \quad 99.9\%$$

Often we wish to find the unavailability (U) of a system, where

$$U = 1 - A \tag{2.74a}$$

$$U = 1 - \frac{\text{uptime}}{\text{uptime} + \text{downtime}} \tag{2.74b}$$

In the above example

$$U = 1 - 0.999$$

$$U = 0.001 \text{ or } 0.1\%$$

With radiolinks we will deal with one-way (e.g., in one direction, say west to east) and two-way availabilities. If the unavailability objective for a two-way channel is 0.02%, and outage probabilities in the two directions are independent then the objective for a one-way channel is 0.01% or about 105 min/year for a two-way system and 53 min/year for an equivalent one-way system.

2.12.2 Contributors to Unavailability

CCIR Rep. 445-2 lists five major contributors to outage on LOS radiolinks:

☐ Equipment failure
☐ Primary power failure
☐ Propagation (fading or excessive ISI* on digital systems)
☐ Maintenance and human error
☐ Unlocated

*ISI = intersymbol interference.

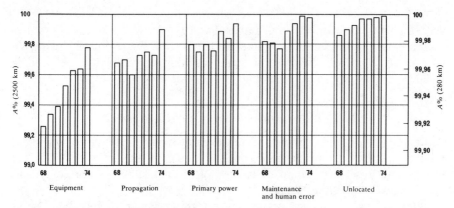

FIGURE 2.43 Availability versus causes of failure for the complete network of the Italian radio-relay system generating in the frequency band 7125–7750 MHz. From CCIR Rep. 445-2. Courtesy of ITU-CCIR, Geneva (Ref. 1).

A typical distribution of these failures on an annual basis is shown in Figure 2.43.

2.12.3 Availability Requirements

The ATT unavailability objective is 0.01% for a one-way channel over a 4000-mile route (Ref. 15). The equivalent availability is 99.99%.

In Canada a tentative objective of 99.97% is used for a 1000-mile one-way radio system. This corresponds to 99.95% availability on a 2500-km base.

The current United Kingdom availability objective for bidirectional transmission is 99.994% per 100 km, which corresponds to 99.84% for a 2500-km circuit. (Ref. CCIR Rep. 445-2) (Ref. 1).

2.12.4 Calculation of Availability of Radiolinks in Tandem

In this subsection we derive per hop availability given a system availability consisting of n hops in tandem with independent outages on different links. Often such a system availability will derive from the CCIR hypothetical reference circuit, which is 2500-km long. Panter (Ref. 20) assumes that such a 2500-km circuit consists of 54 hops each 30 miles (48 km) long.

We describe the procedure by an example. If a one-way circuit requires an availability of 99.95% over a 2500-km reference circuit, what is the required per-hop availability? First calculate the system unavailability, which, in this case, is $1 - 0.9995 = 0.0005$. Divide this unavailability by 54 or $0.0005/54 = 0.000009259$. This is then the unavailability for one hop. Its availability is $1 - 0.000009259 = 99.999074\%$.

ATT uses a 4000-mile reference circuit with a required availability of 99.99% or an unavailability of 0.0001 or 1×10^{-4}. If we assume as above that each hop is 30 miles long, then there are 133 hops in the reference circuit. We now divide 1×10^{-4} by 133 and the resulting unavailability per hop is 0.000075%. The equivalent availability per hop is then 99.999925%.

2.12.4.1 Discussion of Partition of Unavailability

At first glance we could apportion half the outage (unavailability) to equipment failure and half to propagation outage. Thus, in the case of ATT, the unavailability per hop would be 0.000075%/2 or 0.0000375% for equipment and 0.0000375% for propagation outages.

White of GTE Lenkurt argues against this approach (Ref. 21). Let us consider a 1-year of 8760-hr interval. A year has 525,600 min or 31,536,000 sec. What is the annual expected outage when the unavailability is 0.000075%?

$$8760 \text{ hr} \times 0.00000075 = 0.00657 \text{ hr}$$

$$525,600 \text{ min} \times 0.00000075 = 0.3942 \text{ min}$$

$$31,536,000 \text{ sec} \times 0.00000075 = 23.652 \text{ sec}$$

If we assign half of this number to equipment outage and half to propagation outage, we then have 11.8 sec outage per year for each.

The next step is to apply the conventional formula for availability [equation (2.72)] and calculate MTBF in hours:

$$0.000075\% = \frac{\text{MTBF}}{\text{MTBF} + \text{MTTR}}$$

However, we first must assign a reasonable value for MTTR or repair time.

Consider that most LOS radiolink sites are unmanned, and when a failure occurs, a technician must be sent to that site. He/she must be alerted, gather up tool kit and required parts, and travel to that site, possibly 60–100 miles away, and, of course, time must be allowed to carry out the repair. Values for MTTR in the literature for this application are from 2 to 10 hr. We use the worst case, then

$$99.999925\% = \frac{\text{MTBF}}{\text{MTBF} + 10}$$

$$\text{MTBF} = 1.32 \times 10^6 \text{ hr or } 150.68 \text{ years.}$$

If we allotted half the outage to equipment failure and half to propagation, we must double the MTBF, requiring an MTBF of about 301 years!

The argument then follows that propagation outage, say 32 sec a year, might consist of many events (short fades) in 1 year, whereas with equipment reliability we are dealing with one event every 301 years.

It would follow, then, that we treat these two types of outages separately and independently for they are truly not summable. It is like summing 6 apples and 4 oranges resulting in 10 lemons. What is driving us to these large values of MTBF is the large values for MTTR. On sophisticated military radio terminals (non-LOS radiolink), MTTR runs at about 0.3 to 0.5 hr. It is assumed that the technician is on site and on duty.

2.12.4.2 Propagation Availability

We then treat propagation availability separately, but it too requires some special considerations of reasonableness. Again let us turn to an example. We use the Canadian values (Section 2.12.3) of 99.95% for a 2500-km reference circuit and the equivalent unavailability is 0.0005. Assume, again, 54 hops in tandem for the 2500-km reference circuit. The unavailability per hop is then $0.0005/54 = 0.000009259$ or an availability of 99.9990741%. This, of course, assumes a very worst case fading where all hops fade simultaneously but independently. This is unrealistic.

Panter (Ref. 20) reports a more reasonable worst case where we would allow only one-third of the hops to fade at once, $54/3 = 18$; thus, we can say that the unavailability due to propagation (multipath fading) is $0.005/18 = 0.0000278$ or an availability of 99.972 would be required per hop. This would be nearly in keeping with CCIR Rec. 395 para 1.2 or $(50/2500) \times 0.1$ of any month (not year) where the availability required is 99.998%/month to a reference level of 47,500 pWp. We essentially derive the same value for a 2500-km reference circuit where $L = 2500$, then the unavailability is $0.001/54$ or 99.998% availability per hop, following CCIR Rec. 395 to the letter.

Again the assumption is made that all hops in tandem are simultaneously subject to fading. Following Panter's reasoning, the divisor would be 18 rather than 54 or a required availability per hop of 99.994% to the reference level of 47,500 pWp.

The reader's attention is also called to CCIR Rep. 604-1 (Ref. 1).

PROBLEMS AND EXERCISES

1. For a 31-mile radiolink hop determine the additional values in feet to be added to obstacle heights at obstacle locations A, B, and C given the following properties:

        $A = 800$ ft above MSL, 5 miles from east terminal site

        $B =$ midpath at 520 ft above MSL

        $C = 720$ ft, 11 miles from west site

    Assume $K = 1.2$, frequency $= 7.1$ GHz.

2. For an initial planning exercise, determine LOS distance assuming smooth earth, grazing, and $K = \frac{4}{3}$. A structure on one end of the path is 420 ft high, and there is space available on a TV tower at the other end at the 465 ft level.

3. The surface refractivity at a certain location is given as 301, what is the refractivity at this location at 8000 ft altitude?

4. What is the free space loss (FSL) for a radiolink 9 miles long operating at 3.955 GHz?

5. What is the free space loss (FSL) for a radiolink 31 km long operating at 37.05 GHz?

6. What is the free space loss (FSL) to the moon at 6 GHz and at 12 GHz?

7. In reference to question 6, how important is it to state where on the earth's surface we measure from assuming LOS conditions?

8. What is the EIRP of a transmitting subsystem operating at 15 GHz with 4.5 dB of line losses? The output of the transmitter at its flange is 20 mW and the antenna gain is 41 dB.

9. A 21-mile radiolink operating at 6.1 GHz has the following characteristics: at the transmitter 120 ft of EW-64 waveguide (1.7 dB/100 ft), transition loss of 0.2 dB, a 4-ft antenna at the transmit end, 0 dBW output at the transmitter waveguide flange. What is the isotropic receive level at the distant end?

10. From question 9, on the receive end of the link there is 143 ft of EW-64 waveguide, a similar 4-ft antenna, and same transition. What is the unfaded RSL?

11. A low noise amplifier (LNA) has a 3 dB noise figure, operates at room temperature and incorporates a bandpass filter in its front end with 120-MHz passband. What is the thermal noise threshold of the receiver in dBm?

12. The first active stage of a radiolink receiver is a mixer with an 8.5 dB noise figure and B_{rf} of the link is 30 MHz. What is the FM improvement threshold?

13. Calculate the peak deviation of an FM radio transmitter designed to CCIR recommendations carrying 600 VF channels in an FDM/FM configuration? Use a peaking factor of 13 dB.

14. Calculate the unfaded carrier-to-noise ratio (C/N) for an FDM/FM radiolink 34-miles long based on CCIR recommendations where the receiver noise figure is 7.5 dB. The transmitter output power is 2.5 W and the operating frequency is 2.1 GHz. The link carries 1200 VF channels in a standard CCITT FDM configuration. Bandwidth is to be calculated according to Carson's rule. Use convential 4-ft antennas at each end. Line losses are 2 dB at each end.

15. What is the gain of a 2-ft radiolink parabolic reflector antenna assuming the conventional efficiency?

16. What fade margin would be assigned to a radiolink with a path with a 0.003% unavailability due to propagation where Rayleigh type fading is assumed?

17. Using the Barnett method, calculate path unavailability for a 45-km path operating at 4 GHz over very smooth terrain on the U.S. Gulf of Mexico coast with a 36-dB fade margin.

18. A certain radiolink path requires a 38-dB fade margin without diversity. What fade margin will be required (worst case) using frequency diversity with a frequency separation of 2%?

19. When calculating total noise at a radiolink receiver, what are the three types of noise to be considered? We may add these directly if the noise units are measured in ____. When operating near threshold, what type of noise predominates?

20. One hop of a 2500-km radiolink system is 45 miles long. What total noise is permitted at the receiver following CCIR recommendations (pWp0)?

21. Assume an NPR of 55 dB, calculate the noise components to reach the value in question 20. Let preemphasis improvement be 3.5 dB and diversity improvement be 3 dB. The link carriers 600 VF channels in a standard FDM/FM configuration. Feeder distortion is 6 pWp. The link operates at 6 GHz. Set the necessary parameters with the design following CCIR recommendations.

22. An uncooled receiver has a noise figure of 3.7 dB; calculate N_0.

23. The RSL at the input of the first active stage of a digital receiving system is -137 dBW, and the bit rate is 45 Mbps, calculate the value for E_b.

24. Name at least four digital equipment characteristics that give rise to modulation implementation loss.

25. On a certain digital link operating with 8 Mbps with 8-ary PSK modulation and the link BER desired was 1×10^{-8}. The link is 21 miles long and the operating frequency was 4 GHz. Allow 2.1 dB for line losses at each end, a modulation implementation loss of 3.1 dB, transmitter output of 0.5 W, and receiver noise figure of 6 dB. For an unfaded condition with zero margin, calculate reasonable antenna aperture size at each end.

26. After siting a particular link and carrying out the installation, we find that somehow the link did not meet performance objectives by about 12 dB. What "fixes" can be done to bring the link up to specification? List these fixes in reasonable order of costs.

27. The required availability of a link is 99.95%. What is the unavailability?

28. There are two aspects to link availability. Describe methods to improve each.

29. A certain radiolink system has 27 hops, and we desire an availability at the far end of the system (receiving) of 99.93%. Calculate the availability per hop required by the "conservative method" and by the "Panter method."

30. Consider a radiolink where the desired availability is 99.995%. Apportion half to equipment failure and allow a mean time to repair (MTTR) of 2 hr. What equipment MTBF is required?

REFERENCES

1. Recommendations and Reports of the CCIR, 1982, XVth Plenary Assembly, Geneva 1982.

2. "Design Handbook for Line-of-Sight Microwave Communication Systems," MIL-HDBK-416, U.S. Department of Defense, Washington, DC, November 1977.

3. "Naval Shore Electronics Criteria, Line-of-Sight Microwave and Tropospheric Scatter Communication Systems," Navelex 0101, 112, U.S. Department of the Navy, Washington, DC, May 1972.

4. B. R. Bean et al., *A World Atlas of Atmospheric Radio Refractivity*, U.S. Department of Commerce, ESSA, Boulder, CO, 1966.

5. R. L. Freeman, *Telecommunication Transmission Handbook*, 2nd ed., Wiley, New York, 1982.

6. R. L. Marks, et al., "Some Aspects of the Design for Line-of-Sight Microwave and Troposcatter Systems," USAF Rome Air Development Center, Rome, NY, NTIS AD 617-686, Springfield, VA, April 1965.

7. H. Brodhage and W. Hormuth, *Planning and Engineering of Radio Relay Links*, 8th ed., Siemens-Heyden & Son Ltd., London, 1978.

8. *Engineering Considerations for Microwave Communication Systems*, GET-Lenkurt, San Carlos, CA, 1975.

9. H. T. Dougherty, "A Summary of Microwave Fading Mechanisms, Remedies and Applications," U.S. Department of Commerce, ESSA Technical Report ERL69-WPL 4, Boulder, CO, March 1968.

10. *Reference Data for Radio Engineers*, 5th ed., ITT-Howard W. Sams, Indianapolis, IN, 1968.

11. A. P. Barkhausen et al., "Equipment Characteristics and Their Relationship to Performance for Tropospheric Scatter Communication Circuits," Tech. Note 103, U.S. National Bureau of Standards, Boulder, CO, January 1962.

12. "Standard Microwave Transmission Systems," RS-252A, Electronics Industry Association, Washington, DC, September 1972.

13. M. J. Tant, *The White Noise Handbook*, Marconi Instruments Ltd., St. Albans, Herts, 1974.

14. "Telecommunications Systems Bulletin No. 10D—Interference Criteria for Microwave Systems in the Private Radio Services," Electronics Industry Association, Washington, DC, August 1983.

15. *Transmission Systems for Communications*, 5th ed., Bell Telephone Laboratories, Holmdel, NJ, 1982.

16. FCC Rules and Regulations Part 21, Vol. VII, Federal Communications Commission, Washington, DC, September 1982.

17. "Linear Modulation Techniques for Digital Microwave," Harris Corp. RADC-TR-79-56, USAF Rome Air Development Center, Rome, NY, August 1979.

18. E. W. Allen, "The Multipath Phenomenon in Line-of-Sight Digital Transmission Systems," *Microwave Journal*, (May 1984).

19. K. Feher, *Digital Communications Microwave Applications*, Prentice-Hall, Englewood Cliffs, NJ, 1981.

20. P. F. Panter, *Communication Systems Design Line-of-Sight Microwave and Troposcatter Systems*, McGraw-Hill, New York, 1972.

21. R. F. White, *Reliability in Microwave Communications Systems—Prediction and Practice*, GTE-Lenkurt, San Carlos, CA, 1970.

22. R. L. Freeman, *Reference Manual for Telecommunication Engineering*, Wiley, New York, 1985.

23. U.S. Military Standard, "Common Long Haul and Tactical Communication System Technical Standards," MIL-STD-188-100, U.S. Department of Defense, Washington, DC, November 1972.

24. J. D. Oetting, "A Comparison of Modulation Techniques for Digital Radio," *IEEE Trans. Comm.* **Com-27** (Dec., 1979).

OVER-THE-HORIZON RADIOLINKS

3.1 OBJECTIVES AND SCOPE

This chapter deals with the design of radiolinks that operate beyond line-of-sight (LOS). Two transmission modes achieving over-the-horizon communications will be examined:

- ☐ Diffraction
- ☐ Scattering

Diffraction over one or more obstacles can be the predominant transmission mode on most shorter paths (i.e., less than 100 miles) displaying long-term median transmission loss (path loss) on the order of 170–190 dB. On longer paths (i.e., from 100 to about 500 miles or more) scattering off a nonhomogeneous atmosphere will be the predominant transmission mode with long-term median transmission loss of 180–260 dB.

Emphasis will be placed on the design of radiolinks operating in the troposcatter mode. A brief review of the design for knife-edge and smooth earth diffraction links is also presented.

3.2 APPLICATION

Over-the-horizon radiolinks, whether based on the diffraction or scatter transmission modes, use larger installations and are more expensive than their LOS counterparts. The scattering and diffraction phenomena limit the frequency bands of operation from about 250 MHz to under 6 GHz. Table 3.1 provides a useful comparison of LOS and scatter/diffraction systems.

TABLE 3.1. Comparison of Radiolinks: LOS versus Tropo/Diffraction

Item	LOS	Tropo/diffraction
VF channels	Up to 1800/ 2700 per carrier	Up to 240 per carrier
Bit rate	90 Mbps or more per carrier	2400 bps to 4 Mbps per carrier
Path length	1–50 miles	50–500 miles
Transmit power	0.1–10 W	100–50,000 W
Receiver noise figure	4–12 dB	Less than 4 dB
Diversity	None or dual	Dual or quadruple
Antenna aperture size	1–12 ft	6–120 ft

A summary of some advantages of tropospheric scatter and diffraction links is presented below:

☐ Military/tactical for transportability where the total number of installations is minimized for point-to-point area coverage.

☐ Reduces the number of relay stations required to cover a given large distance when compared to LOS radiolinks. Tropospheric scatter/ diffraction systems may require from one-tenth to one-third the number of relay stations as an LOS radiolink system over the same path.

☐ Provides reliable multichannel communication across large stretches of water (e.g., over inland lakes, to offshore islands or oil rigs, between islands) or between areas separated by inaccessible terrain.

☐ May be ideally suited to meet toll-connecting requirements in areas of low population density.

☐ Useful for multichannel connectivity crossing territories of another political administration.

☐ Requires less maintenance staff per route-kilometer than comparable LOS radiolinks over the same route.

☐ Allows multichannel communication with isolated areas, especially when intervening territory limits or prevents the use of repeaters.

3.3 INTRODUCTION TO TROPOSPHERIC SCATTER PROPAGATION

The picture of the tropospheric scatter mechanism is still pretty speculative. One reasonable explanation is expressed in USAF Technical Order 31Z-10-13 (Ref. 1). It states that the ability of the troposphere to act as a refractive medium is based on the variations of refractivity caused by heating and

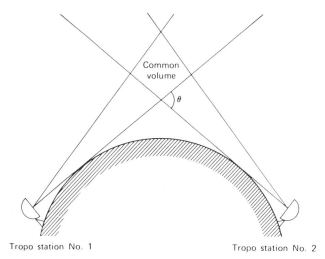

FIGURE 3.1 Tropospheric scatter model; θ is the scatter angle.

cooling of the atmosphere's water content. The variation is inversely proportional to altitude, with the greatest variation taking place nearest the earth's surface. The atmosphere is constantly in motion with respect to a point on earth, and this motion causes small irregularities, in refractivity, colloquially called "blobs." Such blobs are large with respect to the wavelengths used in tropospheric scatter radio systems and present a slightly different index of refraction than the surrounding medium. These relatively abrupt changes in the index of refraction produce a scattering effect of the incident radio beam. However, nearly all of the energy of the ray beam passes through these irregularities continuing onward out through free space, and only a small amount of energy is scattered back toward earth.

Figure 3.1 illustrates the concept of *common volume*, the common area subtended by the transmit and receive ray beams of a tropospheric scatter line. It is in this common volume where, for a particular link, the useful scattering takes place to effect the desired communications.

The scatter angle θ governs the received power level in an inverse relationship, with the receive level falling off rapidly as the scatter angle increases. The scatter angle is the angle between a ray from the receiving antenna and a corresponding ray from the transmit antenna, as shown in Figure 3.1.

Fading is common on scatter/diffraction paths. Two general types of fading are identified: short-term and long-term fading.

Short-term fading is characterized by changes in the signal level around the hourly median value. Short-term fading or fast fading is more prevalent on short hops where the fade rate may be as many as 20 fades per second.

The amplitude of these short-term fades follows a Rayleigh distribution (Figure 3.2). From the figure it will be seen that the median (50 percentile) is exceeded by about 5.5 dB 10% of the time, and 90% of the time the signal does

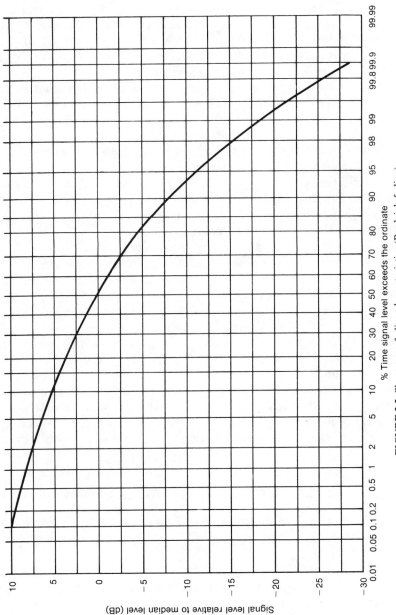

FIGURE 3.2 Short-term fading characteristics (Rayleigh fading).

not drop more than 8 dB below the median value. The difference between the 10% and 90% values is called the fading range, which is a measure of signal swing about the median. Typically, on a scatter hop this range is 13.5 dB if the signal follows completely a Rayleigh distribution. Values of less than 13.5 dB indicate the presence of a nonfading component.

Diversity is universally used on troposcatter links. Diversity tends to keep short-term fading within reasonable bounds, to no greater than 10 dB below the mean value of a single Rayleigh fading signal (Ref. 1), which is usually accommodated in the noise performance objectives of the system.

Long-term fading is another matter, which must be accommodated in the overall link design. Here we are dealing with long-term variations in signal level, which can be defined as variations in the hourly median values of transmission loss (path loss) over periods of hours, days, months, or years. Such long-term fading is caused by slow changes in the general condition of the transmission medium, the troposphere.

Long-term fades are characterized by day-to-day fades and seasonal fades. The day-to-day variations are caused by changes in various properties of the scatter volume such as temperature, density, moisture content, and altitude of the scattering layers. Seasonal variations are due to similar causes, but can be traced to seasonal cycles. The best propagation conditions are usually summer days and the worst case is during winter nights.

Long-term fading generally follows a log-normal distribution, and the fading *decreases* as hop length increases. Whereas the fade range on an 80-km circuit is around 30 dB, for a 500-km circuit it is about 11 dB. Increasing hop length generally requires an increase in scatter angle, raising the altitude of the common volume. The scatter properties of the atmosphere tend to change more at lower elevations and become more uniform at higher elevations.

In the path analysis phase of link design (Section 3.4.2) we will only deal with long-term fading.

3.4 TROPOSPHERIC SCATTER LINK DESIGN

3.4.1 Site Selection, Route Selection, Path Profile, and Field Survey

3.4.1.1 Introduction

Site selection, route selection, path profile, and field survey are carried out in a similar manner as outlined and discussed in Sections 2.2, 2.3, and 2.5. Here, then, we will point out the differences and areas of particular emphasis.

3.4.1.2 Site Selection

Consider these important factors. Whereas on LOS systems several decibels of calculation error may impact hop cost by one or several thousand dollars, impact on over-the-horizon hops may be on the order of hundreds of thousands of dollars or more. Thus, special attention must be paid to accuracy in site position, altitudes and horizon angles, and bearings.

Tropospheric scatter/diffraction sites will be larger, often requiring greater site improvement including freshwater, sanitary systems, living quarters, and more prime power and larger backup power plants. Radiated electromagnetic interference (EMI) is of greater concern. Takeoff angle (θ_{et}, θ_{er}) is critical. For each degree reduction of takeoff angle there is a 12-dB reduction (approximately) in median long-term transmission loss.

3.4.1.3 Route Selection

The route should be selected with first choice to those sites with the most *negative* and last choice to those with the largest positive takeoff angle. The effect of slight variations in path length is negligible for constant takeoff angles. The transmission loss on an over-the-horizon link will vary only slightly for changes in path length of less than about 10 miles (16 km). In a given area, therefore, it is usually best to select the highest feasible site, which also provides adequate shielding from potential interference, even though this may result in a slightly longer path than some location at a lower elevation.

Reference 2 suggests the following formula to estimate takeoff angle, which is valid only for smooth earth (or over water paths):

$$\theta_{et, er} = -0.000686 \left(h_{ts, rs} \right)^{1/2} \tag{3.1}$$

where $\theta_{et, er}$ is in radians; h_{ts} is the height in meters of the transmitting antenna above mean sea level (MSL); h_{rs} is the height in meters of the receiving antenna above MSL. The formula (3.1) is based on an effective earth radius of 4250 km, which is representative of a worst case condition.

3.4.1.4 Path Profile

A path profile of a proposed tropo/diffraction route is carried out in a similar manner as in Section 2.3. Tropo engineers prefer the use of $\frac{4}{3}$ paper. Takeoff angle is a more important parameter than K factor. Thus $\frac{4}{3}$ paper may prove more convenient in the long run. Key obstacles to be plotted are the horizons from each site. The horizon is the first obstacle that the ray beam will graze.

Basic tropospheric scatter path geometry is shown in Figure 3.3. From the path profile we will derive the following:

d = great circle distance between sites (km).

d_{Lt} = distance from transmitter site to transmitter horizon (km).

d_{Lr} = distance from receiver site to receiver horizon (km).

h_{ts} = elevation above MSL of the center of the transmitting antenna (km).

h_{rs} = elevation above MSL of the center of the receiving antenna (km).

h_{Lt} = elevation above MSL of the transmitter horizon point (km).

h_{Lr} = elevation above MSL of the receiver horizon point (km).

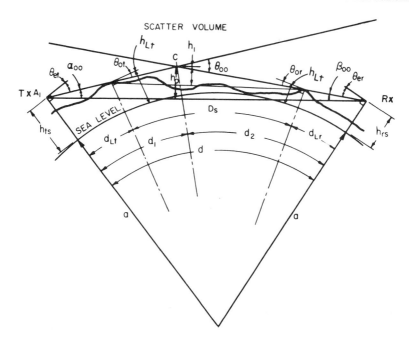

DISTANCES ARE MEASURED IN KILOMETERS ALONG A
GREAT CIRCLE ARC.

$$\theta_{oo} = \frac{D_s}{a} + \theta_{ot} + \theta_{or} = \frac{d}{a} + \theta_{et} + \theta_{er}$$

FIGURE 3.3 Tropospheric scatter path geometry.

3.4.2 Link Performance Calculations

3.4.2.1 Introduction

Free space loss (FSL) is based on theory and on unfaded LOS paths, the calculated receive signal level (RSL) and the measured level will turn out to be within 0.1 and 0.2 dB or less of each other. Transmission loss equations and curves for diffraction and troposcatter paths are empirical, based on hundreds of paths in many parts of the world. The cited references contain empirical methods to calculate tropo transmission loss:

1. NBS (NBS Tech Note 101) (Ref. 4).
2. CCIR (CCIR Rep. 238-2) (Ref. 5).
3. Yeh (Ref. 6).
4. Rider (Ref. 7).
5. Collins (Ref. 8).
6. Longley–Rice (computer model) (Ref. 12).

In our opinion, methods following NBS Tech Note 101 are the most well accepted worldwide. The method to be described is based on NBS 101 as set forth in USAF Tech Order 31Z-10-13 (Ref. 1), which we have simplified. However, the reader is warned that there can be considerable variation between calculated median loss values and measured loss, as much as 6 dB in some cases. It is for this reason that the term service probability must be dealt with, as will be described.

3.4.2.2 Two Definitions

The terms "time availability" and "service probability" must be distinguished and understood. We will first calculate the long-term median basic transmission loss L_{cr}. This is then corrected for a particular climatic region to derive the basic median transmission loss L_n. If the RSL on a link were calculated using this value (i.e., L_n), the RSL would reach or exceed this value only 50% of the time. This is the "time availability" of the link. Of course, we would wish an improved time availability of a link, usually 99% or better. If we design the link for a time availability of 99%, then 1% of the time the RSL will be less than the objective. In section 2.7 we called this the link availability or propagation reliability.

The basic median transmission loss is described by the notation $L_n(0.5, 50)$, or more generally $L_n(Q, q)$. Here the q refers to the time availability and Q refers to the *service probability*.

The service probability concept is used to obtain a measure of prediction uncertainty due to our lack of complete knowledge regarding the propagation mechanism, the semiempirical nature of the prediction formulas, and the uncertainties in equipment performance.

A service probability of 0.5 or 50% tells us that only half the links in a large population given identical input conditions will meet the time availability value. Often we engineer links for a service probability of 0.95 or 95%. Then only 5% of the links will fail to meet the time availability. The last step in calculating transmission loss is to extend the transmission loss to take into account prediction uncertainty, which we call here service probability.

3.4.2.3 The Propagation Mode

There are two possible modes for over-the-horizon transmission: diffraction or troposcatter. In most cases the path profile will tell us the mode. For paths just over LOS, the diffraction mode will predominate. For long paths, the troposcatter mode will predominate. If the basic median transmission loss of one mode is 15 dB more than the other mode, that with the higher loss can be neglected, and the one with the lower loss predominates.

To aid in determining the mode of propagation, the following criteria (Ref. 1) may be used:

☐ The distance (in kilometers) at which diffraction and forward scatter losses are approximately equal is $65(100/f)^{1/3}$, where f is the operating frequency in megahertz. For distances less than this value, diffraction will generally be predominant; for those distances greater than this value, the tropo mode predominates.

☐ For most paths having an angular distance (see Section 3.4.2.6) of at least 20 mrad, the diffraction effects may be neglected and the path can be considered to be operating in the troposcatter mode.

3.4.2.4 Method of Approach

First, we will discuss the equation for calculation of basic long-term tropospheric scatter transmission loss L_{bsr}, thence the geometry and substeps to carry out the calculation. The next step will be to calculate the basic long-term diffraction transmission loss L_{bd} for two selected diffraction modes. Then we discuss the mixed mode case, tropo/diffraction. The next operation is to calculate the reference value of basic transmission loss L_{cr} and extend to the basic transmission loss value $L_n(0.5, 50)$ for a region of interest. This value is then extended for the desired time availability, for the 50% service probability. The last step, if desired, is to again extend the value for an improved service probability, usually 95%.

3.4.2.5 Basic Long-Term Median Tropospheric Scatter Loss, L_{bsr}

$$L_{bsr}(dB) = 30 \log F - 20 \log d + F(\theta d) - F_0 + H_0 + A_a \qquad (3.2)$$

where f = the operating frequency in megahertz

d = greater circle path length in kilometers

$F(\theta d)$ = attenuation function

F_0 = scattering efficiency correction factor

H_0 = frequency gain function

A_a = atmospheric absorption

θd = product of the angular distance and path distance (angular distance in radians, path distance in kilometers)

We disregard the F_0 and H_0 terms. It should be pointed out that the CCIR method (Ref. 5) is similar to the NBS method, and it, too, disregards the scattering efficiency correction factor and the frequency gain function.

A_a, the atmospheric absorption term, is calculated using Figure 3.4 using distance d and the operating frequency f.

The attenuation function $F(\theta d)$ is one of the most significant terms in calculating L_{bsr}. Figures 3.11 through 3.15 plot values of decibel loss versus the product θd for values of surface refractivity and the path asymmetry factor s. Section 3.4.2.6 describes the steps necessary to calculate values of θ, the scatter angle in radians, sometimes called the angular distance.

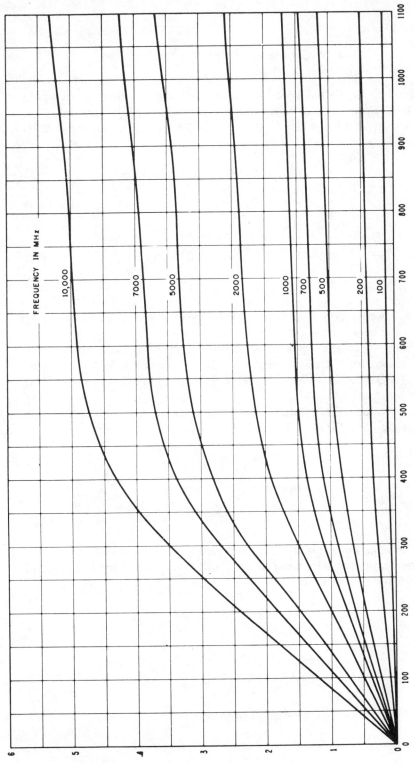

FIGURE 3.4 Estimate of median oxygen and water vapor absorption (dB) (Ref. 1).

147

3.4.2.6 Calculation of θ, the Scatter Angle

Turn to Section 3.4.1.4 and annotate the values taken from the path profile. These are used as the basis for calculating θ or angular distance. Compute values of $H_{st,\,sr}$, the average height of the surface above MSL (km).

Compute h_{st} and h_{sr}, where h_{st} and h_{sr} are the average heights above MSL (km) for the transmit site and receive site:

$$h_{st} = h_{Lt} \quad \text{if } h_{Lt} < h_{ts} + 0.15 \text{ km} \tag{3.3}$$

$$h_{sr} = h_{Lr} \quad \text{if } h_{Lr} < h_{rs} + 0.15 \text{ km} \tag{3.4}$$

or

$$h_{st} = h_{ts} \quad \text{if } h_{Lt} > h_{ts} + 0.15 \text{ km} \tag{3.5}$$

$$h_{sr} = h_{rs} \quad \text{if } h_{Lr} > h_{rs} + 0.15 \text{ km} \tag{3.6}$$

If there are two or more values of h_{Lt} or h_{Lr}, average the values before using the preceding equations.

Determine N_0 for the locations of h_{st} and h_{sr}. N_0 is the surface refractivity reduced to sea level. Use Figure 3.5 to obtain values for N_0 for locations outside of CONUS. Use Figure 3.6 for Continental United States (CONUS) locations.

Compute values of N_{st} and N_{sr}, which are values of refractivity (N) at the surface of the earth for the transmit and receive sites, respectively:

$$N_{st,\,sr} = N_0\exp(-0.1057h_{st,\,sr}) \tag{3.7}$$

Compute N_s, which is the value of N for the total path:

$$N_s = \tfrac{1}{2}(N_{st} + N_{sr}) \tag{3.8}$$

Obtain the value for a, the effective earth radius, from Figure 3.7.

Compute values of the takeoff angles (horizon angles) θ_{et} for transmit site and θ_{er} for the receiving site:

$$\theta_{et} = \frac{h_{Lt} - h_{ts}}{d_{Lt}} - \frac{d_{Lt}}{2a} \tag{3.9}$$

$$\theta_{er} = \frac{h_{Lr} - h_{rs}}{d_{Lr}} - \frac{d_{Lr}}{2a} \tag{3.10}$$

Compute D_s, the distance between horizon points:

$$D_s = d - (d_{Lt} + d_{Lr}) \tag{3.11}$$

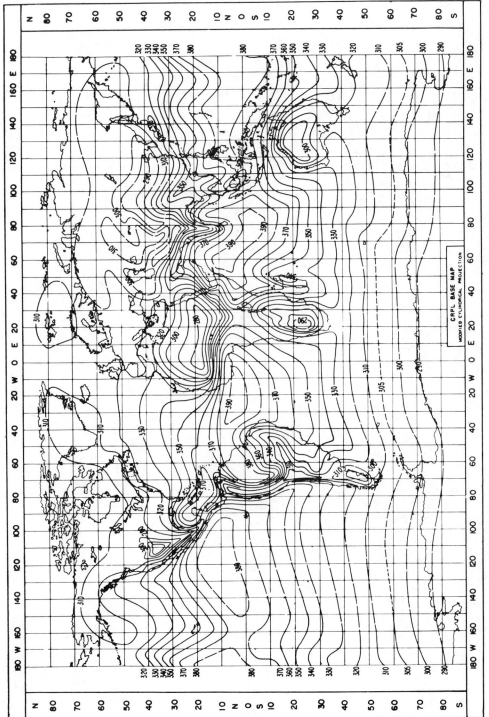

FIGURE 3.5 Minimum monthly surface refractivity values (N_0) referred to MSL (Ref. 1).

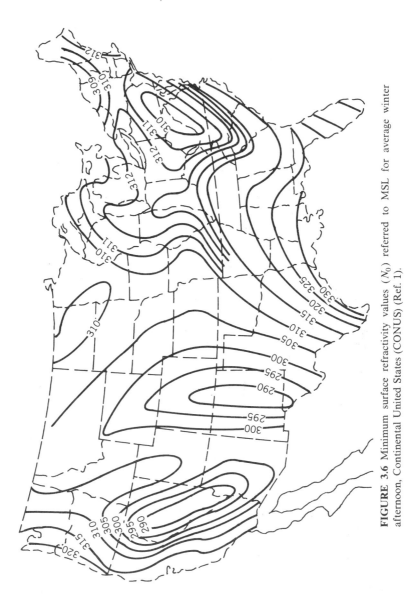

FIGURE 3.6 Minimum surface refractivity values (N_0) referred to MSL for average winter afternoon, Continental United States (CONUS) (Ref. 1).

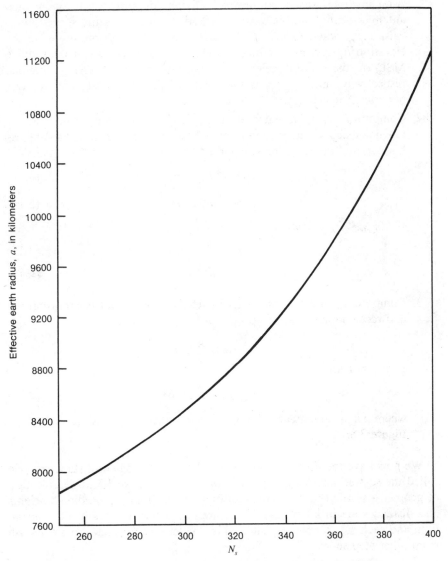

FIGURE 3.7 Effective earth radius versus surface refractivity N_s (Ref. 1).

The following procedures are used to compute the effective height of the transmitting and receiving antennas, respectively, h_{te}, h_{re}.

1. Compute the average height above MSL, $\bar{h}_t$, $\bar{h}_r$, of the central 80% of the horizon between each antenna and its respective point, h_{Lt}, h_{Lr}.

$$\bar{h}_{t,r} = \frac{1}{25} \sum_{i=3}^{27} h_{ti,ri} \text{ (km)} \tag{3.12}$$

All elevations are in kilometers. Divide the distance between each antenna location and its respective horizon into 31 evenly spaced intervals, $d_{ti, ri}$, where $i = 0, 1, 2, 3, \ldots, 30$. From the path profile, write the elevation $h_{ti, ri}$ corresponding to each. h_{t0} and h_{r0} are the heights above MSL of the ground below the transmitting and receiving antenna, respectively, and $h_{t30, r30}$ is the height of the transmitter and receiver horizons, respectively.

2. Compute h_t and h_r, which are the heights of the transmit and receive antennas above intervening terrain between the transmit site and its horizon and the receiving site and its horizon. If $h_{t0, r0} > \bar{h}_{t, r}$, then

$$h_t = h_{ts} - \bar{h}_t \qquad (3.13)$$

$$h_r = h_{rs} - \bar{h}_r \qquad (3.14)$$

If $h_{t0, r0} < \bar{h}_{t, r}$, then

$$h_t = h_{ts} - h_{t0} \qquad (3.15)$$

$$h_r = h_{rs} - h_{r0} \qquad (3.16)$$

3. Compute h_{te} and h_{re}, which are the effective heights of the transmitting and receiving antennas, respectively. If $h_{t, r} < 1$ km, then

$$h_{te, re} = h_{t, r} \qquad (3.17)$$

If $h_{t, r} > 1$ km, then

$$h_{te, re} = h_{t, r} - \Delta h_e \qquad (3.18)$$

where Δh_e is an antenna height correction factor. To find this value, use Figure 3.8.

We now have the necessary inputs to calculate the angular distance θ, also called the scatter angle. The reader should refer to Figure 3.3 for clarification of geometrical references and relationships. All angles are in radians; heights and distances are in kilometers. There are 12 numbered steps in the calculation, and each step number is followed by the letter t (for theta) to distinguish from other step series.

1t. Compute θ_{0t} and θ_{0r}, which are the angular elevations of the horizon ray at the transmitting and receiving points, respectively:

$$\theta_{0t} = \theta_{et} + \frac{d_{Lt}}{a} \qquad (3.19)$$

$$\theta_{0r} = \theta_{er} + \frac{d_{Lr}}{a} \qquad (3.20)$$

where a is the effective earth radius.

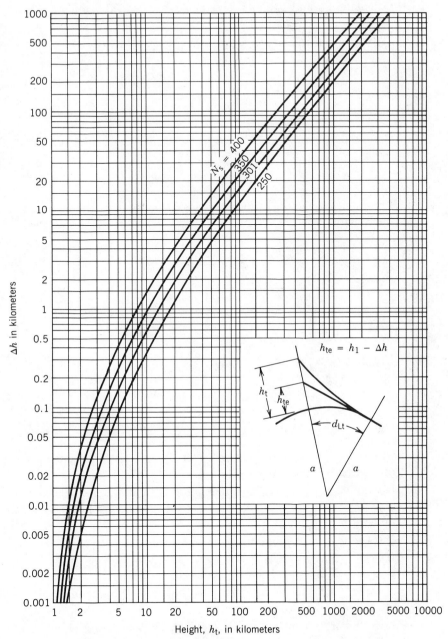

FIGURE 3.8 Reduction of antenna height for a very high antenna (Ref. 1).

2t. Calculate α_{00} and β_{00}, and these angles are defined in Figure 3.3.

$$\alpha_{00} = \frac{d}{2a} + \theta_{et} + \frac{h_{ts} - h_{rs}}{d} \tag{3.21}$$

$$\beta_{00} = \frac{d}{2a} + \theta_{er} + \frac{h_{rs} - h_{ts}}{d} \tag{3.22}$$

3t. Compute θ_{00}, which is the uncorrected scatter angle (radians).

$$\theta_{00} = \alpha_{00} + \beta_{00} \tag{3.23}$$

4t. Compute d_{st} and d_{sr}, which are the distances from the crossover of the horizon rays to the transmitter and receiver points, respectively:

$$d_{st} = \frac{d\alpha_{00}}{\theta_{00}} - d_{Lt} \tag{3.24}$$

$$d_{sr} = \frac{\beta_{00}}{\theta_{00}} - d_{Lr} \tag{3.25}$$

If θ_{0t} or θ_{0r} is negative, compute

$$d'_{st} = d_{st} = |a\theta_{0t}| \quad \text{for negative } \theta_{0t} \tag{3.26}$$

$$d'_{sr} = d_{sr} - |a\theta_{0r}| \quad \text{for negative } \theta_{0r} \tag{3.27}$$

where $d'_{st, sr}$ are the adjusted values of d_{st} and d_{sr} for negative values of θ_{0t} and θ_{0r}, respectively.

5t. Determine $\Delta\alpha_0$ and $\Delta\beta_0$ as a function of θ_{0t} and θ_{0r}; $\Delta\alpha_0$ relates to θ_{0t} and $\Delta\beta_0$ relates to θ_{0r}. If $0 < \theta_{0t, 0r} < 0.1$, use steps 6 and 7 and then proceed to step 10.

If $0.1 < \theta_{0t, 0r} < 0.9$, use step 8 and proceed to step 10.

If $\theta_{0t, 0r}$ is negative, use step 9 and then proceed to step 10. For $\theta_{0t, 0r}$ greater than 0.9, the corresponding value of $\Delta\alpha_0$ or $\Delta\beta_0$ is negligible. $\Delta\alpha_0$ and $\Delta\beta_0$ are parameters used to correct α_{00} and β_{00} for a refractivity N_s of 301.

6t. From Figure 3.9 determine $\Delta\alpha_0$ and $\Delta\beta_0$ for $N_s = 301$. Note in Figure 3.9 that θ_{0t} and θ_{0r} are in milliradians and that the X-axis is valid for both. Also note that the derived values of $\Delta\alpha_0$ and $\Delta\beta_0$ are in milliradians and must be converted to radians.

7t. Determine $\Delta\alpha_0$ and $\Delta\beta_0$ for given values of N_s. Read $C(N_s)$ from Figure 3.10 as a function of N_s. The parameter $C(N_s)$ is used to correct

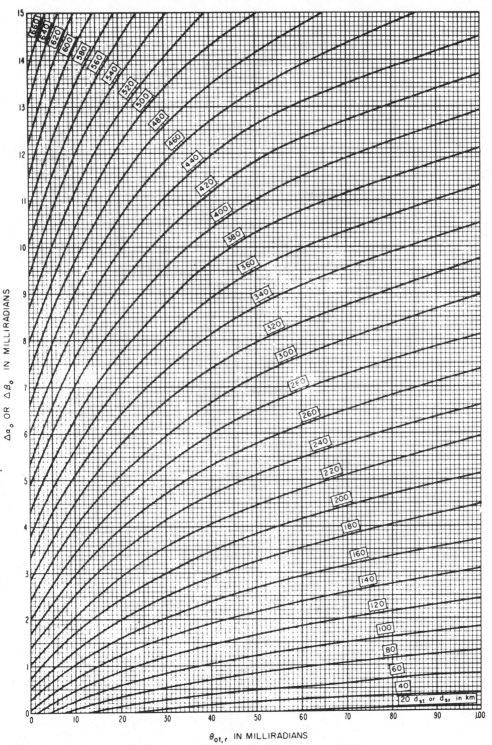

FIGURE 3.9 Correction terms $\Delta\alpha_0\,\Delta\beta_0$ for $N_s = 301$ (Ref. 1).

155

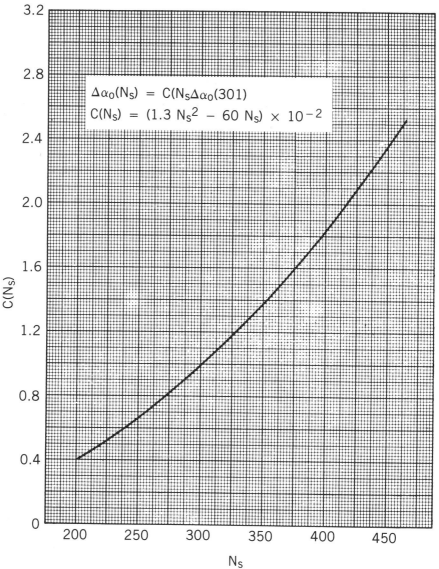

FIGURE 3.10 The coefficient $C(N_s)$ (Ref. 1).

α_{00} and β_{00} for the effects of a nonlinear refractivity index.

$$\Delta\alpha_0(N_s) = C(N_s)\Delta\alpha_0(301) \tag{3.28}$$

$$\Delta\beta_0(N_s) = C(N_s)\Delta\beta_0(301) \tag{3.29}$$

8t. Repeat steps 6 and 7 for $\theta_{0t} = 0.1$ and $\theta_{0r} = 0.1$ and to each add the following factor to the values of step 7. Use Figure 3.9 for $\theta_{0t,0r} = 0.1$. For $\Delta\alpha_0$ use θ_{0t} and d_{st} and for $\Delta\beta_0$ use θ_{0r} and d_{sr}.

$$\text{Factor (8t)} = N_s(9.97 - \cot\theta_{0t,0r})[1 - \exp(-0.05d_{st,sr})](10^{-6}) \tag{3.30}$$

9t. Repeat steps 6 and 7 for $\theta_{0t} = 0$ and $\theta_{0r} = 0$ using Figure 3.9.

10t. Calculate α_0 and β_0, which are the corrected values of α_{00} and β_{00}:

$$\alpha_0 = \alpha_{00} + \Delta\alpha_0 \tag{3.31}$$

$$\beta_0 = \beta_{00} + \Delta\beta_0 \tag{3.32}$$

11t. Calculate the scatter angle (angular distance) θ:

$$\theta = \alpha_0 + \beta_0 \tag{3.33}$$

12t. Calculate the path asymmetry factor s.

$$s = \frac{\alpha_0}{\beta_0} \tag{3.34}$$

3.4.2.7 Calculation of Basic Long-Term Median Tropospheric Scatter Loss, L_{bsr}

Calculate the product θd (i.e., multiply θ times d). θ is the scatter angle (in radians) and d is the great circle distance of the tropo hop in kilometers.

Compute the basic median tropospheric scatter transmission loss L_{bsr}:

$$L_{bsr} = 30\log f - 20\log d + F(\theta d) + A_a \tag{3.35}$$

Remember, in this simplified method, we omit the terms F_0 and H_0. Obtain values for $F(\theta d)$ as follows: if $\theta d \leq 10$, read from Figure 3.11; If $\theta d > 10$, read from Figures 3.12 to 3.15 for given values of N_s. If $s > 1$, read curves s_1, where $s_1 = 1/s$.

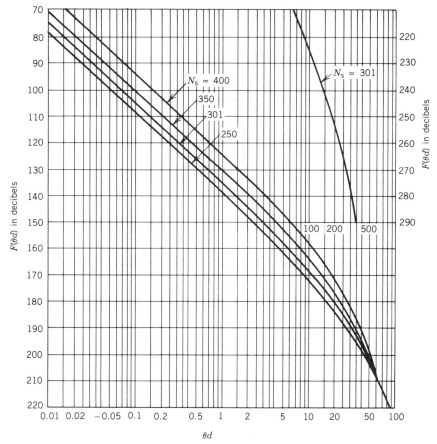

FIGURE 3.11 The attenuation function $F(\theta d)$; d is measured in kilometers and θ is the scatter angle in radians (Ref. 1).

3.4.2.8 Basic Long-Term Median Diffraction Loss, L_{bd}

3.4.2.8.1 Types of Diffraction Paths

Figure 3.16 illustrates six types of diffraction paths. Three of the more common paths are described below: knife-edge diffraction over a single, isolated obstacle (1) with and (2) without reflections (Section 3.4.2.8.2). Step numbers using the letter k; (3) smooth earth diffraction (such as over comparatively flat desert or over water where diffraction is the predominant mode) (Section 3.4.2.8.3). Step numbers using the letter s.

Note: All the following distances and elevations given are in kilometers, all angles are in radians.

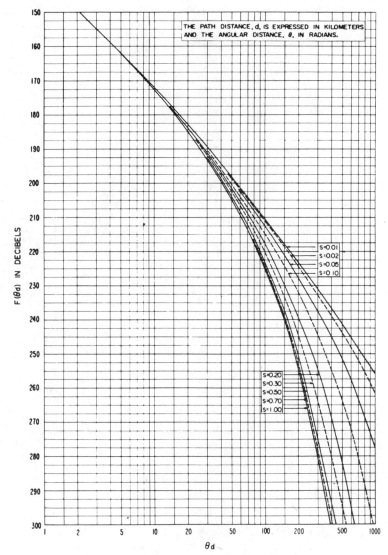

FIGURE 3.12 The function $F(\theta d)$ for $N_s = 250$ (Ref. 1).

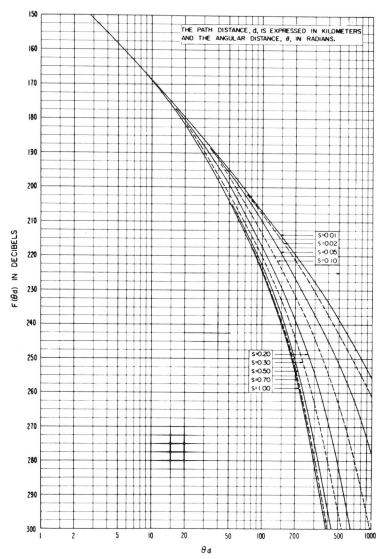

FIGURE 3.13 The function $F(\theta d)$ for $N_s = 301$ (Ref. 1).

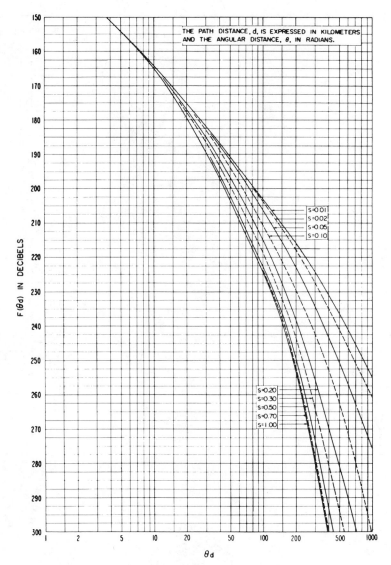

FIGURE 3.14 The function $F(\theta d)$ for $N_s = 350$ (Ref. 1).

161

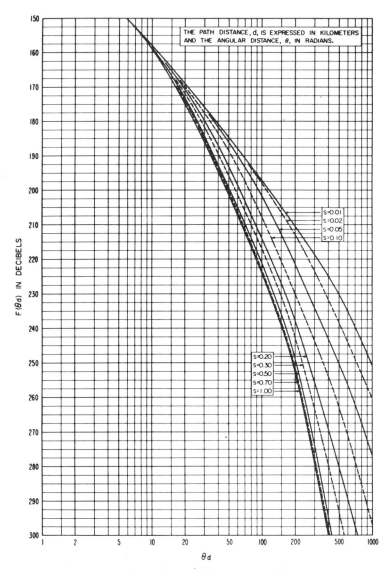

FIGURE 3.15 The function $F(\theta d)$ for $N_s = 400$ (Ref. 1).

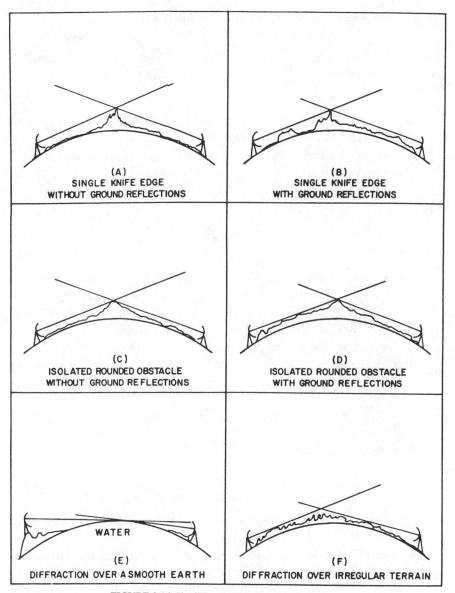

FIGURE 3.16 Six diffraction modes of propagation.

163

3.4.2.8.2 KNIFE-EDGE DIFFRACTION OVER A SINGLE, ISOLATED OBSTACLE

For this diffraction mode a common horizon is required between the transmit and receive antennas. The common horizon is usually a sharp mountain top or sharp ridge.

1k. For an ideal knife-edge without ground reflections, calculate L_{bd} using step 11k and employ the terms $A(v, 0)$, L_{bf}, and A_a.

2k. For an ideal knife-edge with ground reflections again use step 11k and employ terms $A(v, 0)$, L_{bf}, A_a, $G(h_1)$, and $G(h_2)$.

3k. Calculate v:

$$v = 2.583\theta \sqrt{\frac{fd_{Lt}d_{Lr}}{d}} \qquad (3.36)$$

where θ is the scatter angle as calculated in the previous section, step 11t; f is the operating frequency; d_{Lt} and d_{Lr} are from the path profile; and d is the great circle distance taken from the path profile.

4k. If $v < 2.4$, go to step 5k. If $v > 2.4$, go to step 6k.

5k. Determine $A(v, 0)$ from Figure 3.17 or by step 6k.

6k. Calculate $A(v, 0)$, which is the diffraction attenuation relative to free space as a function of v:

$$A(v, 0) = 12.953 + 20 \log v \qquad (3.37)$$

7k. Calculate L_{bf}, which is the free space loss (dB):

$$L_{bf} = 32.45 + 20 \log f + 20 \log d \qquad (3.38)$$

where f is the operating frequency in megahertz and d is the great circle distance of the hop in km.

8k. Steps 9k and 10k are used to compute the additional loss introduced by ground reflections. Reflections may occur in either or both the transmitting or receiving portions of the path. For the applicable portion compute $\bar{h}$ and $G(\bar{h})$ in steps 9k and 10k. If reflections may be neglected in one portion of the path, its corresponding value of $G(\bar{h})$ is then zero.

9k. Calculate $\bar{h}_1$ and $\bar{h}_2$.

$$\bar{h}_{1,2} = 7.23 \sqrt[3]{\frac{f^2}{d_{Lt}^2, d_{Lr}^2}} \left(h_{te, re}\right)^{4/3} \qquad (3.39)$$

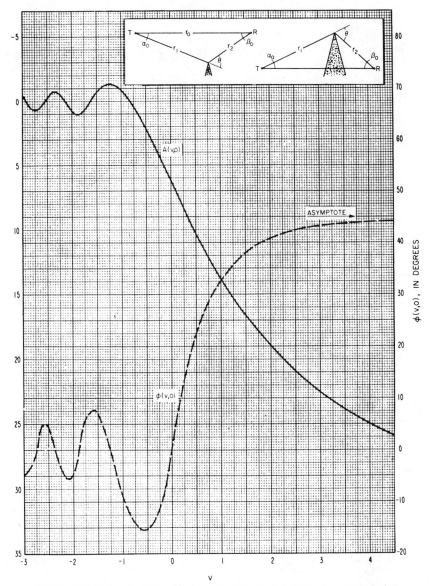

FIGURE 3.17 Knife-edge diffraction loss $A(v, 0)$ in decibels below free space (Ref. 1).

For $\bar{h}_1$ use d_{Lt} and h_{te} and for $\bar{h}_2$ use d_{Lr} and h_{re}, where h_{te} and h_{re} are the effective antenna heights from previous section.

10k. Read value of $G(\bar{h}_{1,2})$ for corresponding values of $\bar{h}_1$, $\bar{h}_2$ from Figure 3.18, where $G(\bar{h}_{1,2})$ is the reflection loss(es) in decibels in the transmitting and receiving segments, respectively.

11k. Compute the basic long-term median diffraction loss L_{bd}.

$$L_{bd} = L_{bf} + A(v,0) + A_a - G(\bar{h}_1) - G(\bar{h}_2) \qquad (3.40)$$

where A_a is obtained from Figure 3.4.

3.4.2.8.3 DIFFRACTION OVER SMOOTH EARTH

1s. Calculate C_0:

$$C_0 = \left(\frac{8497}{a} \right)^{1/3} \qquad (3.41)$$

where a = effective earth radius from Figure 3.7 given N_s (Section 3.4.2.6)

C_0 = a parameter for computing $K(a)$

2s. From Figure 3.19 determine K for an earth radius of 8497 km $[K(a = 8497)]$, where K is a function of the operating frequency.

3s. Determine $b°$ from Figure 3.20, where $b°$ also is a function of the operating frequency.

4s. Calculate $K(a)$, which is a value of K for an effective earth radius other than 8497 km:

$$K(a) = C_0 K(a = 8497 \text{ km}) \qquad (3.42)$$

5s. From Figure 3.21 determine $B(K, b°)$, which is a function of $K(a)$ and $b°$.

6s. Determine $C_1(k, b°)$ from Figure 3.22.

7s. Calculate B_0:

$$B_0 = f^{1/3} C_0^2 B(K, b°) \qquad (3.43)$$

8s. Calculate the following values:

$$x_0 = dB_0 \qquad (3.44a)$$

$$x_1 = d_{Lt} B_0 \qquad (3.44b)$$

$$x_2 = d_{Lr} B_0 \qquad (3.44c)$$

O ≤ K ≤ 0.1   b = 90°, 180°

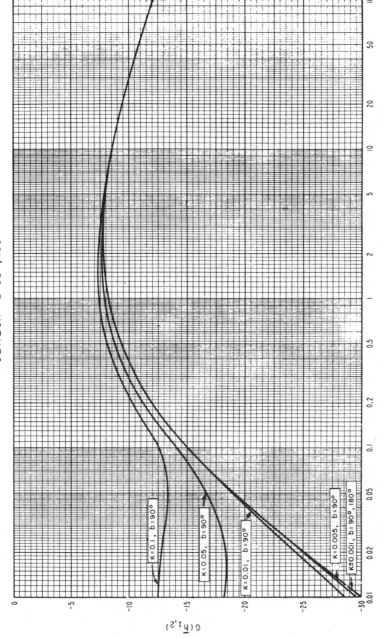

FIGURE 3.18 The residual height gain function, $G(\bar{h}_{1,2})$.

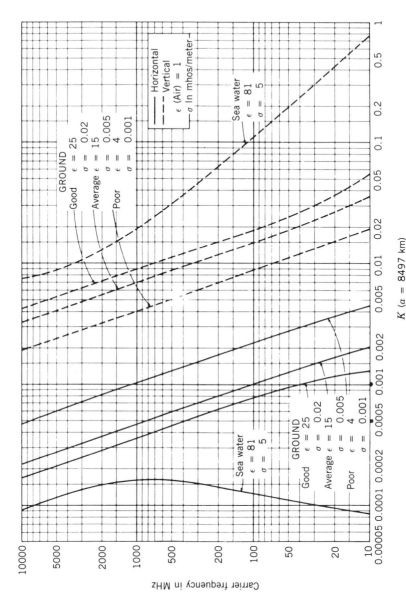

FIGURE 3.19 The parameter K for an effective earth radius $a = 8497$ km (Ref. 1).

168

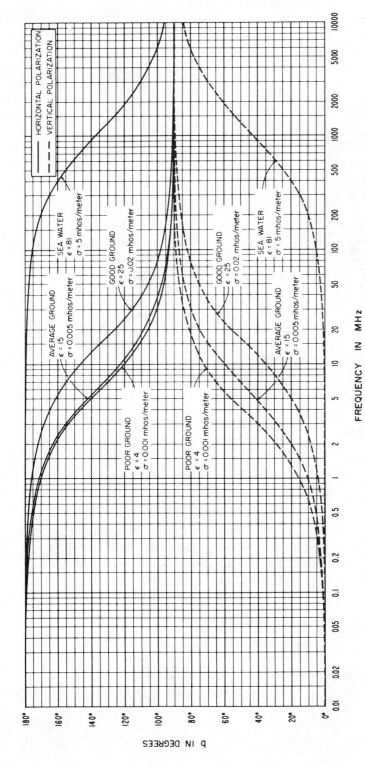

FIGURE 3.20 The parameter b° for ground wave propagation over a spherical earth (Ref. 1).

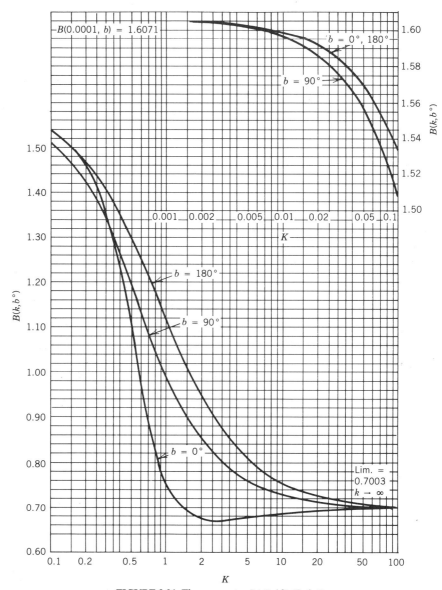

FIGURE 3.21 The parameter $B(K, b°)$ (Ref. 1).

where d is the great circle distance for the path and $d_{Lt, Lr}$ is the distance from the transmitter, receiver terminals to their respective horizons. Derived from path profile.

9s. Determine $G(x_0)$, $F(x_1)$, and $F(x_2)$ from Figure 3.23. These values consider the effect of distance and antenna height on the diffraction loss.

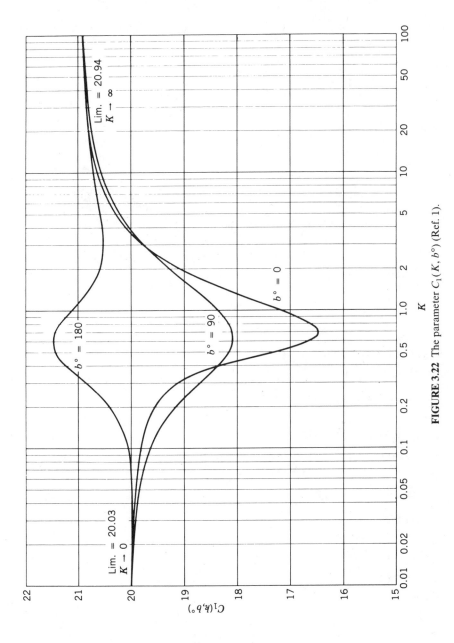

FIGURE 3.22 The parameter $C_1(K, b°)$ (Ref. 1).

171

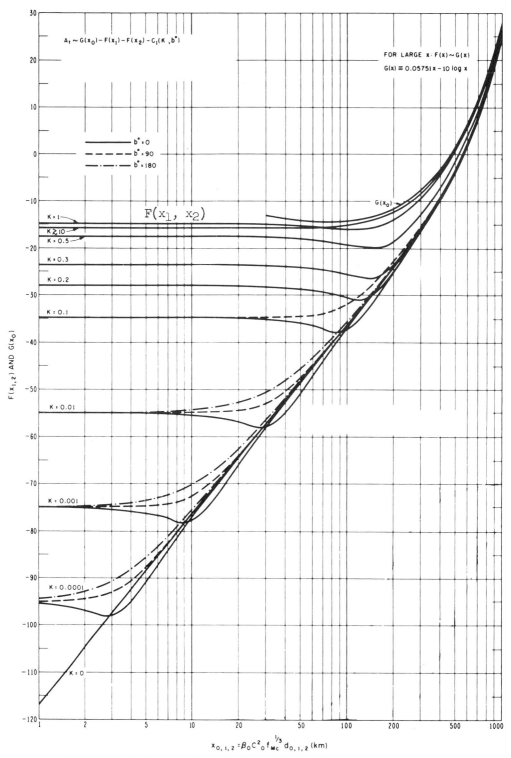

FIGURE 3.23 The functions $F(x_1)$, $F(x_2)$, and $G(x_0)$ for the range $0 \leq K \leq 1$ (Ref. 1).

10s. Calculate A, which is the diffraction attenuation relative to free space:

$$A_{dB} = G(x_0) - F(x_1) - F(x_2) - C_1(K, b°) \qquad (3.45)$$

11s. Calculate L_{bf}, the free space loss:

$$L_{bf} = 32.45 + 20 \log f + 20 \log d \qquad (3.46)$$

where f is the operating frequency in megahertz and d is the great circle distance in kilometers.

12s. Calculate L_{bd}, which is the long-term median value of transmission loss (A_a is the atmospheric gaseous loss from Figure 3.4):

$$L_{bd} = L_{bf} + A + A_a \qquad (3.47)$$

3.4.2.9 Calculation of the Combined Reference Value of Long-Term Basic Transmission Loss, L_{cr}

It is assumed here that a positive identification of the predominant transmission mode has not been made. If the tropo mode displays a transmission loss 15 dB or more greater than the diffraction mode, the diffraction mode predominates and

$$L_{cr} = L_{bd} \qquad (3.48a)$$

If the contrary occurs, where the diffraction mode displays a transmission loss 15 dB or more greater than the tropo mode, then the tropo mode predominates and

$$L_{cr} = L_{bsr} \qquad (3.48b)$$

If the two modes have transmission losses within 15 dB of each other, then

$$L_{cr} = L_{bd} - R(0.5) \qquad (3.48c)$$

The value of $R(0.5)$ is determined from Figure 3.24. The value of L_{cr} will be used in the next section.

3.4.2.10 Calculation of Effective Distance, d_e

Calculate d_{sl}, which is the distance where the diffracted and scatter fields are approximately equal over smooth earth:

$$d_{sl} = 65 \left(\frac{100}{f} \right)^{1/3} \qquad (3.49)$$

where f is the operating frequency in megahertz. Compute d_L, which is the sum of d_{Lt} and d_{Lr} for smooth earth with a radius of 9000 km:

$$d_L = 3\sqrt{2h_{te} \times 10^3} + 3\sqrt{2h_{re} \times 10^3} \qquad (3.50)$$

where h_{te} and h_{re} are the effective antenna heights from Step 3 of Section 3.4.2.6.

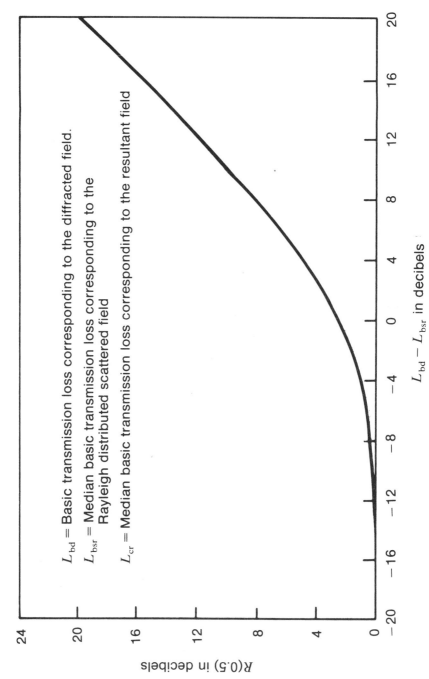

L_{bd} = Basic transmission loss corresponding to the diffracted field.

L_{bsr} = Median basic transmission loss corresponding to the Rayleigh distributed scattered field

L_{cr} = Median basic transmission loss corresponding to the resultant field

FIGURE 3.24 The parameter $R(0.5)$ for determining L_{cr} (Ref. 1).

If $d \leq d_L + d_{sl}$, then

$$d_e = \frac{130d}{d_L + d_{sl}} \qquad (3.50a)$$

If $d > d_L + d_{sl}$, then

$$d_e = 130 + d - d_L - d_{sl} \qquad (3.50b)$$

3.4.2.11 Calculation of the Basic Long-Term Median Transmission Loss, $L_n(0.5, 50)$

$L_n(0.5, 50)$ is the long-term loss adjusted for climatic region and effective distance providing a path time availability of 50% and a service probability of 50%. There are eight climatic regions given in Table 3.2. Select the climatic region from the table appropriate for the path in question. If two climatic regions appear applicable, calculate V_n for both regions and average the value.

TABLE 3.2. Characteristics of Climatic Regions

Climatic Region	Description

1 *Continental Temperature*

This region is characterized by an annual mean N_s of about 320 N-units with an annual range of monthly mean N_s of 20–40 N-units. A continental climate in a large land mass shows extremes of temperature in a "temperate" zone, such as 30°–60° north or south latitude. Pronounced diurnal and seasonal changes in propagation are expected to occur. On the east coast of the United States the annual range of N_s may be as much as 40–50 N-units owing to contrasting effects of arctic or tropical maritime air masses that may move into the area from the north or from the south.

2 *Maritime Temperate Overland*

This region is characterized by an annual mean N_s of about 320 N-units with a rather small annual range of monthly mean N_s of 20–30 N-units. Such climatic regions are usually located from 20° to 50° north or south latitude, near the sea, where prevailing winds, unobstructed by mountains, carry moist maritime air inland. These conditions are typical of the United Kingdom, the west coasts of North America and Europe, and the northwestern coastal areas of Africa.

Although the islands of Japan lie within this range of latitude, the climate differs in showing a much greater annual range of monthly mean N_s, about 60 N-units, the prevailing winds have traversed a large land mass, and the terrain is rugged. One would therefore not expect to find radio propagation conditions similar to those in the United Kingdom, although the annual mean N_s is 310–320 N-units in each location. Climate 1 is probably more appropriate than Climate 2 in this area, but ducting may be important in coastal and oversea areas of Japan as much as 5% of the time in summer.

3 *Maritime Temperate Oversea*

This region is characterized by coastal and oversea areas with the same general characteristics as those for Climate 2. The distinction made is that a radio path with

TABLE 3.2. (*Continued*)

Climatic Region	Description

both horizons on the sea is considered to be an oversea path; otherwise Climate 2 is used. Ducting is rather common for a small fraction of time between the United Kingdom and the European Continent and along the west coast of the United States and Mexico.

4 *Maritime Subtropical Overland*

This region is characterized by an annual mean N_s of about 370 N-units with an annual range of monthly mean N_s of 30–60 N-units. Such climates may be found from about 10°–30° north and south latitude, usually on lowlands near the sea with definite rainy and dry seasons. Where the land area is dry, radio ducts may be present for a considerable part of the year.

5 *Maritime Subtropical Oversea*

This region is characterized by conditions observed in coastal areas with the same range of latitude as Climate 4. The curves for this climate were based on an inadequate amount of data and have been deleted. It is suggested that the curves for Climates 3 or 4 be used, selecting whichever seems more applicable to each specific case.

6 *Desert, Sahara*

This region is characterized by an annual mean N_s of about 280 N-units with year-round semiarid conditions. The annual range of monthly mean N_s may be from 20 to 80 N-units.

7 *Equatorial*

This region is characterized by a maritime climate with an annual mean N_s of about 360 N-units and annual range of 0–30 N-units. Such climates may be observed from 20°N to 20°S latitude and are characterized by monotonous heavy rains and high average summer temperatures. Typical equatorial climates occur along the Ivory Coast and in the Congo of Africa.

8 *Continental Subtropical*

This region is typified by the Sudan and monsoon climates, with an annual mean N_s of about 320 N-units and an annual range of 60–100 N-units. This is a hot climate with seasonal extremes of winter drought and summer rainfall, usually located from 20° to 40°N latitude.

A Continental polar climate, for which no curves are shown, may also be defined. Temperatures are low to moderate all year round. The annual mean N_s is about 310 N-units with an annual range of monthly mean N_s of 10 to 40 N-units. Under polar conditions, which may occur in middle latitudes as well as in polar regions, radio propagation would be expected to show somewhat less variability than in a continental temperature climate. Long-term median values of transmission loss are expected to agree with the reference value L_{cr}.

High mountain areas or plateaus in a continental climate are characterized by low values of N_s and year-round semiarid conditions. The central part of Australia with its hot dry desert climate and an annual range of N_s as much as 50–70 N-units may be intermediate between Climates 1 and 6.

Source. Reference 4.

$$L(0.5) = L_{cr} - V(0.5, d_e) \quad dB$$

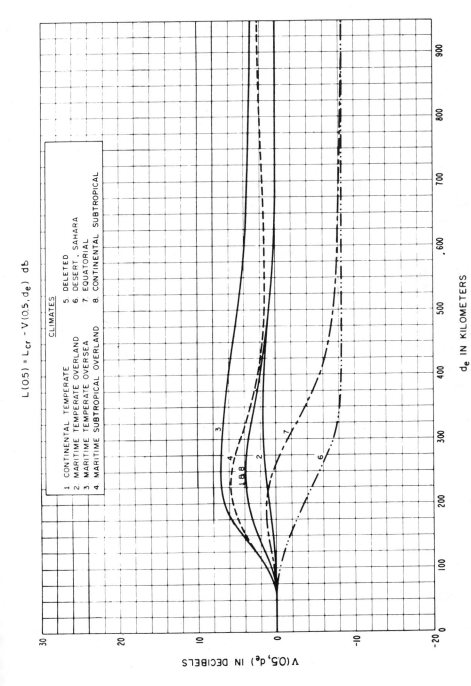

FIGURE 3.25 The function $V_n(0.5, d_e)$ for eight climatic regions (Ref. 1).

177

Note that the small n after the V is indicative of the climatic region. After selecting the climatic region (n), next obtain the appropriate value (in dB) for $V_n(0.5, d_e)$ from Figure 3.25 for the region and effective path distance (d_e) calculated in the previous section. Then calculate $L_n(0.5, 50)$:

$$L_n(0.5, 50) = L_{cr} - V_n(0.5, 50) \tag{3.51}$$

The transmission loss $L_n(0.5, 50)$ now must be extended from its present time availability of 50% to the desired time availability.

3.4.2.12 Extending to the Desired Time Availability, q

To extend the time availability from its median or 50% value, we must first calculate an additional loss, $Y_n(q, d_e, f)$, to be added to the value of L_n calculated above; q is the desired time availability, d_e is the effective distance from Section 3.4.2.10, and f is the operating frequency. The value for Y_n deals with the fading characteristics of the path. Fading is a function of operating frequency, path length, and climatic region (n). If more than one climatic region is indicated (i.e., a particular path has characteristics of two regions), then we calculate Y_n as follows for regions called i and j:

$$Y_n(q, d_e) = \sqrt{0.5 Y_i^2(q, d_e) + 0.5 Y_j^2(q, d_e)} \tag{3.52}$$

Climatic Regions 1, 6, and 8 require special treatment to calculate or derive values for $Y_n(q, d_e)$.

We first treat Climatic Region 1 (continental temperature climate). For this case there are two approaches to calculate values for $Y_1(q, d_e)$. The first approach is to use the curves directly, which are found in Figures 3.28, 3.29, and 3.30. For the second, more refined approach, we derive a value for $Y_1(0.9, d_e)$ for 100 MHz from Figure 3.26 and then multiply the value thus derived by a frequency correction factor $g(0.9, f)$ for the desired frequency f. This factor is taken from Figure 3.27 and is used for operating frequencies 2 GHz and below. The product so derived is then applied to the formulas given in the legend box in Figures 3.28, 3.29, and 3.30.

Example 1. Let $d_e = 200$ km, $f = 900$ MHz. Calculate the value of $Y_1(0.99, 200, 900)$.

From Figure 3.29 we derive -19.9 dB directly, using the first approach.

For the second approach, the function $Y_1(0.9, 200, 100)$ from Figure 3.26 is -8.1 dB. This is the value for Y_1 for a 0.90 time availability, effective distance of 200 km, and frequency of 100 MHz. The frequency correction factor is taken from Figure 3.27, and we see that for 900 MHz (worst case-summer) the value is 1.275. Multiplying these two values together (1.275×8.1 dB), we get -10.33 dB. Apply the formula from the legend in Figure 3.29 for $q = 0.99$ and $Y(0.99) = -10.33 \times 1.82 = -18.8$ dB or 1.1 dB less (loss) than by the first approach.

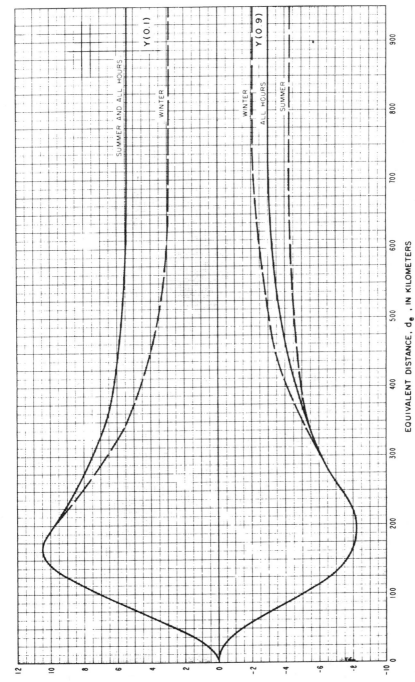

FIGURE 3.26 Long-term power fading function $Y(q, d_e, 100\ \text{MHz})$ for continental temperate climate (Region 1) (Ref. 1).

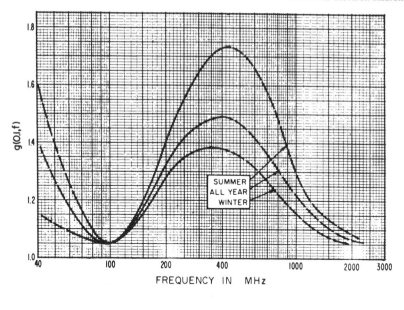

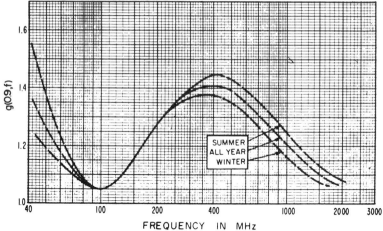

FIGURE 3.27 The parameter $g(q, f)$ for Climatic Region 1 (Ref. 1).

Reference 2 reports that the more refined approach can improve accuracies by more than 2 dB.

A similar approach is used for climatic regions 6 and 8. A frequency correction factor $g(q, f)$ is derived from Figure 3.31 for the specified frequency. Figures 3.32 and 3.33 give values of Y_n for Climatic Regions 6 and 8, respectively, for 1000 MHz. The next step is to determine values for long-term power fading from Figures 3.32 and 3.33 for the appropriate effective distance d_e. This value is then multiplied by the frequency correction factor $g(q, f)$ to determine $Y_n(q, d_e, f)$.

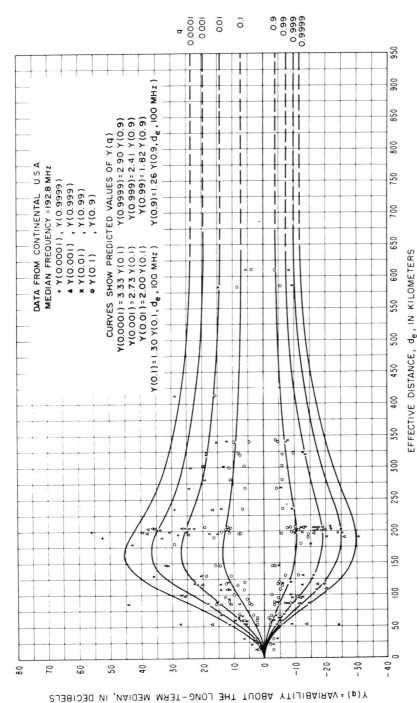

FIGURE 3.28 Long-term power fading, continental temperate climate, 250–450 MHz (Ref. 1).

DATA FROM CONTINENTAL U.S.A.
MEDIAN FREQUENCY = 192.8 MHz

+ Y(0.0001) , Y(0.9999)
▲ Y(0.001) , Y(0.999)
✕ Y(0.01)   , Y(0.99)
○ Y(0.1)   , Y(0.9)

CURVES SHOW PREDICTED VALUES OF Y(q)

Y(0.0001) = 3.33 Y(0.1)   Y(0.9999) = 2.90 Y(0.9)
Y(0.001) = 2.73 Y(0.1)   Y(0.999) = 2.41 Y(0.9)
Y(0.01) = 2.00 Y(0.1)   Y(0.99) = 1.82 Y(0.9)

Y(0.1) = 1.30 Y(0.1, d_e, 100 MHz)   Y(0.9) = 1.26 Y(0.9, d_e, 100 MHz)

EFFECTIVE DISTANCE, d_e, IN KILOMETERS

Y(q) = VARIABILITY ABOUT THE LONG-TERM MEDIAN, IN DECIBELS

q
0.0001
0.001
0.01
0.1
0.9
0.99
0.999
0.9999

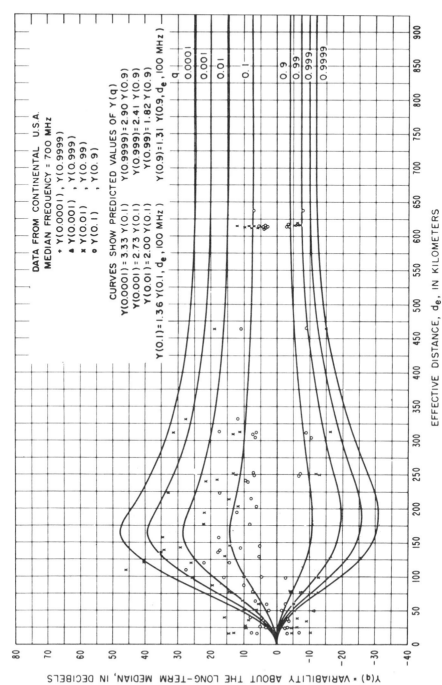

FIGURE 3.29 Long-term power fading, 450 to 1000 MHz, continental temperate climate (Region 1) (Ref. 1).

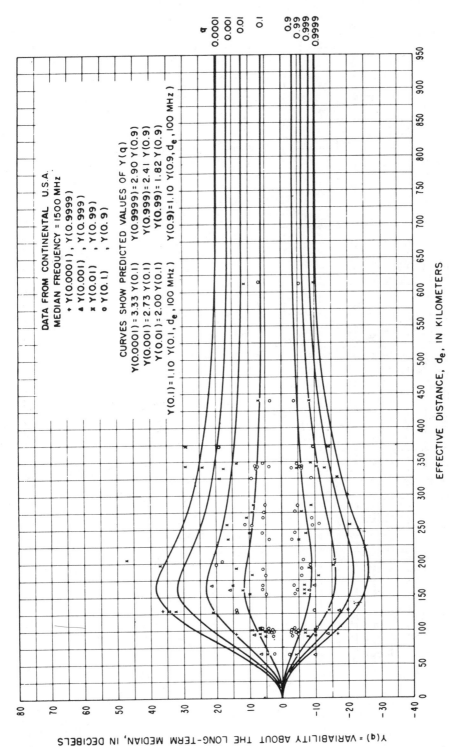

FIGURE 3.30 Long-term power fading, continental temperate climate, frequency > 1000 MHz (Ref. 1).

183

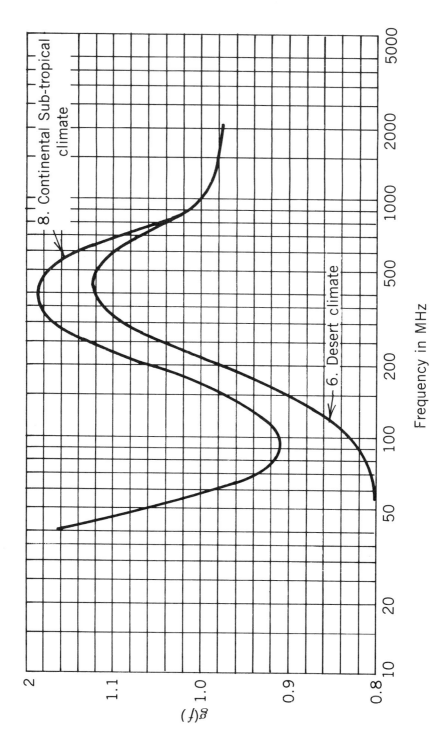

FIGURE 3.31 The frequency correction factor $g(q, f)$ for Climatic Regions 6 and 8 (Ref. 1).

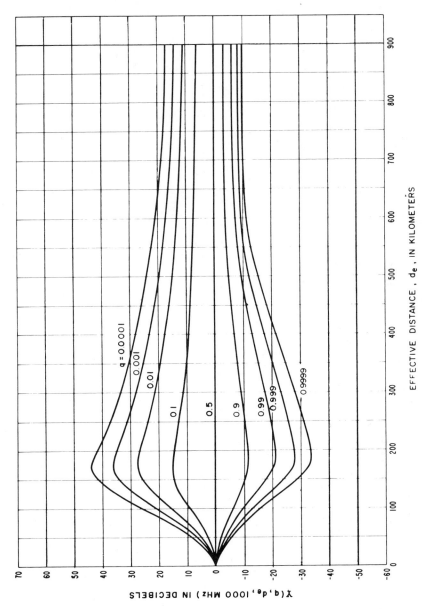

FIGURE 3.32 Long-term power fading function for Climatic Region 6, desert, Sahara (Ref. 1).

185

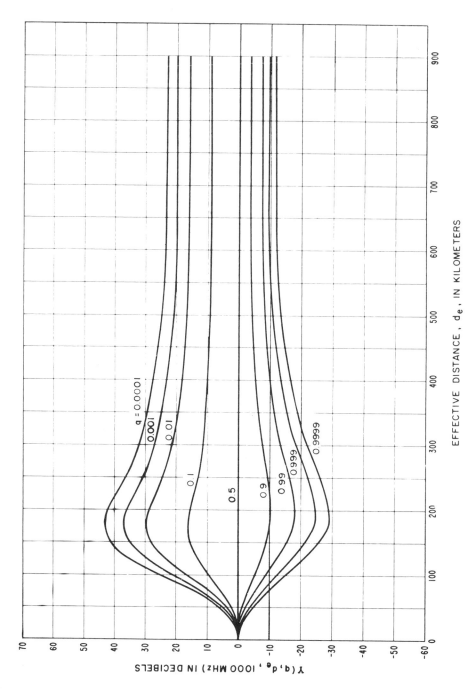

FIGURE 3.33 Long-term power fading function for Climatic Region 8, continental subtropical. (Ref. 1).

186

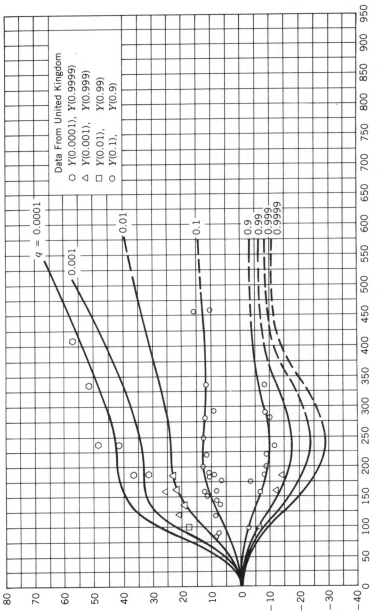

FIGURE 3.34 Long-term power fading, maritime temperate climate, over land, band III (150–250 MHz) (Ref. 1).

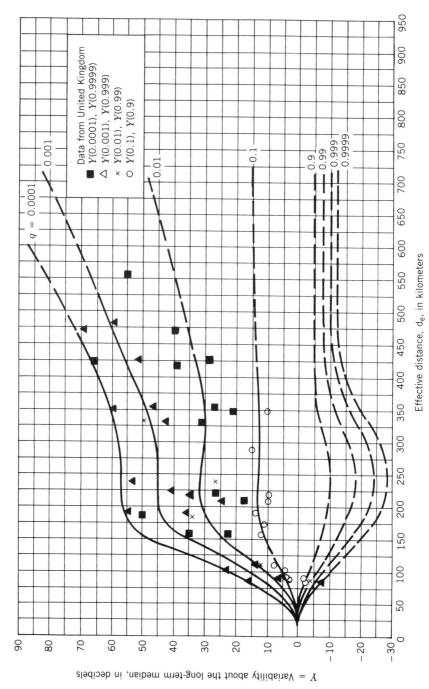

FIGURE 3.35 Long-term power fading, maritime temperate climate, over land, bands IV and V (450–1000 MHz; Climatic Zone 2.) (Ref. 1).

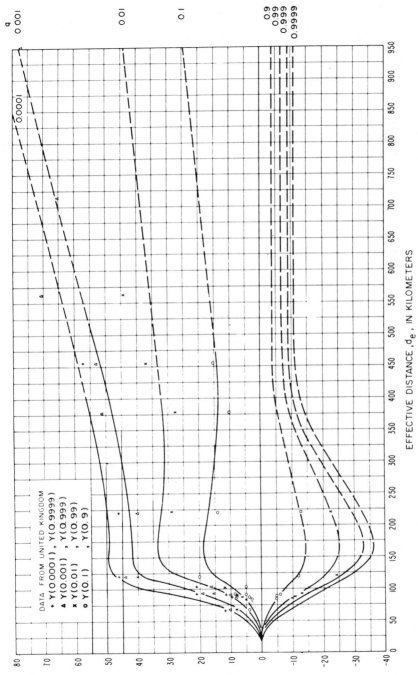

FIGURE 3.36 Long-term power fading, maritime temperate climate, over sea, band III (150–250 MHz). Climatic Region 3 (Ref. 1).

189

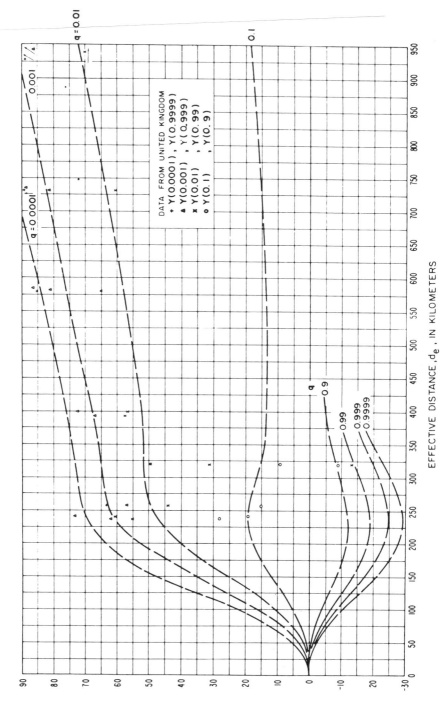

FIGURE 3.37 Long-term power fading, maritime temperate climate, over sea, bands IV and V (450–1000 MHz). Climatic Region 3 (Ref. 1).

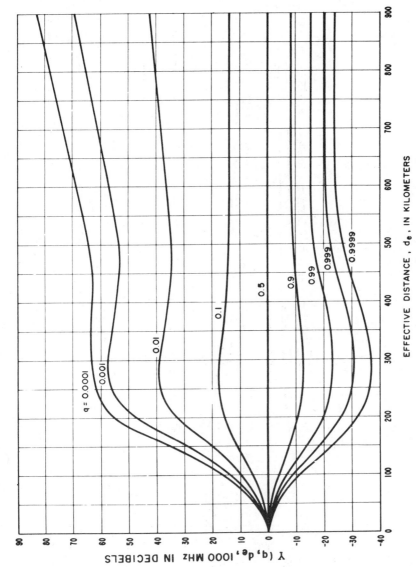

FIGURE 3.38 Long-term power fading function for Climatic Region 4, maritime subtropical, over land (Ref. 1).

191

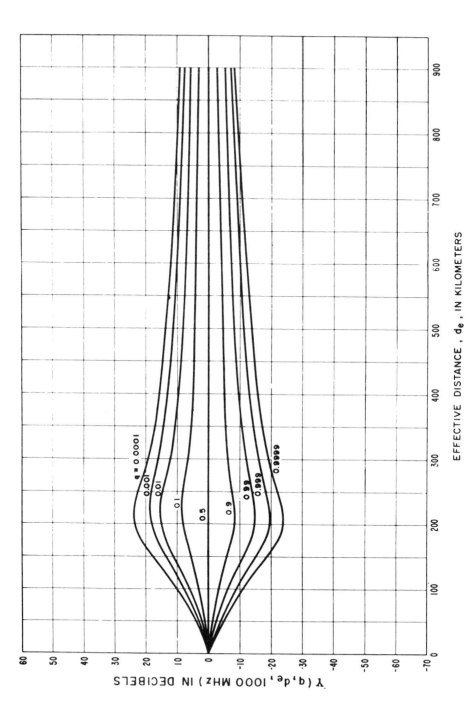

FIGURE 3.39 Long-term power fading function for Climatic Region 7, equatorial (Ref. 1).

Example 2. Determine $Y_n(0.999, 350, 400)$ (i.e., Climatic Region 6, path time availability 0.999, effective distance of 350 km, and operating frequency of 400 MHz). First derive the frequency correction factor from Figure 3.31. Here $g(q, 400) = 1.2$. Next determine the value for $Y_6(0.999, 350, 1000)$ from Figure 3.32. This value is -17 dB. $Y_6(0.999, 350, 400) = -17$ dB $\times 1.2 = -20.4$ dB.

For the remaining climatic regions—2, 3, 4, and 7—we determine $Y_n(q, d_e, f)$ for the applicable climatic region n, path time time availability q, effective distance d_e, and applicable frequency f. Read values of Y_n for the appropriate climatic regions from Figures 3.33–3.39.

To calculate $L_n(q)$, the transmission loss for any time availability q, use the following formula:

$$L_n(q) = L_n(0.5) - Y_n(q) \tag{3.53}$$

where $L_n(q)$ for a service probability of 0.5 (i.e., 50%) is taken from formula (3.51).

3.4.2.13 Extending to the Desired Service Probability

The service probability concept is used to obtain a measure of the prediction uncertainty due to our lack of complete knowledge of the propagation mechanism. The formulas we use are of a semiempirical nature and are based on a finite data sample.

The processes in tropospheric scatter propagation are extremely complex, and it is neither possible nor practical to provide numerical values of all possible parameters and their effects on the time distribution of transmission loss. Consequently, the specific values calculated in accordance with the material in the previous sections must be considered as mean values resulting from an ensemble of propagation paths for which the parameters used in the transmission loss calculations are exactly identical, but which differ from each other in additional respects that cannot be included in the formulation of the models and methods used. It is reasonable to expect that long-term measurements over such an ensemble of paths or links would produce a random (or Gaussian) distribution of transmission loss values for each percentile of time, with the mean of such a distribution identical to the calculated value. The standard deviation of this distribution would then characterize the uncertainty inherent in the prediction or modeling process. The service probability (Q) is the parameter used to specify prediction uncertainty. A common service probability is specified at $Q = 0.95$. This means if we built 100 paths using these same parameters, that, once installed, 95 would equal or exceed specifications and 5 would fail. To calculate the service probability factor, $z_{m0}(Q)\sigma_{rc}(q)$, we first compute $\sigma_c^2(q)$ and $\sigma_{rc}(q)$ for the required values of q, the path time availability:

$$\sigma_c^2(q) = 12.73 + 0.12Y_n^2(q, d_e) \tag{3.54}$$

$$\sigma_{rc}(q) = \sqrt{\sigma_c^2(q) + \sigma_r^2} \tag{3.55}$$

where $\sigma_{rc}(q)$ is the prediction error between the predicted and the observed values of transmission loss for a fraction q of all hours. $\sigma_c^2(q)$ is the path-to-path variance of the difference between observed and predicted long-term median values of transmission loss. σ_r^2 is the variance assigned to errors in estimating equipment parameters; use a value of 4 dB^2 for new circuits and 2 db^2 for established circuits. Y_n was derived in Section 3.4.2.12.

The next step is to determine $z_{m0}(Q)$ from Figure 3.40. Often 0.95 is specified. $z_{m0}(Q)$ is the standard normal deviate.

When calculating the product $z_{m0}(Q)\sigma_{rc}(q)$ for the values of q, it has been found through experience that tables are useful and may be constructed as follows, first for the 50% service probability case:

Time Availability (q) (%)	$Y_n(q)$ (dB)	Transmission Loss (dB)
50		
90		
99		
99.9		
99.99		

A second table is then constructed to extend the 50% service probability case to 95% (i.e., $Q = 0.95$) or to some other desired service probability:

Time Availability (q) (%)	$z_{m0}(Q)\sigma_{rc}(q)$ (dB)	Transmission Loss (dB)
50		
90		
99		
99.9		
99.99		

Example 3. In Climatic Region 1 a tropo path has an effective distance of 325 km and $L_n(0.5) = 218$ dB. Prepare the two tables in accordance with this information. Assume the operating frequency to be 4000 MHz and thus the frequency correction factor equals 1 and may be neglected. Use Figure 3.30 directly to determine $Y_1(q)$.

The calculations are based on formula (3.53):

$$L_n(q) = L_n(0.5) - Y_n(q)$$

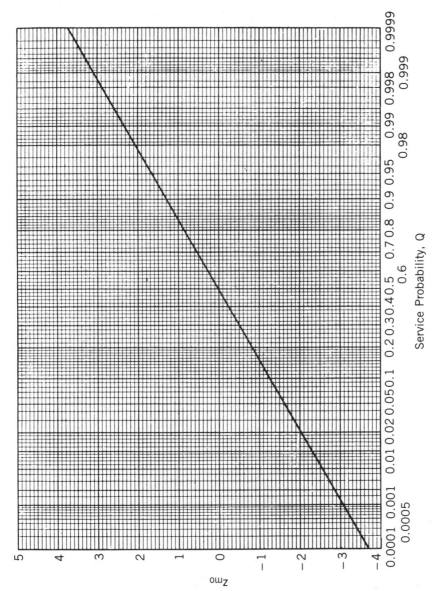

FIGURE 3.40 The standard normal deviate z_{m0} (Ref. 9).

50% SERVICE PROBABILITY CASE

Time Availability (%)	$Y_1(q)$ (dB)	Transmission Loss (dB)
50	0	218
90	-6	224
99	-12	230
99.9	-15	233
99.99	-17.5	235.5

Calculate $\sigma_c^2(0.5)$ (i.e., where $q = 0.50$ or 50%):

$$\sigma_c^2(0.5) = 12.73 + 0$$

$$\sigma_{rc}(0.5) = \sqrt{12.73 + 4}$$

$$= 4.09$$

$$z_{m0}(0.95) = 1.65 \text{ (from Figure 3.40)}$$

$$z_{m0}(0.95) \times \sigma_{rc}(0.5) = 1.65 \times 4.09$$

$$= 6.75 \text{ dB}$$

Calculate $\sigma_c^2(0.90)$:

$$\sigma_c^2(0.90) = 12.73 + 0.12 \times 6^2$$

$$= 12.73 + 4.32$$

$$= 17.05$$

$$\sigma_{rc}(0.90) = \sqrt{17.05 + 4}$$

$$= 4.59$$

$$z_{m0}(0.95) \times \sigma_{rc}(0.90) = 1.65 \times 4.59$$

$$= 7.57 \text{ dB}$$

and so on for the other values of q to complete the table.

EXTENDING TO A 95% SERVICE PROBABILITY

Time Availability (%)	$z_{m0}(0.95) \times \sigma_{rc}(q)$ (dB)	Transmission Loss (dB)
50	6.75	224.75
90	7.57	231.57
99	9.62	239.62
99.9	10.91	243.91
99.99	12.06	248.41

3.4.2.14 Aperture-to-Medium Coupling Loss

3.4.2.14.1 DEFINITION

In chapter 2 antenna gain was treated as free space antenna gain or the ratio of the maximum radiated power density at a point in space to the theoretical maximum radiated power density of an isotropic antenna at the same relative point in space. No degradation factors were included. In tropospheric scatter and diffraction links such degradations cannot be neglected. Whereas LOS radiolink antennas are seldom mounted near ground and antenna apertures are relatively small, in tropo and diffraction links antennas are mounted near ground level and their apertures are comparatively large. This gives rise to two forms of gain degradation:

- ☐ Distortion of wavefront due to antenna position.
- ☐ Radiation pattern distortion due to ground reflections.

These result in a loss of antenna gain when compared to free space gain. We call this *aperture-to-medium coupling loss*. Reference 1 calls this multipath coupling loss and Ref. 2 calls it *loss in path antenna gain*.

In Chapter 2 we could define path antenna gain in decibels as the sum of the transmit and receive antenna gains in decibels or

$$G_p = G_t + G_r \tag{3.56}$$

For the transhorizon systems described in this chapter, we must take into account aperture-to-medium coupling loss, and path antenna gain is defined as

$$G_p = G_t + G_r - L_{gp} \tag{3.57}$$

where L_{gp} is the aperture-to-medium coupling loss or loss in path antenna gain.

There are two ways of treating this parameter:

1. As a loss added to the transmission loss.
2. As a degradation to antenna gains.

We prefer the former. Now we can treat antenna gains as we had previously in Chapter 2. But once we do this, we must add the value of L_{gp} to the path transmission loss.

3.4.2.14.2 CALCULATION OF APERTURE-TO-MEDIUM COUPLING LOSS

One common method of calculation is given in CCIR Rep. 238 (Ref. 5) and assumes each antenna gain does not exceed 55 dB and that there is no significant difference in gains between G_t and G_r. The following formula

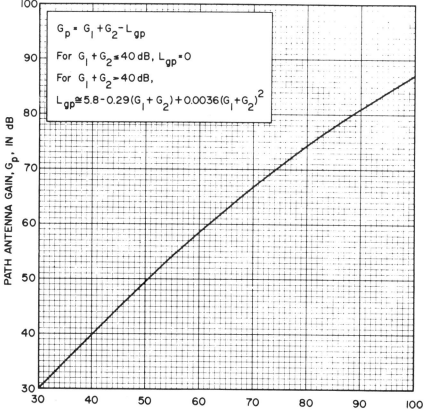

$$G_p = G_1 + G_2 - L_{gp}$$

For $G_1 + G_2 \leq 40\,dB$, $L_{gp} = 0$

For $G_1 + G_2 > 40\,dB$,

$$L_{gp} \cong 5.8 - 0.29(G_1 + G_2) + 0.0036(G_1 + G_2)^2$$

FIGURE 3.41 Path antenna gain G_p for transhorizon links (Ref. 2).

applies to the CCIR approach:

$$L_{gp} = 0.07 \exp[0.055(G_t + G_r)] \qquad (3.58)$$

where G_t and G_r, the gains of the transmitting and receiving antennas, respectively, are given in decibels.

Another method of calculation is given in Ref. 2 and is shown in Figure 3.41. L_{gp} is the aperture-to-medium coupling loss. (Reference 1 can be consulted for still a third method of calculation.)

Example 4. Assume 15-ft parabolic reflector antennas at both ends of a 4-GHz transhorizon link. At 55% efficiency the antenna gains are 43 dB. What is the aperture-to-medium coupling loss?

Method 1 (CCIR):

$$L_{gp} = 0.07 \exp[0.055(43 + 43)]$$

$$= 0.07 \exp(4.73)$$

$$= 7.93 \text{ dB}$$

Method 2 (Ref. 2):

$$L_{gp} = 5.8 - 0.29(43 + 43) + 0.0036(43 + 43)^2$$

$$= 7.49 \text{ dB}$$

3.4.2.15 Sample Problem—Tropospheric Scatter Transmission Loss

A tropospheric scatter path is 283.1 km long; the transmitting frequency is 5000 MHz; the radio refractivity N_0 is 315; the heights above sea level of the antennas are $h_{ts} = 0.2804$ km, $h_{rs} = 0.2439$ km; the elevations above MSL of the horizon points are $h_{Lt} = 0.2195$ km, $h_{Lr} = 0.2743$ km; distances from transmitter and receiver sites to their respective horizons are $d_{Lt} = 39.6$ km, $d_{Lr} = 8.8$ km. Prepare tables of transmission loss values for both 50% and 95% service probabilities and path time availabilities from 90% to 99.99%. Assume Climatic Region 1.

Calculation of basic long-term median tropospheric scatter loss L_{bsr} (Section 3.4.2.5): Compute h_{st} and h_{sr}. Equations (3.3) and (3.4) are valid for this case and then $h_{st} = h_{Lt} = 0.2195$ km and $h_{sr} = h_{Lr} = 0.2743$ km.

Compute values of N_{st} and N_{sr} for h_{st} and h_{sr} [equation (3.7)]:

$$N_{st} = 315 \exp(-0.1057 \times 0.2195)$$

$$N_{st} = 307.8$$

$$N_{sr} = 315 \exp(-0.1057 \times 0.2743)$$

$$N_{sr} = 306$$

Compute N_s [equation (3.8)]:

$$N_s = 306.9$$

Compute effective earth radius a from Figure 3.7 given $N_s = 306.9$: $a = 8580$ km.

Compute the horizon takeoff angles θ_{et} and θ_{er} [equations (3.9) and (3.10)]:

$$\theta_{et} = \frac{0.2195-0.2804}{39.6} - \frac{39.6}{2 \times 8580}$$

$$= -0.00154 - 0.00231$$

$$= -0.00385 \text{ radians}$$

$$\theta_{er} = \frac{0.2743 - 0.2439}{8.8} - \frac{8.8}{2 \times 8580}$$

$$= 0.00345 - 0.00051$$

$$= 0.00294 \text{ radians}$$

Compute the distance between horizon points [equation (3.11)]:

$$D_s = 283.1 - 39.6 - 8.8$$

$$= 234.7 \text{ km}$$

Compute the effective antenna heights h_{te} and h_{re}. Divide the distance to the radio horizon for the transmitter and receiver sites, respectively, into 31 uniform segments. Then compute h_t and h_r [equations (3.13)–(3.16)]:

$$d_{Lt} = 39.6 \text{ km} \quad \text{and} \quad d_{Lr} = 8.8 \text{ km}$$

$$\frac{39.6}{31} = (\text{approx.}) \ 1.3 \text{ km}$$

$$\frac{8.8}{31} = (\text{approx.}) \ 0.28 \text{ km}$$

Build two tables as shown below, starting at segment 3 and finishing at segment 27 showing altitude at each segment. (Normally this is taken from the topo maps used in the path profile.)

	TABLE FOR XMIT				TABLE FOR RCVE	
i	d_{tt} (km)	h_{ti} (m)		i	d_{ri} (km)	h_{ri} (m)
3	3 × 1.3	3.9	270	3	3 × 0.28	0.84	240
4	4 × 1.3	5.2	260	4	4 × 0.28	1.12	250
5	5 × 1.3	6.5	260	5	5 × 0.28	1.4	250
6	6 × 1.3	7.8	250	6	6 × 0.28	1.68	250
7	7 × 1.3	9.1	250	7	7 × 0.28	1.96	250
8	8 × 1.3	10.4	250	8	8 × 0.28	2.24	250
9	9 × 1.3	11.7	250	9	9 × 0.28	2.52	250
10	10 × 1.3	13	250	10	10 × 0.28	2.8	250
11	11 × 1.3	14.3	240	11	11 × 0.28	3.08	260
12	12 × 1.3	15.6	240	12	12 × 0.28	3.36	260
13	13 × 1.3	16.9	240	13	13 × 0.28	3.64	260
14	12 × 1.3	18.2	240	14	14 × 0.28	3.92	260
15	15 × 1.3	19.5	240	15	15 × 0.28	4.2	260
16	16 × 1.3	20.8	240	16	16 × 0.28	4.48	260
17	17 × 1.3	22.1	240	17	17 × 0.28	4.76	260
18	18 × 1.3	23.4	240	18	18 × 0.28	5.04	260
19	19 × 1.3	24.7	250	19	19 × 0.28	5.32	260
20	20 × 1.3	26	250	20	20 × 0.28	5.6	260
21	21 × 1.3	27.3	240	21	21 × 0.28	5.88	260
22	22 × 1.3	28.6	240	22	22 × 0.28	6.16	260
23	23 × 1.3	29.9	240	23	23 × 0.28	6.44	260
24	24 × 1.3	31.2	230	24	24 × 0.28	6.72	260
25	25 × 1.3	32.5	230	25	25 × 0.28	7.0	260
26	26 × 1.3	33.8	230	26	26 × 0.28	7.28	260
27	27 × 1.3	35.1	230	27	27 × 0.28	7.56	270
		Total	6100			Total	6420

$$\bar{h}_{ti} = \frac{6100}{25} = 244 \text{ m} \quad \text{or} \quad 0.244 \text{ km} \qquad \bar{h}_r = \frac{6420}{25} = 256.8 \text{ m} \quad \text{or} \quad 0.2568 \text{ km}$$

In the case of the transmit site, h_{t0} is larger than $\bar{h}_{ti}$, then

$$h_t = h_{ts} - \bar{h}_t \quad [\text{equation}(3.13)]$$

$$h_t = 0.2804 - 0.244 = 0.0364 \text{ km}$$

In the case of the receiver site, the reverse is true, and

$$h_r = h_{rs} - h_{r0}$$

$$= 0.2439 - 0.2429$$

$$= 10 \text{ m or } 0.010 \text{ km}$$

(where h_{r0} is the ground elevation at the receiver site, about 10 m below antenna center, where h_{rs} elevation is measured.)

Because h_{t0} and h_{r0} are less than 1 km, use equation (3.17) and $h_{te} = h_t$ and $h_{re} = h_r$:

$$h_{te} = 0.0364 \text{ km}$$

$$h_{re} = 0.010 \text{ km}$$

Calculate the angular distance θ (steps 1t through 11t):

1t. Compute $\theta_{0t,0r}$ [equations (3.19) and (3.20)]:

$$\theta_{0t} = -0.00385 + \frac{39.5}{8580}$$

$$= -0.00385 + 0.0046$$

$$= 0.00075 \text{ radians}$$

$$\theta_{0r} = 0.00294 + 8.8/8580$$

$$= 0.00396 \text{ radian}$$

2t. Calculate α_{00} and β_{00} [equations (3.21) and (3.22)]:

$$\alpha_{00} = 0.0165 - 0.00385 + \frac{0.2804 - 0.2439}{283.1}$$

$$= 0.01278 \text{ radian}$$

$$\beta_{00} = 0.0165 + 0.00294 + \frac{0.2439 - 0.2804}{283.1}$$

$$= 0.0193 \text{ radian}$$

3t. Compute θ_{00} [equation (3.23)]:

$$\theta_{00} = \alpha_{00} + \beta_{00}$$

$$= 0.01278 + 0.0193$$

$$= 0.03208 \text{ radian}$$

4t. Compute d_{st} and d_{sr} [equations (3.24) and (3.25)]:

$$d_{st} = \frac{283.1 \times 0.01278}{0.03208} - 39.6$$

$$= 73.2 \text{ km}$$

$$d_{sr} = \frac{283.1 \times 0.0193}{0.03208} - 8.8$$

$$= 161.5 \text{ km}$$

5t. Determine $\Delta\alpha_0$ and $\Delta\beta_0$ as a function of θ_{0t} and θ_{0r}. θ_{0t} and θ_{0r} are greater than 0 but less than 0.1; thus proceed to steps 6t, 7t, and 10t.

6t. From Figure 3.9 determine $\Delta\alpha_0$ and $\Delta\beta_0$ for $N_s = 301$:

$$\Delta\alpha_0 = 0$$

$$\Delta\beta_0 = 0.5 \text{ milliradian or } 0.0005 \text{ radian}$$

7t. Read $C(N_s)$ from Figure 3.10; $N_s = 306.9$ from above:

$$C(N_s) = 1.05$$

$$\Delta\alpha_0(N_s) = 0$$

$$\Delta\beta_0(N_s) = 1.05 \times 0.0005$$

$$= 0.000525$$

(Note: Steps 8t and 9t are not required in this example.)

10t. Calculate α_0 and β_0 using equations (3.31) and (3.32) (correction factors are from step 7t):

$$\alpha_0 = \alpha_{00} + 0 = 0.01278 \text{ radians}$$

$$\beta_0 = \beta_{00} + 0.000525 = 0.01982 \text{ radians}$$

11t. Calculate the scatter angle (angular distance) θ:

$$\theta = 0.01278 + 0.01982$$

$$= 0.0326 \text{ radian}$$

12t. Calculate the path asymmetry factor s:

$$s = \frac{0.01278}{0.01982}$$

$$= 0.645$$

Calculate the product θd, where θ is the scatter angle or angular distance and d is the path length. The units must be in radians and kilometers, respectively.

$$\theta d = 0.0326 \times 283.1$$

$$= 9.23 \text{ km-radian}$$

Read $F(\theta d)$ from Figure 3.11 ($N_s = 306.9$):

$$F(\theta d) = 167.5 \text{ dB}$$

Compute the basic median tropospheric scatter transmission loss L_{bsr}:

$$L_{bsr} = 30 \log 5000 - 20 \log 283.5 + 167.5 \text{ dB} + 2.7 \text{ dB}$$

$$L_{bsr} = 232.12 \text{ dB}$$

(The value for A_a is taken from Figure 3.4.)

Section 3.4.2.3 tells us that if the angular distance exceeds 20 milliradians, the path is predominantly tropospheric scatter. This path is 32.6 milliradians, significantly greater than 20 milliradians. Now we can say that $L_{cr} = L_{bsr} = 232.12$ dB.

Calculate the effective distance d_e (Section 3.4.2.10):

$$d_{sl} = 65 \left(\frac{100}{5000} \right)^{1/3} \quad [\text{equation (3.49)}]$$

$$= 17.67 \text{ km}$$

$$d_L = 3\sqrt{2 \times 0.0364 \times 10^3} + 3\sqrt{2 \times 0.010 \times 10^3} \quad [\text{equation (3.50)}]$$

$$= 25.6 + 13.42$$

$$= 39 \text{ km}$$

Because d is greater than $d_{sl} + d_L$,

$$d_e = 130 + 283.1 - 17.67 - 39 \quad [\text{equation (3.50b)}]$$

$$= 356 \text{ km}$$

Calculate the basic long-term median transmission loss $L_n(0.5, 50)$ [equation (3.51)] for Climatic region 1:

$$L_n(0.5, 50) = 232.12 \text{ dB} - 3 \text{ dB} \quad (\text{Figure 3.25})$$

$$= 229.12 \text{ dB}$$

Calculate values for $Y_1(q, 356)$ and arrange them in tabular form for values of q. Use Figure 3.30 directly.

TRANSMISSION LOSS TABLE: 50% SERVICE PROBABILITY

$q(\%)$	Y_1 (dB)	Transmission Loss (dB)
50	0	229.12
90	5.5 dB	234.62
99	10	239.12
99.9	13.5	242.62
99.99	16.5	245.62

Prepare a second table for the 95% service probability case. z_{m0} from Figure 3.40 is 1.65. Use equations (3.54) and (3.55).

TRANSMISSION LOSS TABLE: 95% SERVICE PROBABILITY

$q(\%)$	$z_{m0}(95)\sigma_{r0}(q)$ (dB)	Transmission Loss (dB)
50	6.75	235.8
90	7.44	242.1
99	8.84	248.0
99.9	10.25	252.9
99.99	11.6	257.22

3.5 PATH CALCULATION / LINK ANALYSIS

3.5.1 Introduction

We have shown how to calculate transmission loss for over-the-horizon paths, both tropospheric scatter and diffraction. In either case, the calculation is complex. For LOS paths the calculation is rather simple.

Once we have the transmission loss that includes fade margin (long term), the approach to link analysis follows the same methodology as LOS. There is one exception: multipath.

3.5.2 Path Intermodulation Noise—Analog Systems

For an analog LOS system there were two analog noise components: equipment noise and feeder distortion (waveguide distortion). With tropo and diffraction paths we must consider one more intermodulation (IM) noise component, namely, path IM noise.

Compute which is the maximum delay of the multiecho as compared to the main beam, expressed in seconds:

$$\Delta = 5.21d(\Omega_t + \Omega_r)(4\alpha_0 + 4\beta_0 + \Omega_t + \Omega_r) \times 10^{-8} \qquad (3.59)$$

where d = path length in kilometers

α_0, β_0= main beam angles (radians) of the transmitting and receiving antennas, respectively; these were defined in Section 3.4.2.5

Ω_t, Ω_r = 3-dB beamwidths (radians) of the transmitting and receiving antennas, respectively; read beamwidths from Figure 3.42.

Calculate D, the rms composite deviation, for values of $M = 2, 3, 4, 6, 8,$ and 9; M is the peak deviation ratio:

$$D = \frac{Mf_m}{4.46} \qquad (3.60)$$

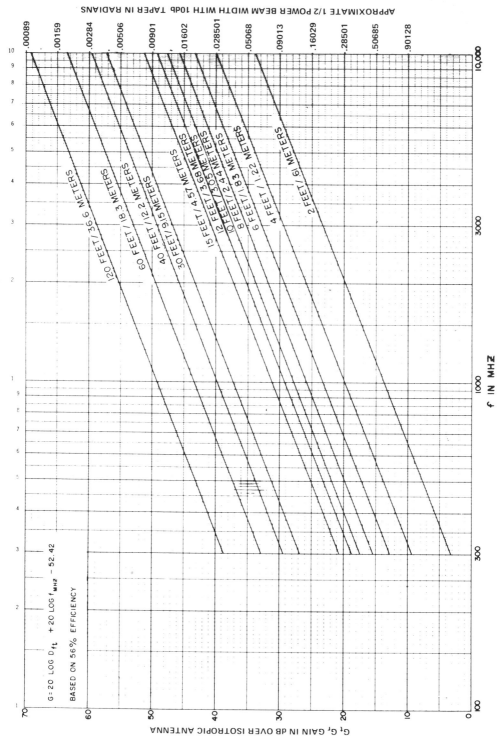

FIGURE 3.42 Antenna gain and beamwidth nomogram. Gain is in decibels and beamwidth is in radians (Ref. 1).

206

where f_m is the maximum baseband frequency. This value may be taken from Table 3.3.

Compute $\gamma_{0.5}$ and $\gamma_{0.01}$, the phase distortion (radians) corresponding to values of D (taken from the previous step) not exceeded more than 50% and 1% of the time:

$$\gamma_{0.5} = 8(\Delta D)^2 \tag{3.61}$$

$$\gamma_{0.01} = 2600(\Delta D)^2 \tag{3.62}$$

Determine $H(\gamma_{0.5})$ and $H(\gamma_{0.01})$ for values of D. Read values from Figure 3.43. Note: for $\gamma < 0.05$, $H(\gamma) = \gamma^2$.

TABLE 3.3. Various System Parameters Calculated as a Function of N, the Number of FDM 4-kHz Voice Channels (Based on CCIR standards)

N^a	Baseband Spectrum (Δf_b) (kHz)^b	Baseband Spectrum (Δf_b) (kHz)^c	BWR (dB)^d	NLR (dBm0)^e	l (N) rms^f	L(N) (dBm0)^g	l(N) peak^h	C(dBp)^i
12	12-60		11.90	3.3	1.46	16.3	6.53	11.10
24	12-108		14.91	4.5	1.68	17.5	7.50	12.91
36	12-156	60-204	16.67	5.2	1.82	18.2	8.13	13.97
48	12-204	60-252	17.92	5.72	1.93	18.72	8.63	14.70
60	12-252	60-300	18.89	6.11	2.02	19.11	9.03	15.30
72	12-300	60-360	19.86	6.43	2.09	19.42	9.46	15.93
84		60-408	20.50	6.7	2.16	19.7	9.66	16.30
96		60-456	21.06	6.92	2.22	19.92	9.91	16.86
108		60-504	21.60	7.1	2.27	20.1	10.11	17.00
120		60-552	22.01	7.3	2.32	20.3	10.35	17.21
180		60-804	23.80	8.02	2.57	21.0	11.2	18.28
240		60-1052	25.05	8.80	2.75	21.80	12.3	18.75
300		60-1300	26.02	9.77	3.08	22.77	13.0	18.75
600		60-2540	29.03	12.78	4.36	25.78	12.46	18.75

$^a N$ = number of 4-kHz voice channels.
^{b} Baseband spectrum beginning at 12 kHz.
^{c} Baseband spectrum beginning at 60 kHz.
^{d} BWR (dB) = bandwidth ratio = 10 log(occupied baseband bandwidth/3.1 kHz).
^{e} NLR (dBm0) = Noise loading ratio
$\qquad = -1 + 4 \log N$ for $12 < N < 240$ (CCIR Standard)
$\qquad = -15 + 10 \log N$ for $N \geq 240$ (CCIR Standard)
$\qquad = -10 + 10 \log N$ for all N (DCA Standard)
$^f l(N)$ rms = antilog(NLR/20)
$^g L(N)$ (dBm0) = NLR + 13(peak loading factor).
$^h l(N)$ peak = antilog $[L(N)/20]$.
$^i C$(dBp) = BWR − NLR + I_w(2.5 dB) (multichannel to single-channel conversion factor).

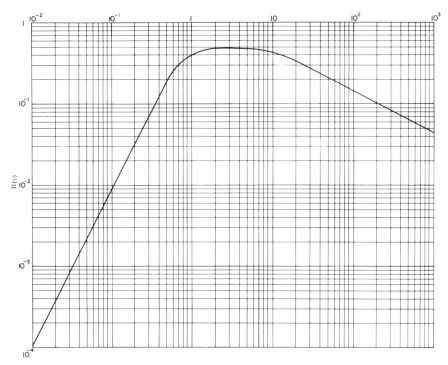

FIGURE 3.43 The parameter $H(\gamma)$ (Ref. 1).

Compute R for values of D, where R is a factor for computing S/N_p:

$$R = 0.288\left(\frac{f_m}{D}\right)^2 \tag{3.63}$$

Compute S/N_p for various values of $H(\gamma)$ and D. S/N_p is the signal-to-path IM noise ratio in dBp:

$$\left(\frac{S}{N}\right)_p = 10\log RH(\gamma) + C + I_d \tag{3.64}$$

where C is multichannel to single-channel conversion factor, which is taken from Table 3.3 and I_d is the diversity improvement factor. For I_d, use one of the following values this is applicable for the equipment to be used on the link: postdetection combining, dual diversity: 3.0 dB; quadruple diversity: 6.0 dB; predetection combining, dual diversity: 2.5 dB; quadruple diversity: 5.0 dB.

Convert S/N_p to N_p (pWp).

$$N_p = \text{antilog} \ \frac{1}{10}\left[90 - \left(\frac{S}{N}\right)_p\right] \qquad (3.65)$$

For the value of N_p for worst-hour performance, use the equivalent value of $H(\gamma_{0.01})$.

Example 5. Assume the previous example in Section 3.4.2.15 and that the antennas have 30-ft apertures, α_0 and β_0 are 0.01278 and 0.01982 radian, respectively, 60 FDM voice channels occupying a baseband of 12–252 kHz (Table 3.3), and let $M = 3$ for simplification.

From Figure 3.42, Ω_t and Ω_r are 0.008 radian each. Then

$$\Delta = 5.2(283.1)(0.008 + 0.008)$$

$$\times (4 \times 0.01278 + 4 \times 0.01982 + 2 \times 0.008) \times 10^{-8}$$

$$= 5.2(283.1)(0.016)(0.146) \times 10^{-8} \quad [\text{equation } (3.59)]$$

$$= 3.44 \times 10^{-8} \text{ sec}$$

Calculate D, the rms composite deviation [equation (3.60)]:

$$D = \frac{3 \times 252 \times 10^3}{4.46}$$

$$D = 169,507 \text{ Hz}$$

Compute

$$\gamma_{0.01} = 2600(3.44 \times 10^{-8} \times 169,507)^2$$

$$= 0.0884 \text{ radian}$$

Determine $H(\gamma_{0.01})$ from Figure 3.43. The value is 7×10^{-3}.
Compute R for the value of D ($D = 169,507$ Hz); use equation (3.63):

$$R = 0.288 \left(\frac{252 \times 10^3}{169,507}\right)^2$$

$$= 0.636$$

Compute S/N_p [equation (3.64)]:

$$\left(\frac{S}{N}\right)_p = 10\log(0.636)(7 \times 10^{-3}) + 15.3 + 6$$

(The value for I_d assumes postdetection combining, quadruple diversity, therefore, $I_d = 6.0$ dB)

$$\left(\frac{S}{N}\right)_p = +235 + 15.3 + 6$$

$$= 44.8 \text{ dB}$$

$$N_p = 33,113 \text{ pWp } 1\% \text{ of the time.}$$

If we calculate it for the median (50% of the time) [equation (3.61)], we arrive at a value of 0.346 pWp. This latter value would be used for total noise calculation.

3.5.3 Sample Link Analysis

Let us treat the link given in Section 3.2.4.15 for two cases using 95% service probability: (1) 50% path availability and (2) 99% path availability. Transmitters are 1 kW, antennas have 30-ft apertures, line losses are 3 dB at each end, quadruple diversity is used, there is a 6-dB diversity improvement, operating frequency is 5000 MHz, 60 voice channels FDM/FM, and receiver noise figure is 3 dB.

calculate EIRP with antenna gain of 51 dB:

$$\text{EIRP} = +30 \text{ dBW} - 3 \text{ dB} + 51 \text{ dB}$$

$$= +78 \text{ dBW}$$

Calculate the aperture-to-medium coupling loss using method 2 in Section 3.4.2.14:

$$L_{gp} = 5.8 - 0.29(51 + 51) + 0.0036(51 + 51)^2$$

$$= 5.8 - 29.58 + 37.45$$

$$= 13.67 \text{ dB}$$

Add this value to the transmission losses for each case.
Compute the total transmission loss for each case:

$$\text{Path time availability } 50\%: 235.8 \text{ dB} + 13.67 \text{ dB} = 249.47 \text{ dB}$$

$$99\%: 248.0 \text{ dB} + 13.67 \text{ dB} = 261.67 \text{ dB}$$

Calculate the isotropic receive level (IRL) for each case:

$$\text{IRL}_{\text{dBW}} = +78 \text{ dBW} - 249.47 \text{ dB} = -171.47 \text{ dBW} \quad (50\%)$$

$$\text{IRL}_{\text{dBW}} = +78 \text{ dBW} - 261.67 \text{ dB} = -183.67 \text{ dBW} \quad (99\%)$$

Calculate the RSL for each case:

$$\text{RSL}_{\text{dBW}} = -171.47 \text{ dBW} - 3 \text{ dB} + 51 \text{ dB}$$

$$= -123.47 \text{ dBW} \quad (50\%)$$

$$\text{RSL}_{\text{dBW}} = -183.67 \text{ dBW} - 3 \text{ dB} + 51 \text{ dB}$$

$$= -135.67 \text{ dBW} \quad (99\%)$$

[Note: RSL = IRL − line losses + antenna gain (See Section 2.6.2)]

Calculate the FM improvement threshold of the far end receiver (see Section 2.9.3):

$$P_{fm(\text{dBW})} = -204 \text{ dBW} + 3 \text{ dB} + 10 \text{ dB} + 10 \log B_{if}$$

Calculate B_{if} in hertz using Carson's rule. (Follow the procedures given in Section 2.6.6.) Table 3.3 gives the highest modulating frequency for a 60-channel configuration as 252 kHz. We must now calculate peak deviation, Δf_p. [equation (2.24)].

Calculate NLR; equation (2.20) is used for 60-channel operation:

$$\text{NLR}_{\text{dB}} = -1 + 4 \log N \quad \text{where } N = 60$$

$$\text{NLR} = = -1 + 7.11$$

$$= 6.11 \text{ dB}$$

Calculate peak deviation:

$$\Delta f_p = 4.47d \left[\log^{-1}(6.11/20) \right]$$

Use 200 kHz per channel from Table 2.5.

$$\Delta f_p = 4.47 \times 200 \times 2.02 \text{ kHz}$$

$$= 1806 \text{ kHz}$$

Apply Carson's rule:

$$B_{if} = 2(1806 + 252) \text{ kHz}$$

$$= 4116 \text{ kHz}$$

$$P_{fm(\text{dBW})} = -204 \text{ dBW} + 3\text{dB} + 10 \text{ dB} + 10 \log(4116 \times 10^3)$$

$$= -124.85 \text{ dBW}$$

The link RSL should be equal or greater than this value for both the 50% and 99% time availability cases. Compare the values:

		FM Improvement Threshold (dBW)	Margin (dB)
RSL (50%)	−123.47	−124.85	+1.38
RSL (99%)	−135.67	−124.85	−10.82

The preceding argument presented is one of the classical methods of carrying out a link analysis. A baseline system has been established. It does not meet requirements as shown in the preceding table: 50% of the time FM improvement threshold is achieved; 1% of the time we are short 10.82 dB. What alternatives are open to us?

1. Accept a reduced time availability. This is probably unacceptable.
2. Increase transmitter output power. Increasing the output power to 10 kW would give an across-the-board link improvement of 10 dB.
3. Add threshold extension. (See Section 3.6 for a description of threshold extension.) With a modulation index of at least 3 ($M = 3$), a 7-dB link improvement can be achieved.
4. Increase antenna aperture. We must remember that as aperture increases, so does aperture-to-medium coupling loss. Thus there is no decibel for decibel improvement.
5. Reduce the number of voice channels, thus reducing bandwidth, which will improve threshold.
6. Resite or reconfigure links to reduce transmission loss.
7. Some marginal improvements can be achieved by lowering receiver noise temperature, using lower-loss transmission lines or higher-efficiency antennas. A thermal noise improvement can be achieved by implementing quadruple diversity rather than dual diversity.

The link design engineer can parametrically compare these alternatives using present worth of annual charges technique for commercial systems or life cycle costs technique for military systems. Present worth of annual charges is described in Ref. 9.

3.6 THRESHOLD EXTENSION

Threshold extension is a method of lowering the FM improvement threshold on an FM link by replacing a conventional FM modulator with a threshold extension demodulator. Assuming a modulation index of 3 ($M = 3$), the amount of improvement that can be expected over a conventional FM demodulator is about 7 dB. In the example given in Section 3.5.3 where the

FM improvement threshold was given as -124.85 dBW for the conventional demodulator, implementing a threshold extension demodulator would lower or extend the threshold to -131.85 dBW.

Threshold extension works on an FM feedback principle, which reduces the equivalent instantaneous deviation, thereby reducing the required bandwidth B_{if}. This, in turn, effectively lowers the receiver noise threshold. A typical receiver with a threshold extension module may employ a tracking filter, which instantaneously tracks the deviation with a steerable bandpass filter having a 3-dB bandwidth of approximately four times the top baseband frequency. The control voltage for the filter is derived by making a phase comparison between the feedback signal and the IF input signal.

3.7 DIGITAL TRANSHORIZON RADIOLINKS

3.7.1 Introduction

Digital tropospheric scatter and diffraction links are being implemented on all new military construction and will be attractive for commercial telephony as we approach the era of an all-digital network. The advantages are similar to those described in the previous chapter. However, on tropo and diffraction paths the problem of dispersion due to multipath can become acute. It can be handled, in fact, it can be turned to advantage with the proper equipment.

The design of a digital tropo or diffraction link in most respects is similar to the design of its analog counterpart. Siting, path profiles, and calculation of transmission loss, including aperture-to-medium coupling loss, use identical procedures as those previously described. However, the approach to link analysis and the selection of modulation scheme differ. These issues will be discussed below.

3.7.2 Digital Link Analysis

Digital tropo/diffraction link analysis can be carried out by a method which is very similar to that described in Section 2.15.5.3:

☐ Calculate the EIRP
☐ Algebraically add the transmission loss (Section 3.4.2)
☐ Add the receiving antenna gain (G_r)
☐ Algebraically add the line losses incurred up to the input of the LNA (low noise amplifier)

The aperture-to-medium coupling loss must be accounted for. As was previously suggested, we can add this value to the transmission loss. Another method is to subtract this loss from the receiving antenna gain (G_r), but aperture-to-medium coupling loss must be accounted for.

One point of guidance: the analysis will follow a single string on a link. We mean here that the EIRP is calculated for a single transmitter and its associated antenna and then down through a single diversity branch on the receiving side as though it were operating alone.

The objective of this exercise is to calculate E_b/N_0. From equation (2.71A).

$$\frac{E_b}{N_0} = \text{RSL}_{\text{dBW}} - 10\log(\text{BR}) - (-204 \text{ dBW} + \text{NF}_{\text{dB}}) \qquad (3.66)$$

[i.e., the first two terms in equation (3.66) represent E_b and the last two terms represent N_0, where the subtraction sign implies logarithmic division]. In equation (3.66) BR is the bit rate in bps and NF_{dB} is the noise figure of the receive chain or the noise figure of the LNA, the latter being sufficient in most cases.

Calculate RSL [similar to equation (2.65)]

$$\text{RSL}_{\text{dBW}} = \text{EIRP}_{\text{dBW}} + T_L + G_r + L_{Lr} \qquad (3.67)$$

where T_L is the transmission loss including aperture-to-medium coupling loss, G_r is the gain of the receiving antenna, and L_{Lr} is the total line losses from the antenna feed to the input of the LNA.

In tropo/diffraction receiving systems, in addition to waveguide loss, there are duplexer losses, preselector filter loss, transition losses, and possibly others.

Example 6. Refer to example in Section 3.5.3. Assume a digital link with a bit rate of 1.544 Mbps operating at 5000 MHz requiring a 99% time availability and 95% service probability; the transmission loss is 248.0 dB, the aperture-to-medium coupling loss = 13.67 dB, and, therefore, $T_L = 261.67$ dB; 30-ft dishes are used on each end of the link; the receiver noise figure is 3 dB; line losses are 3 dB at each end; modulation is QPSK; and transmitter output is 10 kW. Calculate E_b/N_0 and BER using Figure 3.44.

First calculate EIRP [equation (2.6)] and then RSL [equation (3.67)]:

$$\text{EIRP}_{\text{dBW}} = 10\log 10{,}000 - 3 \text{ dB} + 51 \text{ dB}$$

$$= +88 \text{ dBW}$$

$$\text{RSL} = +88 \text{ dBW} - 261.67 \text{ dB} + 51 \text{ dB} - 3 \text{ dB}$$

$$= -125.67 \text{ dBW}$$

Use equation (3.66) to calculate E_b/N_0:

$$E_b/N_0 = -125.67 \text{ dBW} - 10\log 1.544 \times 10^6 + 204 \text{ dBW} - 3 \text{ dB}$$

$$= 13.44 \text{ dB}$$

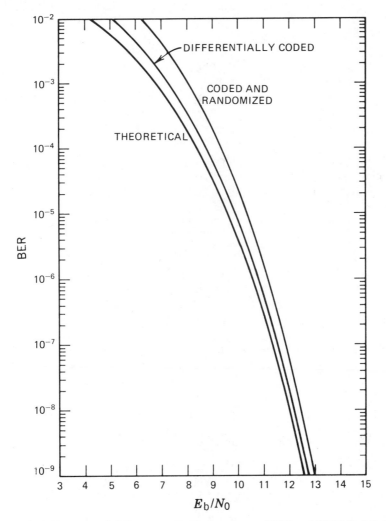

FIGURE 3.44 Bit error probability versus E_b/N_0 for coherent BPSK and QPSK (Ref. 9, Section 9-10.2).

If we allow a 3-dB modulation implementation loss, we are left with a net E_b/N_0 of 10.44 dB. Turning to Figure 3.44 we find that the equivalent BER on the theoretical curve to be 1×10^{-6}. We would expect to have this value BER 99% of the time.

3.7.3 Dispersion

Dispersion is the principal cause of degradation of BER on digital transhorizon links. With conventional waveforms such as BPSK, MPSK, BFSK, and

MFSK, dispersion may be such, on some links, that BER performance is unacceptable.

Dispersion is simply the result of some signal power from an emitted pulse that is delayed, with that power arriving later at the receiver than other power components. The received pulse appears widened or smeared or what we call dispersed. These late arrival components spill over into the time slots of subsequent pulses. The result is intersymbol interference (ISI), which deteriorates bit error rate.

Expected values of dispersion on transhorizon paths vary from 30 to 350 nsec. The cause is multipath. The delay can be calculated from equation (3.59). This equation shows that delay is a function of path length, antenna beamwidth, and the scatter angle components α_0 and β_0.

3.7.4 Some Methods of Overcoming the Effects of Dispersion

One simple method to avoid overlapping pulse energy is to time-gate the transmitted energy, which allows a resting time after each pulse. Suppose we were transmitting a megabit per second and we let the resting time be half a pulse width. Then we would be transmitting pulses of 500 nsec of pulse width, and there would be a 500-nsec resting time after each pulse, time enough to allow the residual delayed energy to subside. The cost in this case is a 3-dB loss of emitted power.

A two-frequency approach taken to reach the same objective in the design of the Raytheon AN/TRC-170 DAR modem, which is the heart of this digital troposcatter radio terminal, is to transmit on two separate frequencies alternatively gating each. The two-frequency pulse waveform is simply the time interleaving of two half-duty cycle pulse waveforms, each on a separate frequency. This technique offers two significant advantages over the one-frequency waveform. First, the two signals (subcarriers) are interleaved in time and are added to produce a composite transmitted signal with nearly constant amplitude, thereby nearly recovering the 3 dB of power lost due to time-gating. The operation of this technique is shown in Figure 3.45.

The second advantage is what is called intrinsic or implicit diversity. This can be seen as achieved in two ways. First, the residual energy of the "smear" can be utilized, whereas in conventional systems it is destructive (i.e., causes intersymbol interference). Second, on lower bit rate transmission, where the bit rate is R, R is placed on each subcarrier, rather than $R/2$ for the higher bit rates. The redundancy at the lower bit rates gives an order of in band diversity. The modulation on the AN/TRC-170 is QPSK on each subcarrier. The maximum data rate is 4.608 Mbps, which includes a digital orderwire and service channel.

The AN/TRC-170 operates in the 4.4–5.0-GHz band with a transmitter output power of 2 kW. The receiver noise figure is 3.1 dB. In its quadruple-

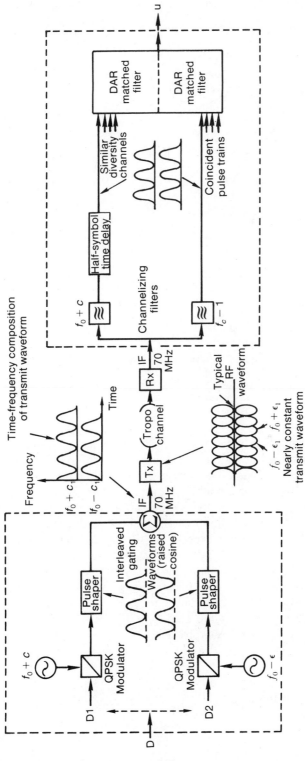

FIGURE 3.45 Operation of the two-frequency AN/TRC-170 modem (Ref. 13).

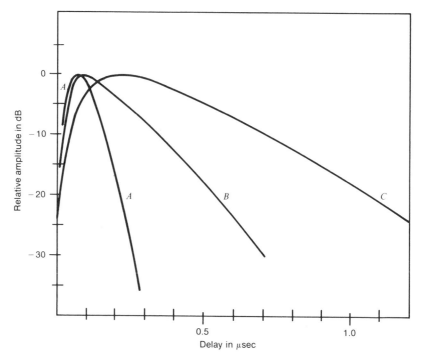

FIGURE 3.46 Characterization of multipath profiles *A*, *B*, and *C* (Ref. 10).

diversity configuration with 9.5-ft antennas and when operating at a trunk bit rate of 1.024 Mbps, the terminal can support a path loss typically of 240 dB (BER = 1×10^{-5}). This value is based on an implicit diversity advantage for a multipath delay spread typical of profile *B* (see Figure 3.46). On a less dispersive path based on profile *A*, the performance would be degraded by about 2.6 dB or a transmission loss of 237.4 dB. With a more dispersive path based on profile *C*, we would expect a 1.1-dB improvement over that of profile *B* (Ref. 10).

The more dispersive the path, the better the equipment operates up to about 1 μsec of rms dispersion. This maximum value would be shifted upward or downward depending on the data rate.

Three multipath profiles are shown in Figure 3.46; rms values for the multipath delay spread of each profile are

Profile *A*:   65 nsec
Profile *B*;   190 nsec
Profile *C*:   380 nsec

PROBLEMS AND EXERCISES

1. Show diagrammatically the "common volume" and "scatter angle."

2. There are two types of fading encountered on tropospheric scatter links. Define and name the types of distributions (statistical distributions of fade events) encountered.

3. Given smooth earth conditions and site elevations of transmitter and receiver are 150 and 250 m above MSL, respectively. What are the takeoff angles at the transmitter and receiver sites?

4. Given sea level refractivity of 310, calculate surface refractivity for a tropo link where the transmitter and receiver sites are 1221 and 1875 m above MSL, respectively.

5. A radiolink 200 km long is to be installed in the eastern United States; the sea level refractivity is 320. The transmitter site is 158 m above MSL and the receiver site is 315 m above MSL. Allow 11 m to antenna centers at each end. The altitude of the transmitter horizon is 330 m and the receiver horizon is 490 m, and the distances to each horizon are 60 and 30 km, respectively. Assume $h_{st} = h_{ts} = 169$ m and $h_{sr} = h_{rs} = 326$ m. Calculate the effective earth radius.

6. Using the information given in question 5, calculate the horizon takeoff angles.

7. As in question 6, calculate the effective antenna heights assuming linear increments in altitude.

8. As in question 6, calculate the angular distance (or scatter angle).

9. Using information developed in questions 5–8, with an operational frequency of 1800 MHz, calculate the basic transmission loss L_{bsr}.

10. From question 9, calculate the long-term median transmission loss L_n.

11. From question 10, the service probability factor is 95%, compute the path loss for a time availability of 99.9%.

12. Assuming 60-ft antennas, calculate the aperture-to-medium coupling loss using both methods given in this chapter.

13. Using information from questions 11 and 12, define the parameters of a baseline system that uses quadruple diversity. Separate the diversity antennas by 400 wavelengths. Select waveguide or coaxial cable transmission line using best judgment. What other losses will be encountered? Calculate values of RSL.

14. What is the FM improvement threshold if the IF bandwidth is 4260 kHz (60 channels of FDM telephony)?

15. Scale the system allowing 3 pWp/km. (How much bigger or smaller than baseline?) Allow an NPR = 55 dB.

16. For a digital modulation scheme, an E_b/N_0 of 13 dB is required 99.9% of the time to achieve an error rate of 10^{-5}. Rescale the system from baseline.

16. What sort of dispersion will be found on this sample path? Give the answer in nanoseconds.

17. If the scatter angle on a certain path can be reduced by 1.5°, will the transmission loss be increased or decreased and by approximately how much?

REFERENCES

1. "General Engineering Beyond-Horizon Radio Communications," USAF T.O. 31Z-10-13, U.S. Department of Defense, Washington, DC, October 1971.

2. "Military Handbook Facility Design for Tropospheric Scatter (Transhorizon Microwave System Design)," MIL-HDBK-417, U.S. Department of Defense, Washington, DC, November 1977.

3. P. F. Panter, *Communication Systems Design Line-of-Sight and Troposcatter Systems*, McGraw-Hill, New York, 1972.

4. P. L. Rice, A. C. Longley, K. A. Norton, and A. P. Barsis, "Transmission Loss Prediction for Tropospheric Communication Circuits, NBS Tech Note 101, U.S. National Bureau of Standards, Boulder, CO, May 1965 (revised January 1967).

5. "Propagation Data Required for Trans-Horizon Radio Relay Systems," CCIR Rep. 238-2, XIth Plenary Assembly, CCIR, Oslo 1966.

6. L. P. Yeh, "Simple Methods for Designing Troposcatter Circuits," *IRE Trans Comm Sys* (September 1960).

7. C. C. Rider, "Median Signal Level Prediction for Tropospheric Scatter," *Marconi Rev* (Third quarter, 1962).

8. "Instruction Manual for Tropospheric Scatter—Principles and Applications," USAEPG-SIG 960-67, U.S. Army Electronic Proving Ground, Ft. Huachuca, AZ, March 1960.

9. R. L. Freeman, *Reference Manual for Telecommunications Engineering*, Wiley, New York, 1985.

10. T. E. Brand, W. J. Connor, and A. J. Sherwood, "AN/TRC-170—Troposcatter Communication System," NATO Conference on Digital Troposcatter, Brussels, March, 1980.

11. R. L. Freeman, *Telecommunication Transmission Handbook*, 2nd ed., Wiley, New York, 1981.

12. (Longley–Rice): G. A. Hufford, A. G. Longley, and W. A. Kissick, "A Guide to the Use of the ITS Irregular Terrain Model in the Areas Prediction Mode," U.S. Department of Commerce, Washington, DC, April 1982. NTIA Report, 82-100.

13. W. J. Connor, "AN/TRC-170—A New Digital Troposcatter Communication System," IEEE ICC '78 Conference Record.

BASIC PRINCIPLES OF SATELLITE COMMUNICATIONS— ANALOG SYSTEMS

4.1 SCOPE AND APPLICATION

Satellite communication has seen remarkable growth over the last 20 years and should see similar growth over the next 20 years. This growth may be tempered by orbital and frequency congestion and accelerating implementation of fiber optic links.

Initially, commercial satellite radiolinks were used on long transoceanic circuits almost exclusively. These links were relayed over the INTELSAT* series of satellites, which are now in phase VI of development (i.e., INTELSAT I through VI). Domestic satellite communication was pioneered by Canada with the ANIK series of satellites. ANIK provides essentially three services: high-capacity trunking among major Canadian population centers, thin-line connectivity to sparsely populated rural areas such as the far north, and TV programming relay.

Ships at sea are now served by INMARSAT† using such satellites as MARISAT and MARECS, and transponders on INTELSAT. Earth stations serving INMARSAT are found in nearly all major maritime countries. The United States and the Soviet Union have a joint venture for search and rescue (SAR) alert and location by satellite.

India, Brazil, Indonesia, the Arab states, and Europe have regional or national domestic satellite systems. The United States, however, is probably the leader in this area. TV programming relay is one area of business activity involving satellites that is mushrooming. TV relay by satellite is being utilized by:

☐ Broadcasters
☐ Cable TV

---

*INTELSAT = International Telecommunication Satellite (Organization).
† INMARSAT = International Marine Satellite (Organization).

☐ Industrial/educational users

☐ Direct to home subscriber

The U.S. armed forces have a number of their own specialized satellite systems such as Defense Satellite Communication System (DSCS), Fleet Satellite (FLTSAT), Satellite Data System (SDS), and MILSTAR. MILSTAR is an advanced digital-processing satellite system operating in the 44/20 GHz bands.

Another major application of domestic/regional satellites is for telephone circuit trunking.

Private industrial networks often lease satellite transponders to provide long-haul network connectivity. Satellite Business Systems (SBS) provides leased or circuit switched service directly to customer premises bypassing local telephone network connectivity.

This chapter, along with Chapters 5 and 6, sets forth methods of design of radiolinks by satellite. Chapter 7 details terminal design. The subsections that follow provide the general principles of satellite link design and analysis for analog transmission. The general intent is to build on the background already developed in previous chapters. The primary thrust is system dimensioning and its rationale. The chapter also provides standard interface information for a number of example systems.

4.2 SATELLITE SYSTEMS—AN INTRODUCTION

4.2.1 Two Broad Categories of Communication Satellites

This text will deal with two broad categories of communication satellites. The first is the repeater satellite, affectionately called the "bent pipe" satellite. The second is the processing satellite, which is used exclusively on digital circuits,

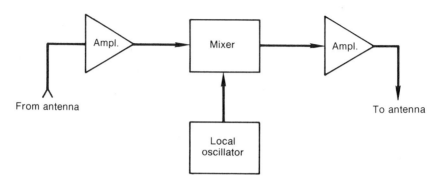

FIGURE 4.1 Simplified functional block diagram of the payload of a conventional translating RF repeater or bent pipe satellite.

where, as a minimum, the satellite demodulates the uplink signal to baseband and regenerates that signal. Analog circuits use exclusively "bent pipe" techniques; digital circuits may use either variety.

The bent pipe satellite is simply a frequency translating RF repeater. Figure 4.1 is a simplified functional block diagram of the payload of such a satellite. Processing satellites are described in Chapter 5.

4.2.2 Satellite Orbits

There are three types of satellite orbits:

- □ Polar
- □ Equatorial
- □ Inclined

The figure a satellite defines in orbit is an ellipse. Of course, a circle is a particular class of ellipse. A Molniya orbit is a highly inclined elliptical orbit.

The discussion here will dwell almost entirely on geostationary satellites. A geostationary satellite has a circular orbit. Its orbital period is one sidereal day (23 h, 56 min, 4.091 sec) or nominally 24 hr. Its inclination is 0°, which means that the satellite is always directly over the equator. It is geostationary. That is to say that it appears stationary over any location on earth, which is within optical view.

Geostationary satellites are conventionally located with respect to the equator (0° latitude) and a subsatellite point, which is given in longitude at the earth's surface. The satellite's range at this point, and only at this point, is 35,784 km. Table 4.1 gives details and parameters of the geostationary satellite.

Table 4.1 outlines several of the advantages and disadvantages of a geostationary satellite. Most of these points are self-explanatory. For satellites not at geosynchronous altitude and not over the equator, there is the appearance of movement. The movement with relation to a point on earth will require some form of automatic tracking on the earth station antenna to keep it always pointed at the satellite. If a satellite system is to have full earth coverage using a constellation of geostationary satellites, a minimum of three satellites would be required to be separated by 120°. As one moves northward or southward from the equator, the elevation angle to a geostationary satellite decreases. (See Section 4.2.3.) Elevation angles below 5° are generally undesirable, as will be discussed subsequently. This is the rationale in Table 4.1 for "area of no coverage." Handover refers to the action taken by a satellite earth station antenna when a nongeostationary (often misnamed "orbiting satellite") disappears below the horizon (or below 5° elevation angle) and the antenna slews to a companion satellite in the system that is just appearing above the opposite horizon. It should be pointed out here that geostationary satellites do have

TABLE 4.1. The Geostationary Satellite Orbit

For the special case of a synchronous orbit—satellite in prograde circular orbit over the equator:	
Altitude	19,322 nautical miles, 22,235 statue miles, 35,784 km
Period	23 hr, 56 min, 4.091 sec (one sidereal day)
Orbit inclination	0°
Velocity	6879 statue miles/hr
Coverage	42.5% of earth's surface (0° elevation)
Number of satellites	Three for global coverage with some areas of overlap (120° apart)
Subsatellite point	On the equator
Area of no coverage	Above 81° north and south latitude
Advantages	Simpler ground station tracking
	No handover problem
	Nearly constant range
	Very small Doppler shift
Disadvantages	Transmission delay
	Range loss (free space loss)
	No polar coverage

Source. Reference 1.

small residual relative motions. Over its subsatellite point, a geostationary satellite carries out a small apparent suborbit in the form of a figure eight because of higher space harmonics of the earth's gravitation and tidal forces from the sun and the moon. The satellite also tends to drift off station because of the gravitational attraction of the sun and the moon as well as solar winds. Without correction the inclination plane drifts roughly 0.86° per year (Ref. 1 Section 13.4.1.9).

4.2.3 Elevation Angle

The elevation angle or "look angle" of a satellite terminal antenna is the angle measured from the horizontal to the point on the center of the main beam of the antenna when the antenna is pointed directly at the satellite. This concept is shown in Figure 4.2. Given the elevation angle of a geostationary satellite, we can define the range. We will need the range, d in Figure 4.2, to calculate the free space loss or spreading loss for the satellite radiolink.

4.2.4 Determination of Range and Elevation Angle of a Geostationary Satellite

Geostationary satellites operate at an altitude of about 35,785 km above sea level. Unless an earth station is directly under a satellite, however, the distance

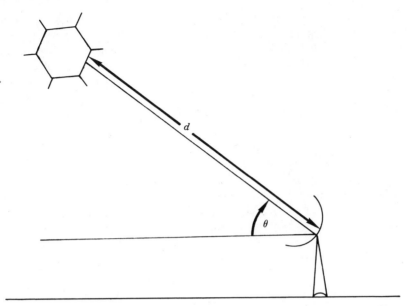

FIGURE 4.2 Definition of elevation angle (θ) or "look angle" and range (D) to satellite.

d of Figure 4.2 will be greater than 35,785 km. The value of d can be established by use of the law of cosines of plane trigonometry. Consider first that the earth station is on the same longitude as the subsatellite point, taken to be at 0° latitude. The subsatellite point is located where a straight line from the satellite to the center of the earth intersects the earth's surface. See Figure 4.3: From the figure we can now state that

$$d^2 = r_0^2 + (h + r_0)^2 - 2r_0(h + r_0)\cos \theta' \qquad (4.1)$$

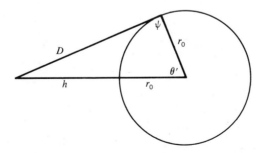

FIGURE 4.3 Geometry for calculation of distance D of a geostationary satellite from an earth terminal on earth's surface.

where θ' is latitude. The equatorial radius of the earth is 6378.16 km, the polar radius is 6356.78 km, and the mean radius is 6371.03 km (Ref. 29). To obtain the most accurate value of d, it would be necessary to take into account the departure of the earth from sphericity, but an approximate value of d can be obtained by taking r_0, the earth is radius, to be 6378 km and h, the height of the satellite above the earth's surface, to be 35,785 km in equation (4.1). Divide all terms in equation (4.1) by $(h + r_0)^2$ or $(42,163)^2$, which gives:

$$\left(\frac{d}{h + r_0} \right)^2 = f^2 + 1 - 2f\cos\theta' \tag{4.2}$$

where $f = r_0/(h + r_0) = 0.1513$. Once d is known, then all three sides of the triangle in Figure 4.3 are known and the angle Ψ can be determined by applying the law of cosines again. The applicable equation is

$$(h + r_0)^2 = d^2 + r_0^2 - 2r_0 d \cos \Psi \tag{4.3}$$

The elevation angle θ measured from the horizontal at the earth terminal is equal to $\psi - 90°$.

For an earth terminal not on the same meridian as the subsatellite point, we can use the equation (from the spherical law of cosines):

$$\cos z = \cos\theta' \cos\phi' \tag{4.4}$$

in equation (4.1) in place of $\cos\theta'$, where ϕ' is the difference in longitude between the subsatellite point and the earth terminal; $\cos Z$ is the angular distance of a great circle path for the special case that one of the end points is at 0° latitude (Figure 4.4). Also, the expression follows from the law of cosines for sides from spherical trigonometry (Ref. 2, Section 44). The azimuth angle α of an earth–space path can be determined by using (Ref. 2, Section 44)

$$\cos\alpha = \tan\theta' \cot Z \tag{4.5}$$

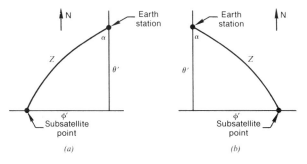

FIGURE 4.4 Projection of right spherical triangles on earth's surface.

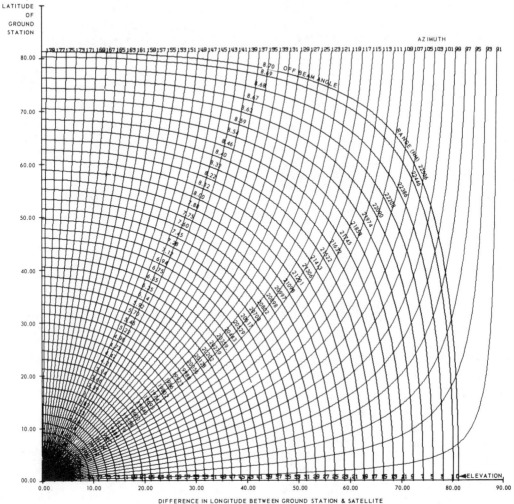

LATITUDE OF GROUND STATION

AZIMUTH

OFF BEAM ANGLE

RANGE (NM)

DIFFERENCE IN LONGITUDE BETWEEN GROUND STATION & SATELLITE

ELEVATION

Note:
Angle off beam center is useful only if satellite has a despun circular beam with beam center at subsatellite point. (Intelsat III)

Satellite direction from earth station	Azimuth Transformation
East & South	None
West & South	Subtract azimuth from 360°
East & North	Subtract azimuth from 180°
West & North	Add 180° to azimuth

Example	Latitude	Longitude
Japan	37° N	141° E
Satellite	0°	176° E
Separation	176°−141°= 35°	
Azimuth	131°	
Elevation	33.5°	
Off Beam Angle	7.26°	
Range	20,680 NM	

FIGURE 4.5 Determination of range to a geostationary satellite, azimuth, and elevation angles. (From Ref. 1, courtesy of COMSAT, Washington, DC.)

227

The angle α is shown in Figure 4.4*a* for an earth terminal located to the east of the subsatellite point. The azimuth angle measured from the north in this case would then be $180° + \alpha$. For an earth terminal located to the west of the subsatellite point (Figure 4.4*b*), the angle from true north is $180° - \alpha$.

Example 1. Calculate azimuth and elevation angles and distance from a point 40°N, 105°W for a satellite where the subsatellite point is located at 119°W.

$$\cos Z = \cos 40° \cos 14° = 0.743$$

From equation 4.1 we find that $d = 37,666$ km and the elevation angle $\theta = 43.73°$; the azimuth angle is $201.2°$.

Range, elevation, and azimuth angles may also be determined by nomogram with sufficient accuracy for link analyses (link budget). See Figure 4.5.

4.3 INTRODUCTION TO LINK ANALYSIS OR LINK BUDGET

4.3.1 Rationale

To size or dimension a satellite terminal correctly, we will want to calculate the receive signal level at the terminal. The methodology is very similar to the methods used in Chapters 2 and 3. There are also certain legal constraints that we should be aware of.

4.3.2 Frequency Bands Available for Satellite Communications

The frequency bands assigned for satellite communications are given in Table 4.2. Generally, these are referred to in band pairs. One of the band pairs, usually of higher frequency than the other, is used for the uplink path (i.e., terminal to satellite), and the other is assigned to the downlink path. In this text we will be dealing with the following commonly used frequency band pairs:

6/4 GHz	5925–6425 MHz	Uplink	Commercial
	3700–4200 MHz	Downlink	
8/7 GHz	7900–8400 MHz	Uplink	Military
	7250–7750 MHz	Downlink	
14/11 GHz	14.0–14.5 GHz	Uplink	Commercial
	11.7–12.2 GHz	Downlink	
30/20 GHz	27.5–30.5 GHz	Uplink	Commercial
	17.7–20.2 GHz	Downlink	
30/20 GHz	30.0–31.0 GHz	Uplink	Military
	20.2–21.2 GHz	Downlink	
44/20 GHz	43.5–45.5 Ghz	Uplink	Military
	20.2–21.2 GHz	Downlink	

TABLE 4.2. WARC 79 Fixed Satellite Service Allocations

Earth-to-Space (GHz)	Region	Bandwidth (MHz)	Space-to-Earth (GHz)	Region	Bandwidth (MHz)
2.655–2.690	$2^b, 3^b$	35	2.50–2.535	$2^b, 3^b$	35
			2.535–2.690	2^b	155
5.725–5.850 ⎫	1	125	3.40–4.20	1, 2, 3	800
5.850–7.075 ⎭	1, 2, 3	1225	4.50–4.80	1, 2, 3	300
7.90–8.40	1, 2, 3	500	7.25–7.75	1, 2, 3	500
12.50–12.7	1	200			
12.70–12.75	1, 2	50			
12.75–13.25 ⎫	1, 2, 3	500			
			10.70–11.70	1, 2, 3	1000
14.00–14.50 ⎭	1, 2, 3	500	11.70–12.30	$2^{b,c}$	600
27.00–27.50	$2, 3^c$	500	12.20–12.50	3^b	300
			12.50–12.75	1, 3	250
27.50–31.00	1, 2, 3	3500	17.70–21.20	1, 2, 3	3500
42.5–43.50 ⎫	1	1000			
47.2–49.20 ⎸	2^a	2000	37.50–40.5	1, 2, 3	3000
49.2–50.2 ⎹	1	1000			
50.40–51.40 ⎭	1	1000			
71.0–74.0	1, 2, 3	3000	81.0–84.0	1, 2, 3	3000
74.0–75.5	1, 2, 3	1500			
92.0–95.0	1, 2, 3	3000	102.0–105.0	1, 2, 3	3000
202.0–217.0	1, 2, 3	15000	149.0–164.0	1, 2, 3	15000
265.0–275.0	1, 2, 3	10000	231.0–241.0	1, 2, 3	10000

[a] Intended for use by, but not limited to, broadcast satellite service feeder links.

[b] Limited to national and subregional services.

[c] Upper band limit (12.3 GHz) may be replaced by a new value in the range 12.1–12.3 GHz at the 1983 RARC for Region 2.

Source. From Reference 4, reprinted with permission.

Tables 4.2 through 4.5 provide more specific frequency assignment information as set forth by the World Administrative Radio Congress of 1979 (WARC-79) (Ref. 3). [*Note.* In Tables 4.2 through 4.5 the reader is urged to consult the latest edition of Radio Regulations (ITU, Geneva), especially for footnotes.]

4.3.3 Free Space Loss or Spreading Loss

Section 1.2 gave the method of calculation of free space loss or, what some call, spreading loss. Equations (1.5), (1.7a), and (1.7b) apply. We just plug

TABLE 4.3. WARC 79 Intersatellite Service Allocations

Band (GHz)	Bandwidth (GHz)
22.55–23.55	1
32.00–33.00	1
54.25–58.20	3.95
59.0–64.0	5
116.0–134.0	18
170.0–182.0	2
185.0–190.0	5

Source. From Reference 4, reprinted with permission.

frequency and distance (range) into these equations, being sure to use the proper units.

Example 2. The elevation angle to a geostationary satellite is 21° and the transmitting frequency is 3.941 GHz. What is the free space loss in decibels?
  Use equation (1.7b), where the distance unit is the nautical mile (nm):

$$L_{dB} = 37.80 + 20 \log 21{,}201 \text{ nm} + 20 \log 3941 \text{ MHz}$$

$$= 196.24 \text{ dB}$$

4.3.4 Isotropic Receive Level—Simplified Model

Consider a downlink at 3941 MHz from a geostationary satellite with a free space loss of 196.24 dB. Let the satellite have an EIRP of $+29$ dBW. What would the isotropic receive level (IRL) be for an earth station with a 21° elevation angle? All other losses are disregarded. The path is line of sight (LOS) by definition. The approach is the same as in Chapter 2 (see Section 2.6.2). Equation (2.7) applies:

$$IRL_{dBW} = EIRP + FSL_{dB}$$

$$IRL_{dBW} = +29 \text{ dBW} + (-196.24) \text{ dB}$$

$$IRL_{dBW} = -167.24 \text{ dBW}$$

The equation has left out a number of small, but when added together, not insignificant losses. Some of these losses are radome loss, pointing losses,

TABLE 4.4. Mobile Satellite Services Allocations

Earth-to-Space (MHz)	Region	Bandwidth (MHz)	Space-to-Earth (MHz)	Region	Footnote
121.45–121.55[a, b, h]	1, 2, 3		—		3572A
242.95–243.05[a, b, h]	1, 2, 3		—		3572A
235.0–322.0[a, h]	1, 2, 3	87			3618
335.4–399.9[a, h]	1, 2, 3	64.4			3618
405.5–406.0[c]	Canada	0.5	—		3533A
406.0–406.1[b]	1, 2, 3	0.1	—		3633A
406.1–410.0[c]	Canada	3.9	—		3634
608.0–614.0[d, e]	2	6.0	—		
806.0–890.0[c, h]	2, 3	3.0			3662C
	Norway				3662CA
	Sweden				3670B
942.0–960.0[c, h]	3, Norway	18			3662C
	Sweden				3662CA
1645.5–1646.5[f]	1, 2, 3	1.0	1554.0–1545.0[f]	1, 2, 3	3695A
(GHz)			(GHz)		
7.90–8.025[a]	1, 2, 3	125	7.250–7.375[a]	1, 2, 3	3764B
14.00–14.50[e, g]	1, 2, 3	500	—		
29.50–30.0[e]	1, 2, 3	500	19.70–20.20[e]	1, 2, 3	
30.00–31.00	1, 2, 3	1000	20.20–21.20	1, 2, 3	
43.50–47.00[h]	1, 2, 3	3500	39.50–40.50	1, 2, 3	3814C
50.40–51.40[e]	1, 2, 3	1000	—		—
66.00–71.00[h]	1, 2, 3	5000			3814C
71.00–74.00	1, 2, 3	3000	81.00–84.00	1, 2, 3	—
95.00–100.00[h]	1, 2, 3	5000			3814C
135.00–142.00[h]		7000			3814C
190.00–200.00[h]		10000			3814C
252.00–265.00[h]		13000			3814C

[a] Footnote allocation.

[b] Emergency position indicating radio beacons only.

[c] Footnote allocation excludes aeronautical mobile-satellite services.

[d] Excludes aeronautical mobile-satellite services.

[e] Secondary allocation.

[f] Distress and safety operations only.

[g] Footnote allocation to land mobile-satellite service only.

[h] No direction specified.

Source. Reference 4, reprinted with permission.

TABLE 4.5. Broadcasting-Satellite Service Allocations

Earth-to-Space (GHz)	Region	Bandwidth (MHz)	Space-to-Earth (GHz)	Region	Bandwidth (MHz)
Feeder links for the broadcast satellite service may, in principle, use any of the fixed satellite service (earth-to-space) bands listed in Table 4.2 with appropriate coordination. However, the following bands were set aside for exclusive or preferential use by such feeder links.					
			$0.62–0.79^a$	1, 2, 3	170
			$2.50–2.69^b$	1, 2, 3	190
10.70–11.70	1	1000	11.70–12.1	1, 3	400
14.50–14.80	$1,^c$ 2, 3	300	12.10–12.2	1, 2, 3	100
17.30–18.1	1, 2, 3	800	12.20–12.5	1, 2	300
			12.50–12.7	$2, 3^b$	200
			12.70–12.75	3^b	50
27.00–27.50	2, 3	500	22.50–23.00	2, 3	500
47.20–49.20	1, 2, 3	2000	40.50–42.50	1, 2, 3	2000
			84.00–86.00	1, 2, 3	2000

^{a} Limited to TV.

^{b} Limited to community reception.

^{c} Excluding Europe.

Source. Reference 4, reprinted with permission.

polarization loss, gaseous absorption loss, and excess attenuation due to rainfall.

4.3.5 Limitation of Flux Density on Earth's Surface

The frequency bands shown in Section 4.3.2 are shared with other services, such as terrestrial point-to-point radiolink microwave. The flux density of satellite signals on the earth's surface must be limited so as not to interfere with terrestrial radio services band-sharing with satellite systems. CCIR recommends the following flux density limits [excerpted from CCIR Rec. 358-2 (Ref. 5), Ref. 1, 13-4.3]:

Maximum Power Flux Density on Earth Surface

UNANIMOUSLY RECOMMENDS

1. that, in frequency bands in the range 1 to 23 GHz shared between systems in the Fixed Satellite Service and line-of-sight radio-relay systems, the maximum power flux-density produced at the surface of the Earth by emissions from a satellite, for all conditions and methods of modulation, should not exceed:

1.1 in the band 1.7 to 2.535 GHz, in any 4 kHz band:

$$
\begin{array}{lll}
-154 & \text{dB(W/m}^2) \text{ for} & \delta \le 5° \\
-154 + (\delta - 5)/2 & \text{dB(W/m}^2) \text{ for } 5° < & \delta \le 25° \\
-144 & \text{dB(W/m}^2) \text{ for } 25° < \delta \le 90°
\end{array}
$$

1.2 in the band 3 to 8 GHz, in any 4 kHz band:

$$
\begin{array}{lll}
-152 & \text{dB(W/m}^2) \text{ for} & \delta \le 5° \\
-152 + (\delta - 5)/2 & \text{dB(W/m}^2) \text{ for } 5° < & \delta \le 25° \\
-142 & \text{dB(W/m}^2) \text{ for } 25° < \delta \le 90°
\end{array}
$$

1.3 in the band 8 to 11.7 GHz, any 4 kHz band:

$$
\begin{array}{lll}
-150 & \text{dB(W/m}^2) \text{ for} & \delta \le 5° \\
-150 + (\delta - 5)/2 & \text{dB(W/m}^2) \text{ for } 5° < & \delta \le 25° \\
-140 & \text{dB(W/m}^2) \text{ for } 25° < \delta \le 90°
\end{array}
$$

1.4 in the band 12.5 to 15.4 GHz, in any 4 kHz band:

$$
\begin{array}{lll}
-148 & \text{dB(W/m}^2) \text{ for} & \delta \le 5° \\
-148 + (\delta - 5)/2 & \text{dB(W/m}^2) \text{ for } 5° < & \delta \le 25° \\
-138 & \text{dB(W/m}^2) \text{ for } 25° < \delta \le 90°
\end{array}
$$

1.5 in the band 15.4 to 23 GHz, in any 1 MHz band:

$$
\begin{array}{lll}
-115 & \text{dB(W/m}^2) \text{ for} & \delta \le 5° \\
-115 + (\delta - 5)/2 & \text{dB(W/m}^2) \text{ for } 5° < & \delta \le 25° \\
-105 & \text{dB(W/m}^2) \text{ for } 25° < \delta \le 90°
\end{array}
$$

where δ is the angle of arrival of the radio-frequency wave (degrees above the horizontal);

2.   that the aforementioned limits relate to the power flux-density and angles of arrival which would be obtained under free-space propagation conditions.

Note.—Definitive limits applicable in shared frequency bands are laid down in Article 7 of the Radio Regulations (Nos. 470N to 470NZB). The CCIR is continuing its study of these problems, which may lead to changes in the recommended limits. At the present time, no changes are proposed to the limits laid down in the Radio Regulations.

4.3.5.1 Calculation of Power Flux Density Levels

The signal level of a wave incident on a satellite antenna is measured by the power flux density S (expressed in watts per square meter) of the approaching wavefront. If we consider the emitter as a point source, the emitted energy spreads uniformly in all directions. The satellite antenna can be considered a point on the surface of a sphere, where the emitter is at the center of the sphere. The radius R_s of the sphere is the distance from the emitter to the satellite antenna. The surface area of the sphere in square meters is given by

$$A_s = 4\pi R_s^2 \tag{4.6}$$

If the distance to the satellite from the emitter was 37,750 km, then

$$A_s = 1.791 \times 10^{16} \text{ m}^2$$

The flux density at the satellite is calculated as if the EIRP of the emitter uniformly covered the total surface A_s. The resulting flux density is given by

$$S = \frac{(k_A)\text{EIRP}}{4\pi R_s^2} = \frac{(k_A)\text{EIRP}}{A_s} \tag{4.7}$$

where k_A is the atmospheric attenuation factor (e.g., gaseous attenuation), which is less than unity. Converting to decibel form gives

$$S(\text{dBW/m}^2) = \text{EIRP}_{\text{dBW}} - 10\log A_s - L_A \tag{4.8}$$

where $L_A = -10\log k_A$, which is the atmospheric attenuation factor in decibels. For a bent pipe satellite we are often given a flux density value (S) to saturate the satellite transponder TWT final amplifier. The earth station EIRP required from a satellite terminal emitter antenna is

$$\text{EIRP} = 10\log A_s + S + L_A \tag{4.9}$$

Example 3. What earth station EIRP is required to saturate a TWT HPA of satellite transponder if the transponder requires -88 dBW/m^2 power flux density for TWT saturation? The distance to the satellite is 23,000 statute miles, and 1 dB is allotted for atmospheric gas attenuation.

First convert 23,000 statute miles to kilometers.

$$23{,}000 \times 1.609 = 37{,}007 \text{ km}$$

Then

$$A_s = 4\pi 37{,}007 \times 37{,}007 \times 10^6$$

$$= 1.3695 \times 10^{15} \times 4 \times 3.14159$$

$$= 1.72098 \times 10^{16}$$

and

$$\text{EIRP} = 162.36 - 88 - 1$$

$$\text{EIRP} = +73.36 \text{ dBW}$$

4.3.6 Thermal Noise Aspects of Low-Noise Systems

We deal with very low signal levels in space communication systems. Downlink signal levels are in the approximate range of -154 to -188 dBW. The objective is to achieve sufficient S/N or E_b/N_0 at demodulator outputs. There are two ways of accomplishing this:

- [ ] By increasing system gain, usually with antenna gain.
- [ ] Reducing system noise.

In this section we will give an introductory treatment of thermal noise analytically, and later the term G/T (figure of merit) will be introduced.

Around noise threshold, thermal noise predominates. To set the stage, we quote from Ref. 6:

> The equipartition law of Boltzman and Maxwell (and the works of Johnson and Nyquist) states that the available power per unit bandwidth of a thermal noise source is

$$p_n(f) = kT \text{ watts/Hz} \tag{4.10}$$

> where k is Boltzmann's constant $(1.3806 \times 10^{-23}$ joule/K) and T is the absolute temperature of the source in kelvins.

Looking at a receiving system, all active components and all components displaying ohmic loss generate noise. In LOS radiolinks, system noise temperatures are in the range of 1500–4000 K, and the noise of the receiver front end is by far the major contributor. In the case of space communications, the receiver front end may contribute less than one-half the total system noise. Total receiving system noise temperatures range from as low as 70 up to 1000 K (for those types of systems considered in this chapter).

In Chapters 2 and 3, receiving system noise was characterized by noise figure expressed in decibels. Here, where we deal often with system noise temperatures of less than 290 K, the conventional reference of basing noise at room temperature is awkward. Therefore, noise figure is not useful at such low noise levels. Instead, it has become common to use effective noise temperature T_e [equation (4.13)].

It can be shown that the available noise power at the output of a network in a band B_w is (Ref. 6)

$$p_n = g_a(f)(T + T_e)B_w \tag{4.11}$$

where g_a is the network power gain at frequency f, T is the noise temperature of the input source, and T_e is the effective input temperature of the network. For an antenna–receiver system, the total effective system noise temperature T_{sys}, conventionally referred to the input of the receiver, is

$$T_{sys} = T_{ant} + T_r \tag{4.12}$$

where T_{ant} is the effective input noise temperature of the antenna subsystem and T_r is the effective input noise temperature of the receiver subsystem. The ohmic loss components from the antenna feed to the receiver input also generate noise. Such components include waveguide or other transmission line, directional couplers, circulators, isolators, waveguide switches, and so forth.

It can be shown that the effective input noise temperature of an attenuator is (Ref. 6)

$$T_e = \frac{p_a}{kB_w g_a} - T_s = \frac{T(1 - g_a)}{g_a} \tag{4.13}$$

where T_s is the effective noise temperature of the source, the lossy elements have a noise temperature T, k is Boltzmann's constant, and g_a is the gain (available loss). p_a is the noise power at the output of the network.

The loss of the attenuator l_a is the inverse of the gain or

$$l_a = \frac{1}{g_a} \tag{4.14}$$

Substituting into Equation (4.13) gives

$$T_e = T(l_a - 1) \tag{4.15}$$

It is accepted practice (Ref. 6) that

$$n_f = l_a + \frac{T_e}{T_0} \tag{4.16}$$

where in Ref. 6 n_f is called the noise figure. Other texts call it the noise factor and

$$\text{NF}_{dB} = 10 \log_{10} n_f \tag{4.17}$$

and T_0 is standard temperature or 290 K. NF_{dB} is the conventional noise figure discussed in Section 2.6.3.1.

From equation (4.15) the noise figure (factor) is

$$n_f = 1 + \frac{(l_a - 1)T}{T_0} \tag{4.18}$$

If the attenuator lossy elements are at standard temperature (e.g., 290 K), the noise figure equals the loss (the noise factor equals the numeric of the loss):

$$n_f = l_a \tag{4.19}$$

or expressed in decibels,

$$\text{NF}_{dB} = 10 \log l_a = L_{a\,(dB)}$$

For low-loss (i.e., ohmic-loss) devices whose loss is less than about 0.5 dB, such as short waveguide runs, which are at standard temperature, equation (4.15) reduces to a helpful approximation

$$T_e \approx 66.8L \qquad (4.20)$$

where L is the loss of the device in decibels.

The noise figure in decibels may be converted to effective noise temperature by

$$\text{NF}_{dB} = 10 \log_{10}\left(1 + \frac{T_e}{290}\right) \qquad (4.21)$$

Example 4. If a noise figure were given as 1.1 dB, what is the effective noise temperature?

$$1.1 \text{ dB} = 10 \log\left(1 + \frac{T_e}{290}\right)$$

$$0.11 = \log\left(1 + \frac{T_e}{290}\right)$$

$$1 + \frac{T_e}{290} = \log^{-1}(0.11)$$

$$1 + \frac{T_e}{290} = 1.29$$

$$T_e = 84.1 \text{ K}$$

4.3.7 Calculation of C/N_0

We present two methods to carry out this calculation. The first method follows the rationale given in Sections 2.12.5.3 and 3.7.2. C/N_0 is measured at the input of the first active stage of the receiving system. For space receiving systems this is the low-noise amplifier (LNA) or other device carrying out a similar function. Figure 4.6 is a simplified functional block diagram of such a receiving system.

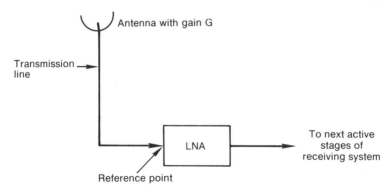

FIGURE 4.6 Simplified block diagram of space receiving system.

C/N_0 is simply the carrier to noise ratio, where N_0 is the noise density in 1 Hz of bandwidth. C is the receive signal level (RSL). Restating equation (4.10):

$$N_0 = kT \qquad (4.22)$$

where k is Boltzmann's constant and T is the effective noise temperature, in this case of the space receiving system. We can now state this identity:

$$C/N_0 = C/kT \qquad (4.23)$$

Turning to Figure 4.6, we see that if we are given the signal level impinging on the antenna, which we call the isotropic receive level (IRL), the receive signal level (RSL or C) at the input to the LNA is the IRL plus the antenna gain minus the line losses, or stated in equation form:

$$C_{dBW} = IRL_{dBW} + G_{ant} - L_{L\,(dB)} \qquad (4.24)$$

where L_L are the line losses in decibels. These losses will be the sum of the waveguide or other transmission line losses, antenna feed losses, and, if used, directional coupler loss, waveguide switch loss, power split loss, bandpass filter loss (if not incorporated in LNA), circulator/isolator losses, and so forth.

To calculate N_0, equation (4.22) can be restated as

$$N_0 = -228.6 \text{ dBW} + 10 \log T_{sys} \qquad (4.25)$$

where -228.6 dBW is the theoretical value of the noise level in dBW for a perfect receiver (noise factor of 1) at absolute zero in 1 Hz of bandwidth. T_{sys} is the receiving system effective noise temperature often called just system noise temperature.

Example 5. Given a system (effective) noise temperature of 84.1 K, what is N_0?

$$N_0 = -228.6 \text{ dBW} + 10 \log 84.1$$

$$= -228.6 + 19.25$$

$$= -209.35 \text{ dBW}$$

To calculate C or RSL, consider the following example.

Example 6. The IRL from a satellite is -155 dBW; the earth station receiving system (space receiving system) has an antenna gain of 47 dB, an antenna feed loss of 0.1 dB, a waveguide loss of 1.5 dB, a directional coupler insertion loss of 0.2 dB, and a bandpass filter loss of 0.3 dB; the system noise temperature (T_{sys}) is 117 K. What is C/N_0?
  Calculate C (or RSL):

$$C = -155 \text{ dBW} + 47 \text{ dB} - 0.1 \text{ dB} - 1.5 \text{ dB} - 0.2 \text{ dB} - 0.3 \text{ dB}$$

$$C = -110.1 \text{ dBW}$$

Calculate N_0:

$$N_0 = -228.6 \text{ dBW} + 10 \log 117 \text{ K}$$

$$= -207.92 \text{ dBW}$$

Thus

$$C/N_0 = C_{dBW} - N_{0 \, (dBW)} \tag{4.26}$$

In this example, substituting:

$$C/N_0 = -110.1 \text{ dBW} - (-207.92 \text{ dBW})$$

$$= 97.82 \text{ dB}$$

The second method of calculating C/N_0 involves G/T, which is discussed in the next section.

4.3.8 Gain-to-Noise Temperature Ratio, G/T

G/T can be called the "figure of merit" of a radio receiving system. It is most commonly used in space communications. It not only gives an experienced engineer a "feel" of a receiving system's capability to receive low level signals effectively, it is also used quite neatly as an algebraically additive factor in space system link budget analysis.

G/T can be expressed by the following identity:

$$\frac{G}{T} = G_{dB} - 10 \log T \tag{4.27}$$

where G is the receiving system antenna gain and T (better expressed as T_{sys}) is the receiving system noise temperature. Now we offer a word of caution. When calculating G/T for a particular receiving system, we must stipulate where the reference plane is. In Figure 4.6 it was called the "reference point." It is at the reference plane where the system gain is measured. In other words, we take the gross antenna gain and subtract all losses (ohmic and others) up to that plane or point. This is the net gain at that plane.

System noise is treated in the same fashion. Equation (4.12) stated

$$T_{sys} = T_{ant} + T_r$$

The antenna noise temperature T_{ant} coming inward in the system, includes all noise contributors, including sky noise, up to the reference plane. Receiver noise T_r includes all noise contributors from the reference plane to the baseband output of the demodulator.

In most commercial space receiving systems, the reference plane is taken at the input to the LNA, as shown in Figure 4.6. In many military systems it is taken at the foot of the antenna pedestal. In one system, it was required to be taken at the feed. It can be shown that G/T will remain constant so long as we are consistent regarding the reference plane.

Calculation of the net gain (G_{net}) to the reference plane is straightforward. It is the gross gain of the antenna minus the sum of all losses up to the reference plane. Calculation of T_{sys} is somewhat more involved. We use equation (4.12). The calculation of T_{ant} is described in section 4.3.8.1 and of T_r in Section 4.3.8.2.

4.3.8.1 Calculation of Antenna Noise Temperature, T_{ant}

The term T_{ant}, or antenna noise, includes all noise contributions up to the reference plane. Let us assume for all further discussion in this chapter that the reference plane coincides with the input to the LNA (Figure 4.6). There are two "basic" contributors of noise: sky noise and noise from ohmic losses.

Sky noise is a catchall for all external noise sources that enter through the antenna, through its main lobe, and through its sidelobes. External noise is largely due to extraterrestrial sources and thermal radiation from the atmosphere and the earth. Cosmic noise is a low level of extraterrestrial radiation that seems to come from all directions.

The sun is an extremely strong source of noise, which can interrupt satellite communications when it passes behind the satellite being used and thus lies in the main lobe of an earth station's antenna receiving pattern. The moon is a much weaker source, which is relatively innocuous to satellite communica-

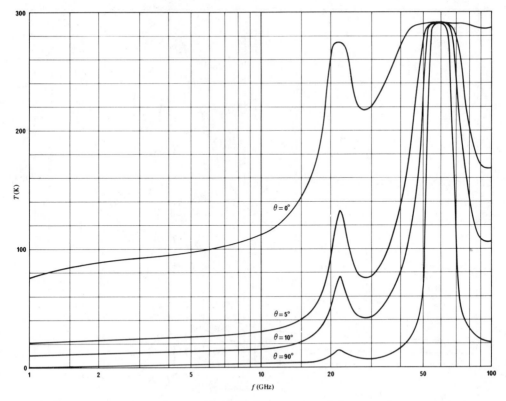

FIGURE 4.7 Sky noise temperature (clear air). Surface pressure is 1 atm; surface temperature is 20°C; surface water-vapor concentration is 3 g/m³; θ is the elevation angle. From CCIR Rep. 720 (Ref. 7). Courtesy of ITU-CCIR, Geneva.

tions. Its radiation is due to its own temperature and reflected radiation from the sun.

The atmosphere affects external noise in two ways. It attenuates noise passing through it, and it generates noise because of the energy of its constituents. Ground radiation, which includes radiation of objects of all kinds in the vicinity of the antenna, is also thermal in nature.

For our discussion we will say that sky noise (T_{sky}) varies with frequency, elevation angle, and surface water-vapor concentration. Figures 4.7, 4.8, and 4.9 give values of sky noise for elevation angles (θ) of 0°, 5°, 10°, and 90° degrees, for water-vapor concentration of 3, 10, and 17 g/m³. These figures do not include ground radiation contributions.

Antenna noise T_{ant} is the total noise contributed to the receiving system by the antenna up to the reference plane. It is calculated by the formula (Ref. 8)

$$T_{ant} = \frac{(l_a - 1)290 + T_{sky}}{l_a} \qquad (4.28)$$

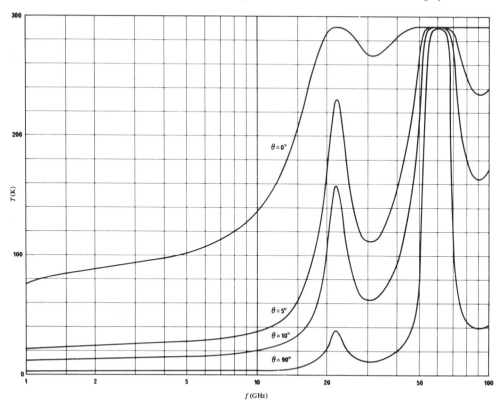

FIGURE 4.8 Sky noise temperature (clear air). Surface pressure is 1 atm; surface temperature is 20°C; surface water-vapor concentration is 10 g/m³; θ is the elevation angle. From CCIR Rep. 720 (Ref. 7). Courtesy of ITU, CCIR,Geneva.

where l_a is the numeric equivalent of the system ohmic losses (in decibels) up to the reference plane. l_a may be expressed, then, as

$$l_a = \log_{10}^{-1} \frac{L_a}{10} \qquad (4.29)$$

where L_a is the sum of the losses to the reference plane.

Example 7. Assume an earth station with an antenna at an elevation angle of 10°, clear sky, 3 g/m³ water-vapor concentration and ohmic losses as follows: waveguide loss of 2 dB, feed loss of 0.1 dB, directional coupler insertion loss of 0.2 dB, and a bandpass filter insertion loss of 0.4 dB. These are the losses up to the reference plane, which is taken as the input to the LNA (Figure 4.10). What is the antenna noise temperature T_{ant}? The operating frequency is 12 GHz.

Determine the sky noise from Figure 4.7. For an elevation angle of 10° and a frequency of 12 GHz, the value is 19 K.

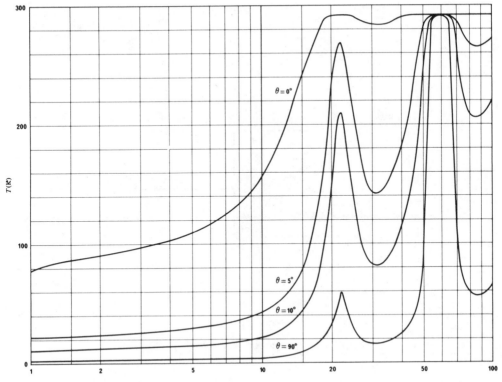

FIGURE 4.9 Sky noise temperature (clear air). Surface pressure is 1 atm; surface temperature is 20°C; surface water-vapor concentration is 17 g/m³; θ = elevation angle. From CCIR Rep. 720 (Ref. 7). Courtesy of ITU-CCIR, Geneva.

Sum the ohmic losses up to the reference plane:

$$L_a = 0.1 \text{ dB} + 2 \text{ dB} + 0.2 \text{ dB} + 0.4 \text{ dB}$$
$$+ 2.7 \text{ dB}$$
$$l_a = \log^{-1}(2.7/10)$$
$$= 1.86$$

Substitute into equation (4.28):

$$T_{ant} = \frac{(1.86 - 1)290 + 19}{1.86}$$
$$= 144.3 \text{ K}$$

4.3.8.2 Calculation of Receiver Noise, T_r

A receiver will probably consist of a number of stages in cascade, as shown in Figure 4.11. The effective noise temperature of the receiving system, which we

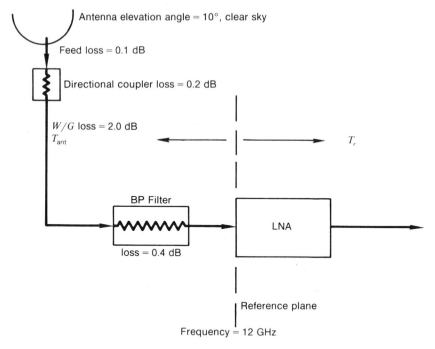

Antenna elevation angle = 10°, clear sky

Feed loss = 0.1 dB

Directional coupler loss = 0.2 dB

W/G loss = 2.0 dB
T_{ant}

T_r

BP Filter

LNA

loss = 0.4 dB

Reference plane

Frequency = 12 GHz

FIGURE 4.10 Example earth station receiving system.

will call T_r, is calculated from the traditional cascade formula:

$$T_r = y_1 + \frac{y_2}{G_1} + \frac{y_3}{G_1 G_2} + \cdots + \frac{y_n}{G_1 G_2 \cdots G_{n-1}} \tag{4.30}$$

where y is the effective noise temperature of each amplifier or device and G is the numeric equivalent of the gain (or loss) of the device.

Example 8. Compute T_r for the first three stages of a receiving system. The first stage is an LNA with a noise figure of 1.1 dB and a gain of 25 dB. The second stage is a lossy transmission line with 2.2-dB loss. The third and final stage is a postamplifier with a 6-dB noise figure and a gain of 30 dB.

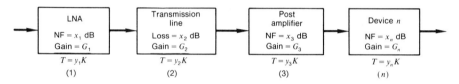

FIGURE 4.11 Generalized cascaded amplifiers/attenuators for noise temperature calculation.

Convert the noise figures to equivalent noise temperatures, using equation (4.21):

$$1.1 \text{ dB} = 10 \log\left(1 + \frac{T_e}{290}\right)$$

$$T_e = 83.6 \text{ K}$$

$$6.0 = 10 \log\left(1 + \frac{T_e}{290}\right)$$

$$T_e = 864.5 \text{ K}$$

Calculate the noise temperature of the lossy transmission line, using equation (4.11). First calculate l_a:

$$l_a = \log^{-1}\left(\frac{L_a}{10}\right)$$

$$= 1.66$$

$$T_e = (1.66 - 1)290$$

$$= 191.3 \text{ K}$$

Calculate T_r, using equation (4.30)

$$T_r = 83.6 + \frac{191.3}{316.2} + \frac{864.5}{316.2 \times 1/1.66}$$

$$= 83.6 + 0.605 + 4.53$$

$$= 88.735 \text{ K}$$

It should be noted that in the second and third terms, we divided by the numeric equivalent of the gain, not the gain in decibels. In the third term, of course, it was not a gain, but a loss (e.g., 1/1.66) where 1.66 is the numeric equivalent of a 2.2-dB loss. It will also be found that in cascaded systems the loss of a lossy device is equivalent to its noise figure.

4.3.8.3 Example Calculation of G/T

A satellite downlink operates at 21.5 GHz. Calculate the G/T of a terminal operating with this satellite. The reference plane is taken at the input to the LNA. The antenna has a 3-ft aperture displaying a 44-dB gross gain. There is 2 ft of waveguide with 0.2 dB/ft of loss. There is a feed loss of 0.1 dB; a bandpass filter has 0.4 dB insertion loss; a radome has a loss of 1.0 dB. The LNA has a noise figure of 5.0 dB and a 30-dB gain. The LNA connects directly to a downconverter/IF amplifier combination with a single sideband noise figure of 13 dB.

Calculate the net gain of the antenna to the reference plane.

$$G_{net} = 44 \text{ dB} - 1.0 \text{ dB} - 0.1 \text{ dB} - 0.4 \text{ dB} - 0.4 \text{ dB}$$

$$= 42.1 \text{ dB}$$

This will be the value for G in the G/T expression.

Determine the sky noise temperature value at the 10° elevation angle, clear sky with dry conditions at 21.5 GHz. Use Figure 4.7. The value is 63 K.

Calculate L_A, the sum of the losses to the reference plane. This will include, of course, the radome loss:

$$L_A = 1.0 \text{ dB} + 0.1 \text{ dB} + 0.4 \text{ dB} + 0.4 \text{ dB} = 1.9 \text{ dB}$$

Calculate l_a, the numeric equivalent of L_A from the 1.9-dB value [equation (4.30)]:

$$l_a = \log^{-1}(1.9/10)$$

$$= 1.55$$

Calculate T_{ant}; use equation (4.28):

$$T_{ant} = \frac{(1.55 - 1)290 + 63}{1.55}$$

$$= 143.55 \text{ K}$$

Calculate T_r. Use equation (4.30). First convert the noise figures to equivalent noise temperatures using equation (4.21). The LNA has a 5.0-dB noise figure, and its equivalent noise temperature is 627 K; and the downconverter/IF amplifier has a noise figure of 13 dB and an equivalent noise temperature of 5496 K:

$$T_r = 627 + 5496/1000$$

$$= 632.5 \text{ K}$$

Calculate T_{sys} using equation (4.12):

$$T_{sys} = 143.55 + 632.5$$

$$= 776.05 \text{ K}$$

Calculate G/T using equation (4.27):

$$\frac{G}{T} = 42.1 \text{ dB} - 10\log(776.05)$$

$$= +13.2 \text{ dB/K}$$

The following discussion, taken from Ref. 4, further clarifies G/T analysis.

Figure 4.12 shows a satellite terminal receiving system and its gain/noise analysis. The notation used is that of the reference document.

The following are some observations of Figure 4.12 (using notation from the reference):

a. The value of T_S is different at every junction
b. The value of G/T (where $T = T_S$) is the same at every junction.

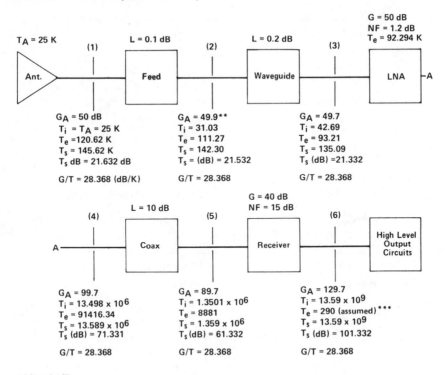

FIGURE 4.12 Example of noise temperature and G/T calculations for cascade of two ports. Note that G/T is independent of position in cascade. From Scientific Atlanta Satellite Communications Symposium 1982, Paper 2A, "Noise Temperature and G/T of Satellite Receiving Systems" (Ref. 4). Reprinted with permission.

c. The system noise temperature at the input to the LNA is influenced largely by the noise temperature of the components that precede the LNA and the LNA itself. The components that follow the LNA have a negligible contribution to the system noise temperature at the LNA input junction (reference plane) if the LNA gain is sufficiently high.

The parameters that have significant influence on G/T are the following:

a. The antenna gain and the antenna noise temperature.

b. The antenna elevation angle. The lower the angle, the higher the sky noise, thus, the higher the antenna noise, and, hence, the lower the G/T for a given antenna gain.

c. Feed and waveguide insertion losses. The lower the insertion loss of these devices, the higher the G/T.

d. LNA. The lower the noise temperature of the LNA, the higher the G/T. The higher the gain of the LNA, the less the noise contribution of the stages following the LNA. For instance, in Figure 4.12, if the gain of the LNA were reduced to 40 dB, the value of T_S would increase to 144.1 K. This means that the value of G/T would be reduced by about 0.26 dB. For an LNA with a gain of only 30 dB, the G/T would then drop by an additional 1.96 dB.

4.3.9 Calculation of C/N_0 Using the Link Budget Technique

The link budget is a tabular method of calculating space communication system parameters. It is a similar approach used in Chapters 2 and 3, where it was called link analysis. In the method presented here the starting point of a link budget is the platform EIRP. The platform can be a terminal or a satellite. In an equation, it would be expressed as follows:

$$\frac{C}{N_0} = \text{EIRP} - \text{FSL}_{\text{dB}} - (\text{other losses}) + G/T_{\text{dB/K}} - k \qquad (4.31)$$

where FSL is the free space loss, k is Boltzmann's constant expressed in dBW or dBm, and the "other losses" may include

☐ Polarization loss
☐ Pointing losses (terminal and satellite)
☐ Off-contour loss
☐ Gaseous absorption losses
☐ Excess attenuation due to rainfall (as applicable)

The off-contour loss refers to spacecraft antennas that are not earth coverage, such as spot beams, zone beams, and MBA (multiple beam antenna), and the contours are flux density contours on the earth's surface. This loss expresses in equivalent decibels the distance the terminal platform is from a contour line. Satellite pointing loss, in this case, expresses contour line error or that the contours are not exactly as specified.

4.3.9.1 Some Link Loss Guidelines

Free space loss is calculated using equations (1.5), (1.7a), and (1.7b). Care should be taken in the use of units for distance (range) and frequency.

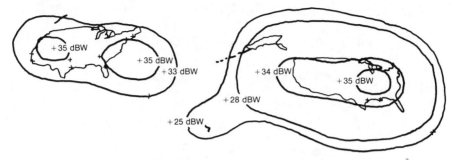

FIGURE 4.13 SP Communications SPACENET 4-GHz EIRP contours for satellite at 70°W (left) and 119°W (right). Courtesy of GTE SpaceNet. (Taken from Ref. 1.)

When no other information is available, use 0.5 dB as an estimate for polarization loss and pointing losses (e.g., 0.5 dB for satellite pointing loss and 0.5 dB for terminal pointing loss).

For off-contour loss, the applicable contour map should be used, placing the prospective satellite terminal in its proposed location on the map and estimating the loss. Figure 4.13 is a typical contour map.

Atmospheric absorption losses are comparatively low for systems operating below 10 GHz. These losses vary with frequency, elevation angle, and altitude. For the 7-GHz downlink, 0.8 dB is appropriate for a 5° elevation angle, dropping to 0.5 dB for 10° and 0.25 dB for 15°; all values are for sea level. For 4 GHz, recommended values are 0.5 dB for 5° elevation angle and 0.25 dB for 10°; all values are for sea level. Atmospheric absorption losses are treated more extensively in Chapter 6.

Excess attenuation due to rainfall is also rigorously treated in Chapter 6. This attenuation also varies with elevation angle and altitude. Suggested estimates are 0.5 dB at 5° elevation angle for 4-GHz downlink band, 0.25 at 10° and 0.15 dB at 15° with similar values for the 6-GHz uplink. For the 7-GHz military band, 3 dB for 5°, 1.5 dB at 10°, and 0.75 dB at 15°, all values at sea level. Use similar values for the 8-GHz uplink.

4.3.9.2 Link Budget Examples

The link budget is used to calculate C/N_0 when other system parameters are given. It is also used when C/N_0 is given when it is desired to calculate one other parameter such as G/T or EIRP of either platform in the link.

Example 9. 4-GHz downlink, FDM/FM, 5° elevation angle. Satellite EIRP + 30 dBW. Range to satellite (geostationary) is 22,208 nautical miles or 25,573 statute miles from Figure 4.5. Terminal $G/T = +20$ dB/K. Calculate C/N_0.

Calculate free space loss from equation (1.7a):

$$\text{FSL}_{dB} = 36.58 + 20\log(4 \times 10^3) + 20\log(25{,}573)$$

$$= 196.78 \text{ dB}$$

EIRP of satellite	+30 dBW
Free space loss	−196.78 dB
Satellite pointing loss	0.5 dB
Off-contour loss	0.5 dB
Atmospheric absorption loss	0.5 dB
Rainfall loss	0.5 dB
Polarization loss	0.5 dB
Terminal pointing loss	0.5 dB
Isotropic receive level	−169.78 dBW
Terminal G/T	+20 dB/K
Sum	−149.78 dBW
Boltzmann's constant	−(−228.6 dBW)
C/N_0	78.82 dB

Example 10. Calculate the required satellite G/T, where the uplink frequency is 6.0 GHz and the terminal EIRP is +70 dBW, and with a 5° terminal elevation angle. The required C/N_0 at the satellite is 102.16 dB (typical for an uplink video link). The free space loss is calculated as in Example 9 but with a frequency of 6.0 GHz. This loss is 200.3 dB. Call the satellite G/T value X.

Terminal EIRP	+70 dBW	
Terminal pointing loss	0.5 dB	
Free space loss	200.3 dB	
Polarization loss	0.5 dB	
Satellite pointing loss	0.5 dB	
Atmospheric absorption loss	0.5 dB	
Off-contour loss	0.5 dB	
Rainfall loss	0.5 dB	
Isotropic receive level at sat.	−133.3 dBW	
G/T of satellite	X dB/K	(initially let $X = 0$ dB/K)
Sum	−133.3 dBW	
Boltzmann's constant	−(−228.6 dBW)	
C/N_0 (as calculated)	95.3 dB	
C/N_0 (required)	102.16 dB	
G/T	+6.86 dB/K	(difference)

This G/T value, when substituted for X, will derive a C/N_0 of 102.16 dB. It is not advisable to design a link without some margin. Margin will compensate for link degradation as well as errors of link budget estimation. The more margin (in decibels) that is added, the more secure we are that the link will work. On the other hand, each decibel of margin costs money. Some compromise between conservation and economic realism should be met. For this link, 4 dB might be such a compromise. If it were all alloted to satellite G/T, then the new G/T value would be $+10.86$ dB/K. Other alternatives to build in a margin would be to increase transmitter power output thereby increasing EIRP, increasing terminal antenna size among other possibilities. The power of using the link budget can now be seen easily. The decibels flow through on a one for one basis, and it is fairly easy to carry out tradeoffs.

It should be noted that when the space platform (satellite) employs an earth coverage (EC) antenna, satellite pointing loss and off-contour loss are disregarded. The beamwidth of an EC antenna is usually accepted as 17° or 18° when the satellite is in geostationary orbit.

4.3.9.3 Calculating System C/N_0

The final C/N_0 value we wish to know is that at the terminal receiver. This must include, as a minimum, the C/N_0 for the uplink and C/N_0 for the downlink. If the satellite transponder is shared (e.g., simultaneous multicarrier operation on one transponder), a value for C/N_0 for satellite intermodulation noise must also be included. The basic equation to calculate C/N_0 for the system is given as

$$\left(\frac{C}{N_0}\right)_{(s)} = \frac{1}{1/(C/N_0)_{(u)} + 1/(C/N_0)_{(d)}} \tag{4.32}$$

Example 11. Consider a bent pipe satellite system where the uplink $C/N_0 = 105$ dB and the downlink $C/N_0 = 95$ dB. What is the system C/N_0?
Convert each value of C/N_0 to its numeric equivalent:

$$\log^{-1}(105/10) = 3.16 \times 10^{10}; \qquad \log^{-1}(95/10) = 0.316 \times 10^{10}$$

Invert each value and add. Invert this value and take 10 log.

$$\left(\frac{C}{N_0}\right)_{(s)} = 94.6 \text{ dB/Hz}$$

Many satellite transponders simultaneously permit multiple carrier access, particularly when operated in the FDMA or SCPC regimes. FDMA (frequency division multiple access) and SCPC (single channel per carrier) systems are described subsequently in this chapter.

When transponders are operated in the multiple-carrier mode, the downlink signal is rich in intermodulation (IM) products. Cumulatively, this is IM noise, which must be tightly controlled, but cannot be eliminated. Such noise must be considered when calculating $(C/N_0)_{(s)}$ when two or more carriers are put through the same transponder simultaneously. The following equation now applies to calculate $(C/N_0)_{(s)}$:

$$\frac{1}{(C/N_0)_{(s)}} = \frac{1}{(C/N_0)_{(u)}^{-1} + (C/N_0)_{(d)}^{-1} + (C/N_0)_{(i)}^{-1}} \qquad (4.33)$$

where the subscripts s, u, d, and i refer to system, uplink, downlink, and intermodulation, respectively.

The principal source of IM noise (products) in a satellite transponder is the final amplifier, often called HPA (high-power amplifier). With present technology, this is usually a TWT (traveling-wave tube) amplifier, although SSAs (solid-state amplifier) are showing increasing implementation on space platforms.

Figure 4.14 shows typical IM curves for a bent pipe type transponder using a traveling wave tube (TWT) transmitter. The lower curve in the figure (curve A) is the IM performance for a large number of equally spaced carriers within a single transponder. The upper curve (curve B) shows the IM characteristics of two carriers.

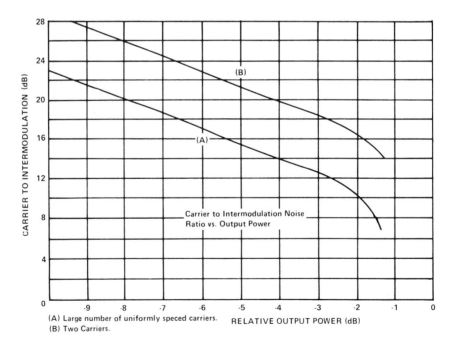

FIGURE 4.14 Output power normalized to single carrier saturation point. Courtesy of Scientific Atlanta (Ref. 4).

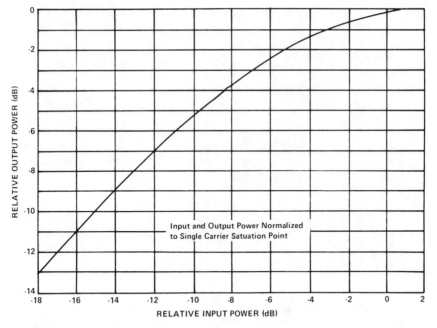

FIGURE 4.15 TWT power transfer characteristics. Courtesy of Scientific-Atlanta (Ref. 4).

In order to bring the IM products to an acceptable level and thus not degrade the system C/N_0 due to poor C/IM performance, the total power of the uplink must be "backed off" or reduced, commonly to a C/IM ratio of 20 dB or better. The result, of course, is a reduction in downlink EIRP. Figure 4.15 shows the effect of input (drive) power reduction versus output power of a TWT. To increase the C/IM ratio to 24 dB for the two-carrier case, we can see from Figure 4.14 that the input power must be backed off by 7 dB. Figure 4.15 shows that this results in a total downlink power reduction of 3 dB. It should also be noted that the resulting downlink power must be shared among the carriers actually being transmitted. For example, if two equal-level carriers share a transponder, the power in each carrier is 3 dB lower than the backed-off value.

4.3.10 Calculating S/N

4.3.10.1 In a VF Channel of an FDM/FM Configuration

To calculate S/N and psophometrically weighted noise (pWp0) in a voice channel when C/N in the IF is given, use the following formula (Ref. 4):

$$\frac{S}{N} = \frac{C}{N} + 20\log\left(\frac{\Delta F_{TT}}{f_{ch}}\right) + 10\log\left(\frac{B_{if}}{B_{ch}}\right) + P + W \qquad (4.34)$$

TABLE 4.6. Global Beam INTELSAT IVA and V Transmission Parameters (Regular FDM / FM Carriers)

Carrier Capacity (No. of Channels) n	Top Baseband Frequency (kHz) f_m	Allocated Satellite Bandwidth Unit (MHz) b_a	Occupied Bandwidth (MHz) b_0	Deviation (rms) for 0 dBm0 Test Tone (kHz) f_r	Multichannel rms Dev. (kHz) f_{mc}	Carrier-to-Total Noise Temperature Ratio at Operating Point (8000 + 200 pW0p from RF Sources) (dBW/K) C/T	Carrier-to-Noise Ratio in Occupied Bandwidth (dB) C/N	Ratio of Unmodulated Carrier Power to Maximum Carrier Power Density Under Full Load Condition (dB/4 kHz)
12^d	60.0	1.25	1.125	109	159	154.7	13.4	20.0
24	108.0	2.5	2.00	164	275	153.0	12.7	22.3
36	156.0	2.5	2.25	168	307	150.0	15.1	22.8
48	204.0	2.5	2.25	151	292	146.7	18.4	22.6
60	252.0	2.5	2.25	136	276	144.0	21.1	22.5
60	252.0	5.0	4.00	270	546	149.9	12.7	25.3
72	300.0	5.0	4.50	294	616	149.1	13.0	25.8
96	408.0	5.0	4.50	263	584	145.5	16.6	25.6
132	552.0	5.0	4.40	223	529	141.4	20.7	24.2^a (×1)
96	408.0	7.5	5.90	360	799	148.2	12.7	27.0
132	552.0	7.5	6.75	376	891	145.9	14.4	27.5
192	804.0	7.5	6.40	297	758	140.6	19.9	25.8^a (×1)

132	552.0	10.0	7.50	430	1020	147.1	12.7	28.0
192	804.0	10.0	9.00	457	1167	144.4	14.7	28.6
252	1052.0	10.0	8.50	358	1009	139.9	19.4	27.0^a (×1)
252	1052.0	15.0	12.40	577	1627	144.1	13.6	30.0
312	1300.0	15.0	13.50	546	1716	141.7	15.6	30.2
432	1796.0	15.0	13.0	401	1479	136.2	21.2	27.6^a (×2)
432^b	1796.0	17.5	15.75	517	1919	138.5	18.2	30.8
432	1796.0	20.0	18.0	616	2276	139.9	16.1	31.5
612	2540.0	20.0	17.8	454	1996	134.2	21.9	78.9^a (×2)
432	1796.0	25.0	20.7	729	2688	141.4	14.1	32.2
792	3284.0	25.0	22.4	499	2494	132.8	22.3	30.0^a (×2)
972	4028.0	36.0	36.0	802	4417	135.2	17.8	34.5
1092^c	4892.0	36.0	36.0	701	4118	132.4	20.7	32.2^a (×2)

[a] This value is X dB lower than the value calculated according to the normal formula used to derive this ratio:

$$-10 \log_{10} \frac{4}{f_{mc}\sqrt{2\pi}}$$

where X is the value in brackets in the last column and f_{mc} is the rms multichannel deviation in kilohertz. The factor is necessary in order to compensate for low modulation index carriers that are not considered to have a Gaussian power density distribution.

[b] Contingency Carrier, used only with INTELSAT IVA.

[c] Not used with INTELSAT V.

[d] Approved for use with INTELSAT V only.

where ΔF_{TT} = rms test tone deviation

$\quad\quad f_{ch}$ = highest baseband frequency

$\quad\quad B_{ch}$ = voice channel bandwidth (3.1 kHz)

$\quad\quad P$ = top VF channel emphasis improvement factor

$\quad\quad W$ = psophometric weighting improvement factor (2.5 dB)

Once the voice channel S/N has been calculated, the noise in the voice channel in picowatts may be determined from

$$\text{noise} = \log^{-1}\{[90 - (S/N)]/10\} \quad (\text{pWp0}) \quad\quad (4.35)$$

Table 4.6 lists the transmission parameters of INTELSAT IVA and V, which we will use below in an example problem. These are typical bent pipe satellite system parameters for communication links.

Example 12. Using Table 4.6 calculate S/N and psophometrically weighted noise in an FDM/FM derived voice channel for a 972-channel system. First use equation (4.34) to calculate S/N and then equation (4.35) to calculate the noise in pWp. Use the value f_r for rms test tone deviation, f_m for the maximum baseband frequency, and B_{if} for the allocated satellite bandwidth. The emphasis improvement may be taken from Figure 2.15. C/N is 17.8 dB in this case:

$$S/N = 17.8 \text{ dB} + 20\log(802 \times 10^3/4028 \times 10^3) + 10\log(36 \times 10^6/3.1 \times 10^3)$$

$$+ 2.5 \text{ dB} + 4.5 \text{ dB}$$

$$= 51.43 \text{ dB}$$

$$\text{noise} = \log^{-1}[(90 - 51.43)/10]$$

$$= 7194.5 \text{ pWp}$$

4.3.10.2 For a Typical Video Channel

As suggested by Ref. 4:

$$\frac{S}{N_v} = \frac{C}{N} + 10\log 3\left(\frac{\Delta f}{f_m}\right)^2 + 10\log\left(\frac{B_{if}}{2B_v}\right) + W + CF \quad\quad (4.36)$$

where S/N_v = peak-to-peak luminance signal-to-noise ratio

$\quad\quad \Delta f$ = peak composite deviation of the video

$\quad\quad f_m$ = highest baseband frequency

$\quad\quad B_v$ = video noise bandwidth (for NTSC systems this is 4.2 MHz)

$\quad\quad B_{if}$ = IF noise bandwidth

$\quad\quad W$ = emphasis plus weighting improvement factor (12.8 dB for NTSC North American systems)

$\quad\quad CF$ = rms to peak-to-peak luminance signal conversion factor (6.0 dB)

For many satellite systems, without frequency reuse, a 500-MHz assigned bandwidth is broken down into 12 36-MHz segments.* Each segment is then assigned to a transponder. For video transmission either a half or full transponder is assigned.

Example 13. A video link is relayed through a 36-MHz (full transponder) transponder bent pipe satellite where the peak composite deviation is 11 MHz and the C/N is 14.6 dB. What is the weighted signal-to-noise ratio? Assume NTSC standards.

Using equation (4.36):

$$\frac{S}{N} = 14.6 \text{ dB} + 10\log\left[3(11/4.2)^2\right] + 10\log(36/8.4) + 12.8 + 6$$

$$= 52.9 \text{ dB}$$

In this case the video noise bandwidth is 4.2 MHz, which is also the highest baseband frequency. European systems would use 5 MHz. The reader should consult CCIR Rec. 421-3 (Ref. 10) or latest revision.

Equation (4.36) is only useful for C/N above 11 dB. Below that C/N value, impulse noise becomes apparent and equation (4.36) is not valid.

4.3.10.3 S/N of the Video-Related Audio or Aural Channel

Conventionally the aural subcarrier is placed above the video in the baseband. To calculate the signal-to-noise ratio of the derived audio, we first calculate the C/N_{sc} or carrier-to-noise ratio of the audio subcarrier when the C/N of the main carrier is given. C/N_{sc} may be calculated from

$$\left(\frac{C}{N}\right)_{sc} = \frac{C}{N} + 10\log\left(\frac{B_{if}}{2B_{sc}}\right) + 10\log\left[\left(\frac{\Delta F_c}{f_{sc}}\right)^2\right] \qquad (4.37)$$

where ΔF_c = peak deviation of the main carrier by the subcarrier
    f_{sc} = subcarrier frequency
    B_{sc} = subcarrier filter noise bandwidth
    B_{if} = IF noise bandwidth

Now we can calculate the signal-to-noise ratio of the aural channel by the following equation:

$$\left(\frac{S}{N}\right)_a = \left(\frac{C}{N}\right)_{sc} + 10\log\left[3\left(\frac{\Delta F_{sc}}{f_m}\right)^2\right] + 10\log\left(\frac{B_{sc}}{2B_a}\right) + E \qquad (4.38)$$

---

*Many new systems coming on line use wider bandwidths.

where f_m = maximum audio frequency

B_a = audio noise bandwidth

ΔF_{sc} = peak subcarrier deviation

E = audio pre/deemphasis advantage, which usually is given the value of 12 dB

Example 14. Calculate the S/N of the aural channel where the aural subcarrier frequency is 6.8 MHz, the peak carrier deviation is 2 MHz; the peak subcarrier deviation is 75 kHz; and the aural channel bandwidth is 15 kHz. Assume C/N is 14 dB, and B_{sc} is 600 kHz, and the audio noise bandwidth is 30 kHz.

First calculate C/N_{sc} using equation (4.37):

$$\left(\frac{C}{N}\right)_{sc} = 14 \text{ dB} + 10\log\left[36 \times 10^6/(2)(600 \times 10^3)\right]$$

$$+ 10\log\left[\left(2 \times 10^6/6.8 \times 10^6\right)^2\right]$$

$$\left(\frac{C}{N}\right)_{sc} = 14 + 14.77 - 10.63$$

$$= 18.14 \text{ dB}$$

Calculate S/N_a by equation (4.38):

$$\left(\frac{S}{N}\right)_a = 18.14 \text{ dB} + 10\log\left[3(75 \times 10^3/15 \times 10^3)^2\right]$$

$$+ 10\log(600 \times 10^3/30 \times 10^3) + 12 \text{ dB}$$

$$= 18.14 + 18.75 + 13 + 12$$

$$= 61.89 \text{ dB}$$

4.3.10.4 Calculation of S/N for Analog Single Channel (SCPC) Systems

SCPC is a method of accommodating multiple carriers on a single satellite transponder. It is a satellite-access method, similar to frequency division multiple access (FDMA), where each carrier is modulated by a single channel, either a voice channel or a program channel. Access methods are discussed in Section 4.4. SCPC modulation techniques may be analog (FM) or digital. In this subsection only the analog (FM) technique is discussed. Companding, weighting, and emphasis are used.

Channel signal-to-noise ratio may be calculated when C/N is given by

$$\frac{S}{N} = \frac{C}{N} + 10\log\left[3(\Delta F/f_m)^2\right] + 10\log\left(\frac{B_{if}}{2B_a}\right) + W + C \qquad (4.39)$$

where ΔF = peak deviation
  f_m = highest modulating frequency (baseband frequency)
  B_a = audio noise bandwidth
  B_{if} = IF bandwidth
  W = emphasis plus weighting improvement factor
  C = companding advantage

Example 15. Calculate the signal-to-noise ratio of an FM SCPC VF channel where the C/N is 10 dB; the peak deviation is 7.3 kHz; the IF noise bandwidth is 25 kHz; the weighting/emphasis advantage is 7 dB; and the companding advantage is 17 dB.

Use formula (4.39):

$$\frac{S}{N} = 10\ \text{dB} + 10\log\left[3\left(7.3 \times 10^3 / 3.4 \times 10^3\right)^2\right]$$

$$+ 10\log\left(25 \times 10^3 / 6.8 \times 10^3\right) + 7\ \text{dB} + 17\ \text{dB}$$

$$= 51.1\ \text{dB}$$

The voice channel bandwidth in this example has a 3.4-kHz bandwidth.

4.3.10.5 System Performance Parameters

Table 4.7 gives some typical performance parameters for video, aural channel, and FDM/FM for a 1200-channel configuration.

4.4 ACCESS TECHNIQUES

4.4.1 Introduction

Access refers to the way a communication system uses a satellite transponder. There are three basic access techniques:

  □ FDMA (frequency division multiple access)
  □ TDMA (time division multiple access)
  □ CDMA (code division multiple access)

With FDMA a satellite transponder is divided into frequency band segments where each segment is assigned to a user. The number of segments can vary from one, where an entire transponder is assigned to a single user, to literally hundreds of segments, which is typical of SCPC operation. For analog telephony operation, each segment is operated in an FDM/FM mode. In this case FDM group(s) and/or supergroup(s) are assigned for distinct distant location connectivity.

TABLE 4.7. Some Typical Satellite Link Performance Parameters

System Parameters	Units	FDM-FM	Video
TV Video			
C/N	dB	—	14.6
Maximum video frequency	MHz	—	4.2
Overdeviation	dB	—	—
Peak operating deviation	MHz	—	10.7
FM improvement	dB	—	13.2
BW improvement	dB	—	6.3
Weighting/emphasis improvement	dB	—	12.8
P-rms Conversion Factor	dB	—	6.0
Total improvement	dB	—	38.3
S/N (peak-to-peak/rms-luminance signal)	dB	—	52.9
TV program channel (subcarrier)			
Peak carrier deviation	MHz	—	2.0
Subcarrier frequency	MHz	—	7.5
FM improvement	dB	—	11.5
BW improvement	dB	—	14.0
Total improvement	dB	—	2.5
C/N_{sc} (subcarrier)	dB	—	16.9
Peak subcarrier deviation	kHz	—	75
Maximum audio frequency	kHz	—	15
FM improvement	dB	—	18.8
BW improvement	dB	—	13.8
Emphasis improvement	dB	—	12.0
Total improvement	dB	—	44.6
S/N (audio)	dB	—	59.2
FDM/FM			
Number channels	—	1200	—
Test tone deviation (rms)	kHz	650	—
Top baseband frequency	kHz	5260	—
FM improvement	dB	18.2	—
BW improvement	dB	40.7	—
Weighting improvement	dB	2.5	—
Emphasis (top slot) improvement	dB	4.0	—
Total improvement	dB	29.0	—
TT/N (test tone to noise ratio)	dB	49.0	—
Noise	pWp0	12,589	—

Source. Scientific-Atlanta Satellite Communications Symposium '82. Courtesy of Scientific-Atlanta (Ref. 4).

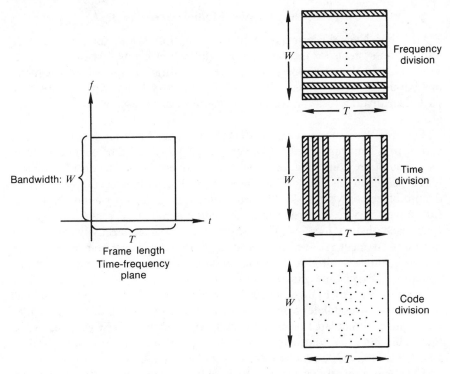

FIGURE 4.16 The three basic multiple access techniques depicted in a time–frequency plane.

TDMA works in the time domain. Only one user appears on the transponder at any given time. Each user is assigned a time slot to access the satellite. System timing is crucial with TDMA. It lends itself only to digital transmission, typically PCM.

CDMA is particularly attractive to military users due to its antijam and low probability of intercept (LPI) properties. With CDMA, the transmitted signal is spread over part or all of the available transponder bandwidth in a time/frequency relationship by a code transformation. Typically, the modulated RF bandwidth is ten to hundreds of times greater than the information bandwidth.

CDMA had previously only been attractive to the military user because of its antijam properties. Since 1980 some interest in CDMA has been shown in the commercial sector for demand access for large populations of data circuit/network users with bursty requirements in order to improve spectral utilization. For further discussion of spread spectrum systems refer to Ref. 28.

Figure 4.16 depicts the three basic types of satellite multiple access. The horizontal axis represents time and the vertical axis represents spectral bandwidth.

4.4.2 Frequency Division Multiple Access (FDMA)

FDMA has been the primary method of multiple access for satellite com-munictaion systems for the past 20 years and will probably remain so for quite some time. In the telephone service, it is most attractive for permanent direct or high-usage (HU) routes, where, during the busy hour, trunk groups to a particular distinct location require 5 or more Erlangs (e.g., one or more FDM groups).

For this most basic form of FDMA, a single earth station transmits a carrier to a bent pipe satellite. The carrier contains an FDM configuration in its modulation consisting of groups and supergroups for distinct distant locations. Each distant location receives and demodulates the carrier, but demultiplexes only those FDM groups and supergroups pertaining to it. For full duplex service, a particular earth station receives, in return, one carrier from each distant location with which it provides connectivity. So, on its downlink, it must receive, and select by filtering, carriers from each distant location, and demultiplex only that portion of each derived baseband that contains FDM channelization destined for it.

Another form of FDMA is SCPC, where each individual telephone channel independently modulates a separate radio-frequency carrier. Each carrier may be frequency modulated or digitally modulated, often PSK. SCPC is useful on low-traffic-intensity routes (e.g., less than 5 Erlangs) or for overflow from FDM/FDMA. SCPC will be discussed in Section 4.4.2.1.

FDMA is a mature technology. Its implementation is fairly straightforward. Several constraints must be considered in system design. Many of these constraints center around the use of TWT as HPA in satellite transponders. Depending on the method of modulation employed, amplitude and phase nonlinearities must be considered to minimize IM products, intelligible cross-talk, and other interfering effects by taking into account the number and size (bandwidth) of carriers expected to access a transponder. These impairments are maintained at acceptable levels by operating the transponder TWT at the minimum input backoff necessary to ensure that performance objectives are met. However, this method of operation results in less available channel capacity when compared to a single access mode.

TWT amplifiers operate most efficiently when they are driven to an operat-ing point at or near saturation. When two or more carriers drive the TWT near its saturation point, excessive IM products are developed. These can be reduced to acceptable levels by reducing the drive, which, in turn, reduces the amplifier efficiency. This reduction in drive power is called backoff.

In early satellite communication systems, IM products created by TWT amplitude nonlinearity was the dominant factor in limiting system operation. However, as greater power became available and narrow bandwidth trans-ponders capable of operation with only a few carriers near saturation became practical, maximum capacity became dependent on a carefully evaluated tradeoff analysis of a number of parameters, which include the following:

1. Satellite TWT impairments including in-band IM products produced by both amplitude and phase nonlinearities and intelligible crosstalk caused by FM-AM-PM conversion during multicarrier operation.
2. FM transmission impairments not directly associated with the satellite transponder TWT such as adjacent carrier interference caused by frequency spectrum overlap between adjacent carriers which gives rise to convolution and impulse noise in the baseband; dual path distortion between transponders; interference due to adjacent transponder IM earth station out-of-band emission and frequency reuse cochannel interference.
3. General constraints including available power and allocated bandwidth; uplink power control; frequency coordination; and general vulnerability to interference.

Backoff was discussed briefly in Section 4.3.9.3. From Figures 4.14 and 4.15 we can derive a rough rule-of-thumb that tells us that for approximately every decibel of backoff, IM products drop 2 dB for the multicarrier case (i.e., more than three carriers on a transponder). Also, for every decibel of backoff on TWT driving power, TWT output power drops 1 dB. Of course, this causes inefficiency in the use of the TWT. As the number of carriers are increased on a transponder, the utilization of the available bandwidth becomes less efficient. If we assume 100% efficient utilization of a transponder with only 1 carrier, then with 2 carriers the efficiency drops to about 90%, with 4 to 60%, with 8 to about 50%, and with 14 carriers to about 40%.

Crosstalk is a significant impairment in FDMA systems. It can result from a sequence of two phenomena:

☐ An amplitude response that varies with frequency-producing amplitude modulation coherent with the original frequency modulation of an RF carrier or FM/AM transfer, and
☐ A coherent amplitude modulation that phase-modulates all carriers occupying a common TWT amplifier due to AM/PM conversion.

As a carrier passes through a TWT amplifier, it may produce amplitude modulation from gain–slope anomalies in the transmission path. Another carrier passing through the same TWT will vary in phase at the same rate as the AM component and thereby pick up intelligible crosstalk from any carrier sharing the transponder. Provisions should be made in specifying TWT amplifier characteristics to ensure that AM/PM conversion and gain–slope variation meet system requirements. Intelligible crosstalk should be 58 dB down or better, and with modern equipment, this goal can be met quite easily.

Out-of-band RF emission from an earth station is another issue; 500 pWp0 is commonly assigned in a communication satellite system voice channel noise budget for earth station RF out-of-band emission. The problem centers on the earth station HPA TWT. When a high-power, wide-band TWT is operated at saturation, its full output power can be realized, but it can also produce severe IM RF products to the up-path of other carriers in an FDMA system. To limit such unwanted RF emission, we must again turn to the backoff technique of the RF drive to the TWT. Some systems use as much as 7-dB backoff. For example, a 12-kW HPA with 7-dB backoff is operated at about 2.4 kW. This is one good reason for overbuilding earth station TWTs and accepting the inefficiency of use.

Uplink power control is an important requirement for FDM/FM/FDMA systems. Sufficient power levels must be maintained on the uplink to meet signal-to-noise ratio requirements on the derived downlinks for bent pipe transponders. On the other hand, uplink power on each carrier accessing a particular transponder must be limited to maintain IM product generation (IM noise) within specifications. This, of course, is the backoff discussed above. Close control is required of the power level of each carrier to keep transmission impairments within the total noise budget.

One method of meeting these objectives is to study each transponder configuration on a case-by-case basis before the system is actually implemented on the satellite. Once the proper operating values have been determined, each earth station is requested to provide those values of uplink carrier levels. These values can be further refined by monitoring stations that precisely measure resulting downlink power of each carrier. The theoretical power levels are then compared to both the reported uplink and the measured downlink levels.

Energy dispersal is yet another factor required in the design of a bent pipe satellite system. Energy dispersal is used to reduce the spectral energy density of satellite signals during periods of light loading (e.g., off busy hour). The reduction of maximum energy density will also facilitate efficient use of the geostationary satellite orbit by minimizing the orbital separation needed between satellites using the same frequency band and multiple-carrier operation of broadband transponders. The objective is to maintain spectral energy density the same for conditions of light loading as for busy hour loading. Several methods of implementing energy dispersal are described in CCIR Report 384-3 (Ref. 12).

One method of increasing satellite transponder capacity is by *frequency reuse*. As the term implies, an assigned frequency band is used more than once. The problem is to minimize mutual interference between carriers operating on the same frequency but accessing distinct transponders. There are two ways of avoiding interference in a channel that is used more than once:

1. By orthogonal polarization.
2. By multiple-exclusive spot beam antennas.

The use of opposite-hand circular or crossed-linear polarizations may be used to effect an increase in bandwidth by a factor of 2. Whether the potential increase in capacity can be realized depends on the amount of cross-polarization discrimination that can be achieved for the cochannel operation. Cross-polarization discrimination is a function of the quality of the antenna systems and the effect of the propagation medium on polarization of the transmitted signal. The amount of polarization "twisting" or distortion is a function of the elevation angle for a given earth station. The lower the angle, the more the twist.

Isolation between cochannel transponders should be greater than 25 dB. INTELSAT V specifies 27-dB minimum isolation between polarizations.

Whereas a satellite operating in a 500-MHz bandwidth might have only 12 transponders (36- or 40-MHz bandwidth each), with frequency reuse, 24 transponders can be accommodated with the same bandwidth. Thus, the capacity has been doubled.

Table 4.6 shows a typical transponder allocation for FDMA operation.

4.4.2.1 Single Channel Per Carrier (SCPC) FDMA

A number of SCPC systems have been implemented and are now in operation, and many more are coming on line. There are essentially two types of SCPC systems: preassigned and demand assigned (DAMA). The latter requires some form of control system. Some systems occupy an entire transponder, whereas others share a transponder with video service leaving a fairly large guardband between the video portion of the transponder passband and the SCPC portion. Such transponder sharing is illustrated in Figure 4.17.

Channel spacings on a transponder vary depending on the system. INTELSAT systems commonly use 45-kHz spacing. Others use 22.5, 30, and 36 kHz as well as 45 kHz. Modulation is FM or BPSK/QPSK. The latter lends itself well to digital systems, whether PCM or CVSD (continuous variable slope delta modulation).

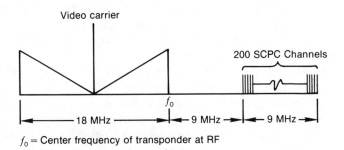

f_0 = Center frequency of transponder at RF

FIGURE 4.17 Satellite transponder frequency plan, video plus SCPC.

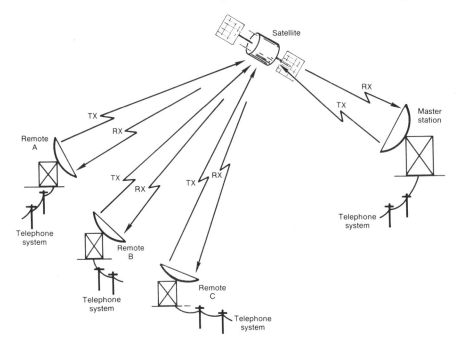

FIGURE 4.18 A typical FM SCPC system.

If we were to divide a 36-MHz bandwidth on a transponder into uniform 45-kHz segments, the total voice channel capacity of the transponder is 800 VF channels. These are better termed half-channels, because, for telephony, we always measure channel capacity as full duplex channels. Thus 400 of the 800 channels would be used in the "go" direction, and the other 400 would be used in the reverse direction. An SCPC system is generally designed for a 40% activity factor. Thus, statistically, only 320 of the channels can be expected to be active at any instant during the busy hour. This has no effect on the bandwidth, only on the loading. Most systems use voice-activated service. In this case, carriers appear on the transponder for the activated channels only. This provides probably the worst case for IM noise for all conventional bent pipe systems during periods of full activation. A simplified drawing of an SCPC system is shown in Figure 4.18.

SCPC channels can be preassigned or DAMA. The preassigned technique is only economically feasible in situations where source/sink locations have very low traffic density during the busy hour (e.g., less than several Erlangs).

DAMA SCPC systems are efficient for source/sink locations of comparatively low traffic intensity, especially under situations with multiple location community of interest (e.g., trunks to many distinct locations) but under 12 channels to any one location. This demarcation line at 12 channels, of course, is where the designer should look seriously at establishing fixed-assignment FDMA/FDM channel group allocation.

The potential improvement in communication system capacity and efficiency is the primary motivation for DAMA systems. Our primary consideration in this section is voice traffic. Call durations average several minutes. The overhead time to connect and disconnect is 1 or 2 sec in an efficient channel assignment system. Some of the issues of importance in demand assignment systems are:

☐ User requirements such as traffic intensity in Erlangs or ccs, number of destinations and grade of service (probability of blocked calls).

☐ Capacity improvements due to implementation of DAMA scheme.

☐ Assignment algorithms (centralized versus distributed control).

☐ Equipment cost and complexity.

Consider an example to demonstrate the potential improvement due to demand assignment. (See Table 4.8.) An earth station is required to communicate with 40 destinations, and the traffic intensity to each destination is 0.5 Erlang with a blocking probability of $P_B = 0.01$. Assume that the call arrivals have a Poisson distribution and the call holding times have an exponential distribution. Then for a trunk group with n channels in which blocked calls are cleared, based on an Erlang B distribution (Ref. 1, Section 1-2), A, in Table 4.8, gives the traffic intensity in Erlangs. If the system is designed for preassignment, each destination will require four channels to achieve $P_B = 0.01$. For 40 destinations, 160 channels will be required for the preassigned case. In the DAMA case, the total traffic is considered since any channel can be

TABLE 4.8. Comparison of Preassigned versus DAMA Channel Requirements Based on a Grade of Service of $P_B = 0.01$

	Channel Requirements[a]					
Number of	$A = 0.1$		$A = 0.5$		$A = 1.0$	
Destinations	Preassigned	DAMA	Preassigned	DAMA	Preassigned	DAMA
1	2	2	4	4	5	5
2	4	3	8	5	10	7
4	8	3	16	7	20	10
8	16	4	32	10	40	15
10	20	5	40	11	50	18
20	40	7	80	18	100	30
40	80	10	160	30	200	53

[a] A = Erlang traffic intensity

Source. "Demand Assignment," Subsection 3.6.5 in H. L. Van Trees *Satellite Communications*, IEEE Press, 1979 (Ref. 14). Reprinted with permission IEEE Press.

assigned to any destination. The total traffic load is 20 Erlangs (40 × 0.5) and the Erlang B formula gives a requirement of only 30 channels. The efficiency of the system with demand assignment is improved over the preassigned by a factor of 5.3 (160/30). Table 4.8 further expands on this comparison using various traffic loadings and numbers of destinations. Reference 13, Chapter 1, provides good introductory information on traffic engineering.

There are essentially three methods for controlling a DAMA system:

- ☐ Polling
- ☐ Random-access central control
- ☐ Random-access distributed control

The polling method is fairly self-explanatory. A master station (Figure 4.18) "polls" all other stations in the system sequentially. When a positive reply is received, a channel is assigned accordingly. As the number of earth stations in the system increases, the polling interval becomes longer, and the system tends to become unwieldy because of excessive postdial delay as a call attempt waits for the polling interval to run its course.

With random-access central control, the status of channels is coordinated by a central computer located at the master station. Call requests (called *call attempts* in switching) are passed to the central computer via a digital orderwire (i.e., digitally over a radio service channel), and a channel is assigned, if available. Once the call is completed and the subscriber goes on-hook, the speech path is taken down and the channel used is returned to the demand-access pool of idle channels. According to system design, there are a number of methods to handle blocked calls [all trunks busy (ATB) in telephone switching], such as queueing and second attempts.

The distributed control random-access method utilizes a processor controller at each earth station accessing the system. All earth stations in the network monitor the status of all channels via continuous updating of channel status information by means of the digital orderwire circuit. When an idle channel is seized, all users are informed of the fact, and the circuit is removed from the pool. Similar information is transmitted to all users when circuits are returned to the idle condition. One problem, of course, is the possibility of double seizure. Also the same problems arise regarding blockage (ATB) as in the central-control system. Distributed control is more costly and complex, particularly in large systems with many users. It is attractive in the international environment because it eliminates the "politics" of a master station.

4.4.3 Brief Overview of Time Division Multiple Access (TDMA)

TDMA operates in the time domain and is applicable only to digital systems because information storage is required. To make Section 4.4 complete, a brief overview of TDMA is presented.

With TDMA, use of a satellite transponder is on a time-sharing basis. At any given moment in time, only one earth station accesses the satellite. Individual time slots are assigned to earth stations operating with that transponder in a sequential order. Each earth station has full and exclusive use of the transponder bandwidth during its time-assigned segment. Depending on bandwidth of the transponder and type of modulation used, bit rates of 10–100 Mbps are used.

If only one carrier appears on the transponder at any time, then intercarrier IM products cannot be developed and the TWT-based HPA of the transponder may be operated at full power. This means that the TWT requires no backoff in drive and may be run at or near saturation or maximum efficiency. With multicarrier FDMA systems, backoffs of 5–10 dB are not uncommon.

TDMA utilizes the transponder bandwidth more efficiently, too. Reference 15 compares approximate channel capacities of an INTELSAT IV global beam transponder operating with a normal INTELSAT Standard A (30-m antenna) earth stations using FDMA and TDMA, respectively. Assuming 10

TABLE 4.9. FDMA versus TDMA

Advantages	Disadvantages
FDMA	
Mature technology	IM in satellite transponder output
No network timing	Requires careful uplink power control
Easy FDM interface	Inflexible to traffic load
TDMA	
Maximum use of transponder power	Still emerging technology
No careful uplink power control	Network timing Complex control
Flexible to dynamic traffic loading	Major digital buffer considerations
Straightforward interface with digital network	Difficult to interface with FDM
More efficient transponder bandwidth usage	
Digital format compatible with Forward error control Source coding Demand access algorithms	
Applicable to switched satellite service	

accesses, typical capacity using FM/FDMA is approximately 450 one-way voice channels. With TDMA, using standard 64-kbps PCM voice channels, the capacity of the same transponder is 900 voice channels. If digital speech interpolation (DSI) is now implemented on the TDMA system, the voice channel capacity is increased to approximately 1800 channels. DSI is discussed in Chapter 5.

Still another advantage of TDMA is flexibility. Flexibility is not only a significant benefit to large systems, but is often the key to system viability in smaller systems. Nonuniform accesses pose no problem in TDMA because time-slot assignments are easy to adjust. This applies to initial network configuration, assignments, reassignments, and demand assignments. This is ideal for a long haul system where traffic adjustments can be made dynamically as the busy hour shifts from east to west. Changes can also be made for growth or additional services.

Disadvantages of TDMA are timing requirements and complexity. Accesses may not overlap. Obviously overlapping causes interference between sequential accesses. Guard times between accesses must be made minimum for efficient operation. The longer the guard times, the shorter the burst length and/or number of accesses. Typically 5–15 accesses can be accommodated per transponder with guard times on the order of 50–200 nsec.

As the world's telecommunication network evolves from analog to digital and as frequency congestion increases, there will be more and more demand to shift to TDMA operation in satellite communications.

Table 4.9 compares FDMA with TDMA.

4.5 INTELSAT SYSTEMS

4.5.1 Introduction

INTELSAT is probably, overall, the best known of the families of communication satellite systems. The INTELSAT family, starting with INTELSAT III of the mid-1960s, has been in operation for the longest period. The key word is international. Its original concept was to provide international trunk connectivity, particularly transoceanic, for telephone administrations. INTELSAT satellites are bent pipe FDMA/FDM/FM. Provision has been made for TDMA, and in mid-1985 TDMA was finally implemented. Current operational satellites are INTELSAT IVA, V, and VI. Frequency reuse is common practice. Transponders operate in the 6/4 and 14/11/12 GHz bands. INTELSAT VI transponder frequency plan is shown in Figure 4.19. Table 4.10 summarizes some of the key characteristics of the INTELSAT space segment for INTELSAT V and VI.

INTELSAT specifies certain requirements for earth stations of member countries to operate with INTELSAT satellites. Earth stations are defined by class: INTELSAT Standard A, B, C, D, E, and F. Table 4.11 summarizes

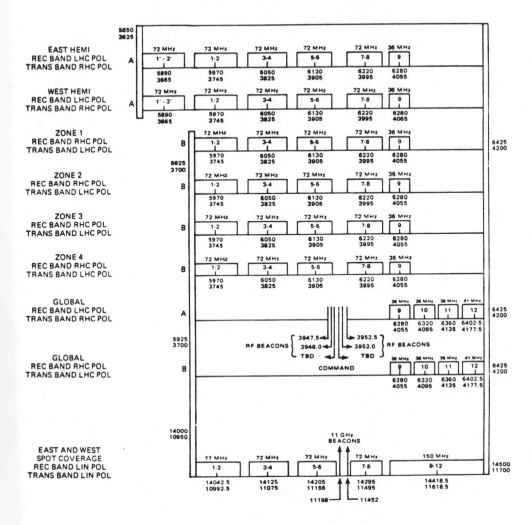

FIGURE 4.19 INTELSAT VI transponder frequency plan. From Ref. 1, Sect. 13-4.10.9. Courtesy of INTELSAT, Washington, DC.

TABLE 4.10. Summary of Some Key Characteristics of INTELSAT V and VI

Characteristic[c]		INTELSAT V		INTELSAT VI
G/T (dB/K)				
6 GHz	EC	−18.6		−14.0
	HC	−11.6		−9.2
				−9.5 (1)
	ZC	−8.6[a]	Z2, Z4	−7.0/−7.5[b]
			Z1, Z3	−2.0/−3.0[b]
14 GHz				
	East spot	0.0		+1.0
	West spot	+3.3		+5.0
EIRP (dBW)				
4 GHz				
	EC	+26.5(7-9)		+23.5 (9)
		+23.5		+26.5
	EH, WH			+28.0 (9)
	EH, WH			+31.0
	Others	+29.0		
	Hemi (9)	+26.0		
	Z1-4			+28.0 (9)
	Z1-4			+31.0
11 GHz				
	East spot	+41.4		+41.1
	West spot	+44.4		+44.4
POLARIZATION				
4 GHz				
	EC, HC	RHC		
	EC (A)			RHC
	EC (B)			LHC
	HC			RHC
	Z1, Z2	LHC		
	Z1-4			LHC
14 GHz	(east orthogonal to west)	Linear		Linear
11 GHz	(east orthogonal to west)	Linear		Linear
(opposite orthogonality to 14-GHz counterpart)				
6 GHz				
	EC	LHC		
	EC (A)			LHC
	EC (B)			RHC
	HC	LHC		LHC
	Z1, Z2	RHC		
	Z1-4			RHC

[a]Assumes minimum antenna gain within a zone coverage is 3 dB higher than that with a hemi coverage (HC). If difference is greater or less than 3 dB, the G/T for zone coverage (ZC) should be changed accordingly, but shall not be less than −11.6 dB/K.

[b]Denotes high and low levels.

[c]EC = earth coverage, called global coverage in INTELSAT VI. HC = hemispherical coverage. ZC = zone coverage. EH = east hemispherical. WH = west hemispherical. INTELSAT V has two zone antennas; INTELSAT VI has four. RHC, LHC: right-hand circular polarization and left-hand circular polarization, respectively. Spot refers to spot-beam antennas.

TABLE 4.11. Comparison of Some Key Parameters of INTELSAT Standard Earth Stations

| Parameter | INTELSAT Earth Stations by Standard Class | | | | | |
	A	B	C	D	E	F
Service	FDM/FM, TV/FM, SCPC/QPSK (PCM)	TV/FM FDM/FM for TV aural SCPC/QPSK (PCM)	FDM/FM only	LDTS[a] CFM[b] (SCPC)	Business SVC IBS[c]	Business SVC (data with FEC) (CQPSK)
Frequency bands (GHz)	6/4	6/4	14/11	6/4	14/11	6/4
Minimum G/T (dB/k)	40.7	31.7	39.0	D1 = 22.7 D2 = 31.7	E1 = 25.0 E2 = 29.0 E3 = 34.0	F1 = 22.7 F2 = 27.0 F3 = 29.0
Minimum antenna gain (dBi)	—	53.2 (at 6 GHz)	—	46.6 (D1 at 6 GHz)		F1 = 47.7 F2 = 51.6 F3 = 53.0 (at 6 GHz)
EIRP TV (max) (dBW)	+88	+85				
EIRP FDM (max) (dBW)	+98.6	+74.7 (TV aural)	+91.8	D1 = +56.1 D2 = +52.7 + 82.1 to F1 + 82.8		to F1 = +91.3 to F2 = +87.3 to E1 = +87.7 to E2 = +85.2
EIRP SCPC (max) (dBW)		+69.8				
Minimum elevation angle (deg)	5	5	10	5	10	5
Tracking	Automatic open and closed loop	Same	Same	Same D2; manual D1	Same E3; manual E1, E2	Same F3; automatic or manual F1, F2

[a]LDTS = low density telephone service.
[b]CFM = companded FM.
[c]IBS = international business service.
Source. INTELSAT: A = BG-28-72E M/6/77; B = BG-28-74E M/6/77; C = BG-28-73E M/6/77; D = BG-56-72E W/9/83; E = BG-53-86 Rev 1 June 1984; F = BG-58-95E W/3/84.

several key characteristics on a comparison basis. The following subsections provide discussion of each INTELSAT class.

4.5.2 INTELSAT Type A Standard Earth Stations*

The Type A earth station is primarily used for intercontinental direct trunk groups and television. The driving parameter is the required G/T of

*See Section 4.5.8.

+40.7 dB/K at 4 GHz. The receiving system must accommodate the entire 500-MHz satellite bandwidth. To achieve the G/T, generally 30-m antennas are required and receiving system noise temperatures are in the range of 70 K. To reach the nearly +100 dBW EIRP, 10-kW transmitters are required. If TV is to be transmitted simultaneously with high-density FDM, often two 10-kW transmitters are used, one for TV and one for FDM, and the outputs of the two transmitters are power combined.

SCPC is cost effective on these large installations for overflow during busy hour and for low-density traffic relations, less than about 5 Erlangs.

The large antennas used on INTELSAT Standard A earth stations produce very narrow beam widths, on the order of 0.1°. Because of satellite suborbital motion and drift, and with such narrow beamwidths, automatic tracking is required, usually active tracking, monopulse of conscan. Further discussion of earth station design may be found in Chapter 7.

4.5.2.1 Noise on a Telephone Channel for an INTELSAT Link

INTELSAT allows 10,000 pWp of noise power in the receive telephone channel. It will be noted that this value corresponds to the noise power value on a 2500-km CCIR hypothetical reference circuit. This 10,000-pWp noise power value is budgeted by INTELSAT as follows:

Space segment	8,000 pWp
Earth stations	1,000 pWp
Terrestrial interference	1,000 pWp
Total	10,000 pWp

Earth station noise contribution is broken down as follows:

Earth station transmitter noise excluding multicarrier IM noise and group delay noise	250 pWp
Noise due to total group delay after necessary equalization	500 pWp
Other earth station receiver noise	250pWp
Total	1000 pWp

The INTELSAT space segment allocates 8000 pWp to up- and downlink thermal noise, transponder IM noise, earth station out-of-band emission, cochannel interference within the operating satellite, and interference from adjacent satellite networks. Within the 8000-pWp allocation, an allowance of 500 pWp is reserved for earth station RF out-of-band emission caused by multicarrier IM from other earth stations in the system.

All noise power values shown are referenced to the zero-test-level point in the overall system.

4.5.3 INTELSAT Standard B Earth Stations

In general, the mandatory performance characteristics specified for the Standard A earth stations apply to Standard B earth stations with exceptions relating only to antenna aperture and a more limited range of transmission capabilities. INTELSAT document BG-28-74E M/6/77 (Ref. 17) states the following:

At the present time the modulation methods shown below are approved for Standard B earth stations which operate in the band 6/4 GHz. Carriers will be assigned to each satellite in a frequency division multiple access (FDMA) basis using one of the following:

1. Frequency division multiplex/Frequency modulation (FDM/FM). For TV-associated audio carriers only.
2. Preassigned single-channel-per-carrier/Pulse Code Modulation/phase shift keying (4-phase) (SCPC/PCM (4ϕ)). For voice activated telephony.
3. Preassigned single-channel-per-carrier/Phase shift keying (4-phase) (SCPC/PSK (4ϕ)). For data service.
4. Frequency modulation television (TV). For television.

Standard B earth stations can communicate through all satellite types (INTELSAT) with standard B earth stations and with standard A earth stations ($G/T = +40.7$ dB/K) by means of using the above modulation methods. At the present time standard B earth stations will not be used to communicate with standard C earth stations.

It should be assumed that the preceding will hold for standard D, E, and F earth stations as well.

The minimum G/T of the standard B earth station at 4 GHz is $+31.7$ dB/K at a 5° elevation angle under clear sky conditions.

The maximum EIRP per carrier for SCPC/PSK, when interworking with a Standard A earth station, is $+63$ dBW, and then working another Standard B earth station, it is $+69.8$ dBW. For full transponder TV, the maximum EIRP is $+81.8$ dBW interworking a Standard A earth station and $+85$ dBW with a Standard B earth station at the distant end; 2.5-MHz FDM/FM for TV aural requires an EIRP of $+74.7$. These values are for a 5° elevation angle.

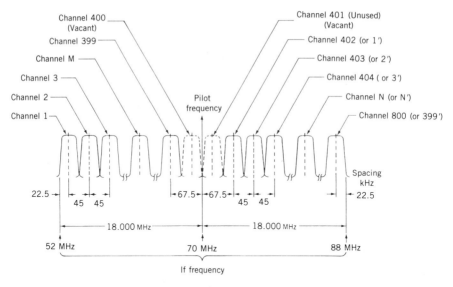

FIGURE 4.20 SCPC frequency plan at a 70-MHz IF for full transponder operation. From INTELSAT BG-28-73E M/6/77 (Ref. 17).

Figure 4.20 shows the SCPC frequency plan utilizing a 36-MHz transponder, which is fully dedicated to this service. It has a capacity of 399 full duplex, preassigned channels.

4.5.4 INTELSAT Standard C Earth Stations*

The INTELSAT Standard C earth station operates in the 14/11 GHz band with INTELSAT V and VI satellites. Operation is FDMA with FDM/FM. As discussed in Chapter 6, excess attenuation due to rainfall and gaseous absorption become major considerations in system design above 10 GHz. These considerations are aptly taken into account in the standard.

4.5.4.1 Gain-to-Noise Temperature Ratio

The following has been taken from INTELSAT BG-28-73E M/6/77 (Ref. 17):

> Approval of an earth station in the category of Standard "C" will only be obtained if these conditions are met for operation in the direction of the satellite and for the required polarization angle.

---

*See Section 4.5.8.

TABLE 4.12

Parameter	INTELSAT V Coverage in which Earth Station is Located	
	"West spot"	"East spot"
A	39.0	39.0
B	29.5	32.5
P_1	10.0	10.0
P_2	0.017	0.017

For any frequency in the band 10.95 to 11.20 and 11.45 to 11.70 GHz, the values of G/T_i must not be less than the values given below for the various criteria.

1. The value of $G/T_1 - L_1$ shall be no less than $A + 20\log_{10}(f/11.2)$ dB/K for all but P_1% of the time*:
2. The value of $G/T_2 - L_2$ shall be no less than $B + 20\log_{10}(f/11.2)$ dB/K for all but P_2% of the time*;
3. In the case of diversity earth stations the quantity which shall comply with the above specifications is the one which, at any given time, is equal to the value of $G/T - L$ of the operating terminal.

The following notes apply to these specification:

Note 1: The term G denotes the receive antenna gain referenced to the input of the receiving amplifier at the frequency of interest.

Note 2: The term L_i ($i = 1, 2$) is the predicted attenuation relative to a clear sky, at the frequency of interest, exceeded for no more than P_i ($i = 1, 2$) percent of the time* along the path to the satellite(s) with which operation is desired. The value L_i shall be that predicted on the basis of statistical distribution of mean attenuation within periods of one minute;

Note 3: The term T_i ($i = 1, 2$) is the receiving system noise temperature, including noise contributions from the atmosphere, referred to the input of the low noise amplifier at the frequency of interest, when the attenuation L_i prevails;

Note 4: The terms G/T_i and L_i are assumed to be given in dB notation; f is the frequency of interest expressed in GHz;

Note 5: The values of A, B, P_1 and P_2 in Table 4.12 shall apply.

---

* The time period to which this percentage applies shall be that period for which statistics are available, preferably a minimum of 5 years.

4.5.4.2. Equivalent Isotropic Radiated Power (EIRP)
for FDM / FM Carriers (Clear Sky)

The following has been taken from INTELSAT document BG-28-73E M/6/77 (Ref. 18):

The EIRP in the direction of the satellite shall, under clear sky conditions and light wind, be maintainable to ± 0.5 dB of the value assigned by INTELSAT.

TABLE 4.13(A). East Up (14 → 4 GHz Cross-Strapped Links)
Required Clear Sky EIRP for FDM / FM Carriers[a]

Bandwidth Unit (MHz)	Regular Carriers	
	Carrier Size (Channels)	EIRP (dBW)
1.25	12	66.5
2.5	24	68.8
	36	71.2
	48	74.6
	60	77.2
5.0	60	71.8
	72	72.1
	96	75.7
	132	79.8
7.5	96	73.5
	132	75.2
	192	80.7
10.0	132	73.8
	192	75.8
	252	80.5
15.0	252	77.5
	312	79.5
	432	85.1
20.0	432	81.2
	612	87.1
25.0	792	88.4
36.0	972	85.3

[a]Note: The gain stability of INTELSAT V allows for ± 2 dB variation during its operating lifetime, and, therefore, 2 dB of earth station EIRP on all carriers is reserved to provide compensation to this level if needed. A 1 dB mandatory margin has also been included in the above levels.

Source. (Reference 18)

TABLE 4.13(B). East Up (14 → 11 GHz Direct Links)
Required Clear Sky EIRP for FDM / FM Carriers[a]

Bandwidth Unit (MHz)	Regular Carriers	
	Carrier Size (Channels)	EIRP (dBW)
2.5	24	67.5
	36	69.9
	48	73.2
	60	75.9
5.0	60	70.5
	72	70.8
	96	74.4
	132	78.9
7.5	96	72.2
	132	73.9
	192	79.8
10.0	132	72.5
	192	75.5
	252	80.2
15.0	252	75.2
	312	78.2
	432	83.8
20.0	432	79.9
	612	85.7
25.0	432	78.9
	792	87.1
36.0	972	84.0

[a]Note: The gain stability of INTELSAT V allows for ±2 dB variation during its operating lifetime, and, therefore, 2 dB of earth station EIRP on all carriers is reserved to provide compensation to this level if needed. A 1 dB mandatory margin has also been included in the above levels.

Source. Reference 18.

This clear sky EIRP will be that necessary to develop the required satellite EIRP assigned to each link. The maximum clear sky EIRP values for FDM/FM carriers are provided in Tables 4.13 (a) to 4.13(d). This tolerance includes all factors contributing to power variation, for example, transmitter power level variations, antenna beam pointing error, and amplitude variations caused by beam scanning.

In order to meet the short-term CCIR performance criterion it is mandatory that means be provided to prevent the power flux density at the satellite from falling

TABLE 4.13(C). West Up (14 → 4 GHz Cross-Strapped Links)
Required Clear Sky EIRP for FDM / FM Carriers[a]

Bandwidth Unit (MHz)	Regular Carriers	
	Carrier Size (Channels)	EIRP (dBW)
1.25	12	63.4
2.5	24	65.7
	36	68.1
	48	72.4
	60	74.1
5.0	60	68.7
	72	69.0
	96	72.6
	132	76.7
7.5	96	70.4
	132	72.1
	192	77.6
10.0	132	71.7
	192	73.7
	252	78.4
15.0	252	74.4
	312	76.4
	432	82.0
20.0	432	78.1
	612	83.9
25.0	432	77.1
	792	85.2
36.0	972	82.2

[a]Note: The gain stability of INTELSAT V allows for ± 2 dB variation during its operating lifetime, and, therefore, 2 dB of earth station EIRP on all carriers is reserved to provide compensation to this level if needed. A 1 dB mandatory margin has also been included in the above levels.

Source. Reference 18.

by more than M dB below the nominal clear sky value for more than K percent of the time in a year. The values of M and K are given in Table 4.14.

Regardless of which method is used to maintain the flux density at the satellite, the flux density level obtained for clear sky EIRP shall not be exceeded by more than 0.5 dB prior to or following cessation of precipitation except for the brief interval following recovery from propagation conditions. This recovery period is

**TABLE 4.13(D). West Up (14 → 11 GHz Direct Links)
Required Clear Sky EIRP for FDM / FM Carriers[a]**

Bandwidth Unit (MHz)	Regular Carriers	
	Carrier Size (Channels)	EIRP (dBW)
2.5	24	64.4
	36	66.8
	48	70.1
	60	72.8
5.0	60	67.4
	72	67.7
	96	71.3
	132	75.4
7.5	96	69.1
	132	70.8
	192	76.3
10.0	132	70.4
	192	72.4
	252	76.1
15.0	252	73.1
	312	75.1
	432	80.7
20.0	432	76.5
	612	82.3
25.0	432	75.5
	792	84.0
36.0	972	80.8

[a] Note: The gain stability of INTELSAT V allows for ±2 dB variation during its operating lifetime, and therefore, 2 dB of earth station EIRP on all carriers is reserved to provide compensation to this level if needed. A 1 dB mandatory margin has also been included in the above levels.

Source. Reference 18.

determined by the power control circuitry and shall be in the range of 1 to 2 sec.*

A means shall be available for changes in EIRP levels to be effected expeditiously with a capability to maintain the new level. Means shall be provided whereby the level of each transmitted carrier can be monitored.

* Provisional, under review.

TABLE 4.14. Atlantic Region 14-GHz Up Path Margins for Power Control Determination

	"M" Up Path Margin (dB)	"K" Criterion (%)[a]
West up		
14 4-GHz links	6.0	0.017
14 11-GHz links (80 MHz)	6.5	0.010
14 11-GHz links (240 MHz)	12.5	0.010
East up		
14 4-GHz links	6.0	0.017
14 11-GHz links (80 MHz)	8.5	0.010
14 11-GHz links (240 MHz)	9.0	0.010

[a] Based on up path rain criterion for this percentage of the time in one year.

Source. Reference 18.

4.5.5 INTELSAT Standard D Earth Stations

The Standard D earth station provides more modest capability in the 6/4 GHz bands and supports what INTELSAT calls "Low Density Telephone Service" (LDTS). There are two subcategories in the Standard D class: D-1, which supports just one telephone channel, and D-2, which supports a greater number of voice channels. These earth stations will operate with INTELSAT V and VI satellites. Satellite access is FDMA/SCPC using companded FM or CFM using preassigned frequencies on a particular transponder.

The gain-to-noise temperature ratios (G/T) required for Class D service are:

$$\frac{G}{T} \geq 22.7 + 20\log_{10}(f/4) \quad \text{dB/K for Standard D-1} \tag{4.40a}$$

$$\frac{G}{T} \geq 31.7 + 20\log_{10}(f/4) \quad \text{dB/K for Standard D-2} \tag{4.40b}$$

where f is the frequency in GHz and G is the gain of the antenna system measured at the input of the LNA. For Class D-1 operation, the minimum gain at 6 GHz is 46.6 dB at a 30° elevation angle, and the gain relationship for other angles is $46.6 - 0.06 (\alpha - 30)$ dBi, where α is the elevation angle in degrees. Assuming about a 60% efficiency, a 13-ft antenna would suffice. The beamwidth (3 dB) of the antenna at 6 GHz is about 0.85°, which probably is sufficiently wide such that autotracking will not be required, assuming that periodic manual trim-up of the pointing will be carried out by the D-1 station operator. INTELSAT recommends autotracking for Class D-2 earth stations.

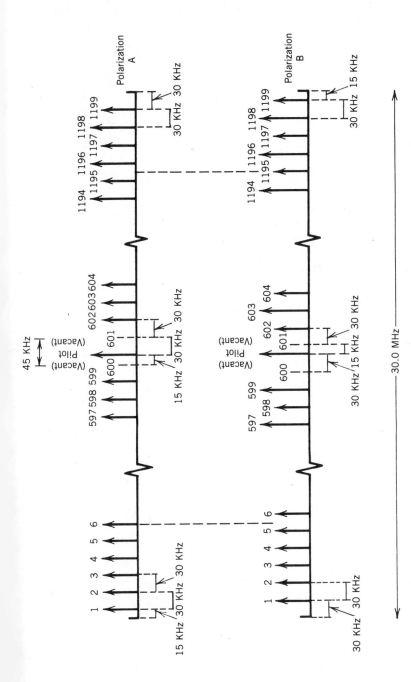

FIGURE 4.21 SCPC/CFM (LDTS) frequency plan for full transponder operation. From INTELSAT BG-56-72E W/9/83 (Ref. 19).

Note: For other than full transponder operation, the pilot may be assigned a frequency which is not at center of the satellite transponder. For those cases the vacant slots will not be MOS. 600 and 601, but will be those adjacent to the pilot. Similarly the 45 KHz offset may be introduced between slots other than mos. 600 and 601.

283

The EIRP required for SCPC operation with global beams of INTELSAT V, VA, and VI at 30° elevation angle is +56.6 dBW for Class D-1 and +52.7 dBW for Class D-2. These values, of course, are expressed as dBW per carrier. For elevation angles other than 30° the specified EIRP for 30° must be adjusted as follows: $-0.06 (\alpha - 30)$ dB, where α is the earth station elevation angle in degrees. The EIRP values given above include a 1.0 dB mandatory margin. Figure 4.21 shows the frequency plan for LDTS and Table 4.15 gives the SCPC/CFM characteristics and transmission parameters. Initially, LDTS will be provided through global beam transponders. In the future provision of hemispheric and zone beams may also be considered. In INTELSAT V satellites the global beam transponders do not make use of dual polarization for frequency reuse. For INTELSAT VA and VI, however, dual polarization is used in the global coverage. Therefore, LDTS stations shall be capable of operating in any circular polarization sense as follows:

Polarization A: uplink is LHCP; downlink is RHCP.

Polarization B: uplink is RHCP; downlink is LHCP.

TABLE 4.15. SCPC / CFM characteristics and Transmission Parameters

Parameter	Requirement
Audio channel input bandwidth	300–3400 Hz
Modulation	FM with companding
Companding	2 : 1 syllabic
Carrier control	Voice-Activated
Channel spacing	30.0 kHz[a]
Channel bandwidth	30.0 kHz
IF noise bandwidth	25.0 kHz
rms test tone deviation for 0 dBm0 at 1 kHz	5.1 kHz
C/N per channel at nominal operating point	10.2 dB
C/N_0 at nominal operating point	54.2 dB Hz
C/N per channel at threshold	6.2 dB
C/N_0 in IF bandwidth at threshold	50.2 dB Hz

[a] The spacing between the two carrier slots adjacent to the center of the frequency band will be 45 kHz.

Source. Reference 19.

However, simultaneous operation in both senses of polarization is not required, neither for the uplink nor for the downlink.

To achieve sufficient S/N in the voice channel both pre-emphasis/de-emphasis and companding are used. The companders that should be used are of the $2:1$ syllabic type in accordance with CCITT Rec. G-162 (Ref. 20).

4.5.6 INTELSAT Standard E Earth Stations

The Standard E earth station is designed for the "business service" and operates in the nominal 14/11 GHz band pair utilizing INTELSAT V, VA, and VI satellites.

The business service basically provides users with data connectivity. Thus, we are dealing with digital transmission, which is covered in Chapter 5, essentially a generic discussion. For completeness of INTELSAT offerings, Standard E (and in Section 4.5.7 Standard F) INTELSAT earth stations are described. A review of the appropriate subsections of Chapter 5 may be helpful in understanding concepts and terminology.

For Standard E business service, three general categories of system usage are envisioned by INTELSAT: (1) a closed network, (2) an open network, and (3) a network that can be interconnected with a public terrestrial digital network. These are described below:

(a) A closed network is intended to provide freedom to the user in selecting the digital system required for his or her particular needs. The performance characteristics for this type of service do not require specifications related to interconnection with other users and can be defined in terms of RF transmission characteristics as defined in INTELSAT document BG-53-86 (Rev. 1) (Ref. 16) and briefly reviewed in this subsection.

(b) An open network requires a certain degree of common terminal features to be defined in order for one user's network to interface with another. INTELSAT intends to expand the Standard E earth station document to include these features at a later date.

(c) The type of network usage which can be interconnected with a public terrestrial digital data network is also contemplated by INTELSAT and is under study by that organization.

INTELSAT (BG-53-86 Rev. 1) states that the satellite data link performance is a bit error rate (BER) equal to or better than 1×10^{-8} with clear sky conditions and 1×10^{-6} under degraded sky conditions for all but 1% of a year. These BERs are attained by the specified EIRP and by user-implemented forward error correction (FEC) techniques using convolutional coding, Viterbi decoding, rate $\frac{1}{2}$ or rate $\frac{3}{4}$. (See Chapter 5.) At the present time, the only approved modulation method for Standard E earth stations is coherent QPSK.

Carriers will be given preassigned frequencies within a given transponder. QPSK/TDMA/FDMA, TDM/QPSK/FDMA, and QPSK/FDMA terminals may be employed for specialized networks using these preassigned frequencies, after review and approval has been obtained from INTELSAT. Minimum elevation angle for Class E earth stations is 10°.

There are three subclasses of the Standard E earth stations—E-1, E-2, and E-3—and the class selected will depend on user requirements with respect to site location, traffic requirements, and desired channel performance. The mandatory G/T values for the several classes are defined below:

$$\text{Standard E-1:}\ \frac{G}{T} \geq 25.0 + 20\log_{10}(f/11)\quad \text{dB/K} \qquad (4.41a)$$

$$\text{Standard E-2:}\ \frac{G}{T} \geq 29.0 + 20\log_{10}(f/11)\quad \text{dB/K} \qquad (4.41b)$$

$$\text{Standard E-3:}\ \frac{G}{T} \geq 34.0 + 20\log_{10}(f/11)\quad \text{dB/K} \qquad (4.41c)$$

where F is the frequency in GHz, G is the antenna gain measured at the input of the LNA in dBi, and T is the receiving system noise temperature at the same reference point measured in K/Hz. Table 4.16 gives the minimum bandwidth requirements for Standard E earth stations. Table 4.17 provides the minimum tracking requirements. Digital carrier reference parameters are shown in Tables 4.18(a) and 4.18(b).

TABLE 4.16. Minimum Bandwidth Requirements for Standard E Earth Stations

Satellite	ITU Region	Earth Station Transmit Frequency Band (GHz)	Earth Station Receive Frequency Band (GHz)	Available Transponders
V, VA, and VI	All	14.00–14.25	10.95–11.2	1–2, 3–4,[d] 5–6
VB[c]	2[b]	14.00–14.25[a]	11.7–11.95	1–2, 3–4, 5–6, 7–8[a]
	1 and 3[b]	14.00–14.25[a]	12.5–12.75	1–2, 3–4, 5–6, 7–8[a]

[a]Standard E earth stations intending to operate in the spot-to-hemi multibeam mode available in transponder 7-8 on INTELSAT VB should have this transmit bandwidth extended to 14.35 GHz.

[b]In INTELSAT VB, the receive band segments of 11.7–11.95 and 12.5–12.75 GHz are interchangeable between the east and west spot beams, so that this spacecraft series can be operated in any Ocean Region.

[c]If INTELSAT VB satellites are configured to operate at 11 GHz (as opposed to 12 GHz), the earth station receive frequency band will be the same as for INTELSAT V, VA, and VI satellites.

[d]Transponder 3-4 is only available in INTELSAT VI.

Source. Reference 16.

TABLE 4.17. Minimum Tracking Requirements for Standard E Earth Stations

Earth Station Standard	INTELSAT V, VA, and VB ($\pm 0.1°$ N/S and $\pm 0.1°$ E/W)	INTELSAT VI[c] ($\pm 0.02°$ N/S and $\pm 0.06°$ E/W)
E-1	Manual, E/W only (weekly peaking)	Fixed antenna[b]
E-2	Manual, E/W and N/S (peaking every 3–4 hr)	Fixed antenna[b]
E-3	Automatic track[a]	Manual, E/W only (weekly peaking)

[a] Program tracking, which would operate from pointing information supplied by INTELSAT, is recommended due to the uncertainty of step track operation in a K band environment which may be subject to severe fading and scintillations.

[b] "Fixed" antenna mounts will still require the capability to be steered from one satellite position to another, as dictated by operational requirements (typically once or twice every 2–3 years). These antennas should also be capable of being steered at least over a range of $\pm 5°$ from beam center for the purpose of verifying that the antenna pointing is correctly set toward the satellite, and for providing a means of verifying the sidelobe characteristics in this range.

[c] Antenna tracking requirements are based on provisional INTELSAT VI stationkeeping tolerances. These tolerances will be reviewed after operational experience is gained with this new series of spacecraft.

Source. Reference 16.

INTELSAT points out that the user may wish to consider the use of different FEC (forward error correction; see Chapter 5) schemes, or no FEC implementation at all, to achieve some other BER performance than that mentioned previously. This is acceptable, according to INTELSAT (BG-53-86 Rev. 1), provided that the transmission rate and bandwidth units given in Tables 4.18(a) and 4.18(b) are not exceeded. If a more robust FEC implementation is required, the information data rate may be reduced for a given transmission rate.

The necessary EIRP per carrier during clear sky conditions is a function of satellite G/T, outage margins, and optimized loading of the entire transponder in question. The nominal uplink EIRP will, therefore, vary from time-to-time and will be established in coordination with INTELSAT's TOCC (Technical Operations Control Center). For the purpose of Class E earth station planning, the maximum clear sky uplink EIRP levels that must be available, if requested, are given in Tables 4.19(a), 4.19(b) and 4.19(c).

The EIRP in the direction of the satellite shall, except under adverse weather conditions, be maintained to within ± 1.5 dB. This tolerance includes all earth station factors that contribute to EIRP variations (e.g., HPA output instability, antenna pointing, and tracking errors, added on a root sum square basis).

Two unmodulated beacons are provided on INTELSAT V, VA, VB, and VI. It is assumed that measurement of their attenuation (relative levels from

TABLE 4.18(A). IBS Reference Transmission Parameters for Direct 14 / 11 or 12 GHz and Cross-Strapped Links[a] (INTELSAT V, VA, and VI; Rate 3 / 4 FEC)

Maximum Information Rate (bit/sec)	Transmission Rate (bit/sec)	Occupied Bandwidth Unit (hertz)	Allocated Bandwidth Unit (hertz)	C/T (dBW/K) (10^{-8})	C/N_0 (dB/Hz) (10^{-8})	C/N (dB) (10^{-8})
64 k	94 k	56 k	67.5 k	−171.0	57.6	
128 k	188 k	112 k	135.0 k	−168.8	60.6	
256 k	376 k	224 k	270 k	−165.0	63.6	
384 k	564 k	336 k	405.0 k	−163.2	65.3	10.1
512 k	752 k	448 k	540.0 k	−162.0	66.6	
768 k	1128 k	672 k	800.0 k	−160.2	68.4	
1.544 M	⌐ 2.3 M	1.38 M	1.625 M	−157.1	71.5	
2.048 M	3.0 M	1.80 M	2.125 M	−155.9	72.7	
4.096 M	6.0 M	3.60 M	4.250 M	−152.9	75.7	
6.312 M	9.3 M	5.58 M	6.500 M	−151.1	77.5	
8.448 M	12.4 M	7.44 M	8.750 M	−149.8	78.8	
Notes 3 and 6	Note 4		Note 5			

[a] For all links, the BER will be better than or equal to 10^{-6} for all but 1% of the year. This means that under clear sky conditions these carriers will be operated with enough margin to ensure that the 1% criterion is met. Depending on the actual transponder and link conditions, INTELSAT may establish the clear sky setting of the link at a C/N better than or equal to 10.1 dB in order to ensure the margins identified are provided. The C/N, C/T, and C/N_0 values for 10^{-6} are 1.1 dB less than those shown for 10^{-8}.

Notes:

1. Carrier sizes not conforming to the reference units are acceptable under the conditions stated in BG-53-86.

2. C/T, C/N_0, and C/N values have been calculated with rate 3/4 FEC as a baseline. The assumed information rate E_b/N_0 is 9.2 dB for a BER of 10^{-8}.

3. The maximum information rate applies to the data rate before 10% overhead and encoding. It is assumed that earth station owners may add up to 10% overhead bits to their information rates for such functions as signaling, alarms, earth station control, encryption key control, or voice channels accompanying teleconferencing transmissions. In some cases the information rate may already include inband overhead channels.

4. Transmission rate = information rate plus 10% overhead $\times \frac{4}{3}$.

5. For carrier sizes from 64 kbit/sec to 768 bit/sec the bandwidth allocated to the carrier in the satellite transponder is a multiple of 22.5 kHz. For carrier sizes of 1.544 Mbit/sec and above the bandwidth allocated to the carrier is a multiple of 125 kHz.

6. The technical parameters for the 4.096, 6.312, and 8.448 Mbit/sec carrier sizes are shown here since they may find future application. Initial service is expected with carrier sizes from 64 kbit/sec to 2.048 Mbit/sec.

Source. Reference 16.

TABLE 4.18(B). IBS Reference Transmission Parameters for Direct 14 / 11 or 12 GHz and Cross-Strapped Links[a] (INTELSAT VB, rate 1 / 2 FEC)

Maximum Information Rate (bit/sec)	Transmission Rate (bit/sec)	Occupied Bandwidth Unit (hertz)	Allocated Bandwidth Unit (hertz)	C/T (dBW/K) (10^{-8})	C/N_0 (dB/Hz) $(10^{-8}$	C/N (dB) (10^{-8})
64 k	141 k	85 k	112.5 k	−172.4	56.2	
128 k	282 k	170 k	202.5 k	−169.4	59.2	
256 k	564 k	340 k	405.0 k	−166.4	62.2	
384 k	846 k	510 k	607.5 k	−164.7	63.9	6.8
512 k	1128 k	680 k	900.0 k	−163.4	65.2	
768 k	1692 k	1020 k	1350.0 k	−161.6	67.0	
1.544 M	3.4 M	2.0 M	2.500 M	−158.7	69.9	
2.048 M	4.5 M	2.7 M	3.250 M	−157.4	71.2	
4.096 M	9.0 M	5.4 M	6.375 M	−154.4	74.2	
6.312 M	13.9 M	8.34 M	9.875 M	−152.5	76.1	
8.448 M	18.6 M	11.16 M	13.000 M	−151.3	77.3	
Notes 3 and 6	Note 3		Note 5			

[a] For all links, the BER will be better than or equal to 10^{-6} for all but 1% of the year. This means that under clear sky conditions these carriers will be operated with enough margin to ensure that the 1% criterion is met. Depending on the actual transponder and link conditions, INTELSAT may establish the clear sky setting of the link at a C/N better than or equal to 10.1 dB in order to ensure the margins identified are provided. The C/N, C/T, and C/N_0 values for 10^{-6} are 1.1 dB less than those shown for 10^{-8}.

Notes:

1. Carrier sizes not conforming to the reference units are acceptable under the conditions stated in BG-53-86.

2. C/T, C/N_0, and C/N values have been calculated with rate 1/2 FEC as a baseline. The assumed information rate E_b/N_0 is 7.6 dB for a BER of 10^{-8}.

3. The maximum information rate applies to the data rate before 10% overhead and encoding. It is assumed that earth station owners may add up to 10% overhead bits to their information rates for such functions as signaling, alarms, earth station control, encryption key control, or voice channels accompanying teleconferencing transmissions. In some cases the information rate may already include inband overhead channels.

4. Transmission rate = information rate plus 10% overhead × 2.

5. For carrier sizes from 64 kbit/sec to 768 bit/sec the bandwidth allocated to the carrier in the satellite transponder is a multiple of 22.5 kHz. For carrier sizes of 1.544 Mbit/s and above the bandwidth allocated to the carrier is a multiple of 125 kHz.

6. The technical parameters for the 4.096, 6.312, and 8.448 Mbit/sec carrier sizes are shown here since they may find future application. Initial service is expected with carrier sizes from 64 kbit/sec to 2.048 Mbit/sec.

Source. Reference 16.

TABLE 4.19(A). Maximum EIRP[a]
(14 to 11 / 12 GHz Direct Links, 10° Elevation Angle)

Receiving Station	EIRP															
	E-1				E-2				E-3				C			
S/C Category	(1)	(2)	(3)	(4)	(1)	(2)	(3)	(4)	(1)	(2)	(3)	(4)	(1)	(2)	(3)	(4)
Information Rate (bit/sec)																
64 k	56.9	59.4	58.0	60.9	54.1	56.0	55.6	57.5	51.0	50.6	49.5	52.1	49.8	—	45.7	49.9
128 k	59.9	62.4	61.0	63.9	57.1	59.0	58.6	60.5	54.0	53.6	52.5	55.1	52.8	—	48.7	52.9
256 k	62.9	65.4	64.0	66.9	60.1	62.0	61.6	63.5	57.0	56.6	55.5	58.1	55.8	—	51.7	55.9
384 k	64.7	67.2	65.8	68.7	61.9	63.8	63.4	65.3	58.8	58.4	57.3	59.9	57.6	—	53.5	57.7
512 k	65.9	68.4	67.0	69.9	63.1	65.0	64.6	66.5	60.0	59.6	58.5	61.1	58.8	—	54.7	58.9
768 k	67.7	70.2	68.8	71.7	64.9	66.8	66.4	68.3	61.8	61.4	60.3	62.9	60.6	—	56.5	60.7
1.544 M	70.8	73.3	71.9	74.8	68.0	69.9	69.5	71.4	64.9	64.5	63.4	66.0	63.7	—	59.6	63.8
2.048 M	71.9	74.4	73.0	75.9	69.1	71.0	70.6	72.5	66.0	65.6	64.5	67.1	64.8	—	60.7	64.9
4.096 M	75.0	77.5	76.1	79.0	72.2	74.1	73.7	75.6	69.1	68.7	67.6	70.2	67.9	—	63.8	68.0
6.312 M	76.9	79.4	78.0	80.9	74.1	76.0	75.6	77.5	71.0	70.6	69.5	72.1	69.8	—	65.7	69.9
8.448 M	78.1	80.6	79.2	82.1	75.3	77.2	76.8	78.7	72.2	71.8	70.7	73.3	71.0	—	66.9	71.1

[a] The EIRP values listed in this table apply to the location of the transmit earth station for all possible beam connections except those shown below where an EIRP adjustment is needed as indicated under "delta EIRP." In the case of spacecraft categories (2) and (4), when the transmit earth station is located between the inner and outer beam edge contours of the west spot beam, it may be necessary to increase the EIRP by an amount up to the maximum delta. The actual adjustment will be supplied by INTELSAT when the location of the earth station is known.

Spacecraft Category	FEC Code Rate	Frequency	Beam Connection	Delta EIRP
(1) V (F1-F9) and VA (F10-F12)	3/4	14/11	West-to-west East-to-east	−3 dB +3 dB
(2) VB (F13-F15)	1/2	14/12	West outer to any beam	Up to +5 dB
(3) VI (F1-F3)	3/4	14/11	West-to-west East-to-east	−3 dB +3 dB
(4) VI (F4-F5)	3/4	14/11	West outer to any beam	Up to +3 dB
VB (F13-15)	3/4	14/11	West outer to any beam	Up to +5 dB

Source. Reference 16.

TABLE 4.19(B). Standard E Maximum EIRP[a]
(14 to 4 GHz Cross-Strapped Links, 10° Elevation Angle)

Receiving Station									EIRP											
	F-1				F-2				F-3				B				A			
S/C Category	(1)	(2)	(3)	(4)	(1)	(2)	(3)	(4)	(1)	(2)	(3)	(4)	(1)	(2)	(3)	(4)	(1)	(2)	(3)	(4)
Information Rate (bit/sec)																				
64 K	61.3	61.2	60.5	61.6	57.5	57.4	57.2	58.0	55.6	55.5	55.8	56.4	53.8	53.1	51.4	53.3	49.4	47.5	45.2	47.9
128 K	64.3	64.2	63.5	64.6	60.5	60.4	60.2	61.0	58.6	58.5	58.8	59.4	56.8	56.1	54.4	56.3	52.4	50.5	48.2	50.9
256 K	67.3	67.2	66.5	67.6	63.5	63.4	63.2	64.0	61.6	61.5	61.8	62.4	59.8	59.1	57.4	59.3	55.4	53.5	51.2	53.9
384 K	69.1	69.0	68.3	69.4	65.5	65.2	65.0	65.8	63.4	63.3	63.6	64.2	61.6	60.9	59.2	61.1	57.2	55.3	53.0	55.7
512 K	70.3	70.2	69.5	70.6	66.5	66.4	66.2	67.0	64.6	64.5	64.8	65.4	62.8	62.1	60.4	62.3	58.4	56.5	54.2	56.9
768 K	72.1	72.0	71.3	72.4	68.3	68.2	68.0	68.8	66.4	66.3	66.6	67.2	64.6	63.9	62.2	64.1	60.2	58.3	56.0	58.7
1.544 M	75.2	75.1	74.4	75.5	71.4	71.3	71.1	71.9	69.5	69.4	69.7	70.3	67.7	67.0	65.3	67.2	63.3	61.4	59.1	61.8
2.048 M	76.3	76.2	75.5	76.6	72.5	72.4	72.2	73.0	70.6	70.5	70.8	71.4	68.8	68.1	66.4	68.3	64.4	62.5	60.2	62.9
4.096 M	79.4	79.3	78.6	79.7	75.6	75.5	75.3	76.1	73.7	73.6	73.9	74.5	71.9	71.2	69.5	71.4	67.5	65.6	63.3	66.0
6.312 M	81.3	81.2	80.5	81.6	77.5	77.4	77.2	78.0	75.6	75.5	75.8	76.4	73.8	73.1	71.4	73.3	69.4	67.5	65.2	67.9
8.448 M	82.5	82.4	81.7	82.8	78.7	78.6	78.4	79.2	76.8	76.7	77.0	77.6	75.0	74.3	72.6	74.5	70.6	68.7	66.4	69.1

[a]The EIRP values listed in this table apply to the location of transmit earth station for all possible beam connections except those shown below where an EIRP adjustment is needed as indicated under "delta EIRP." In the case of spacecraft categories (2) and (4), when the transmit earth station is located between the inner and outer beam edge contours of the west spot beam, it may be necessary to increase the EIRP by an amount up to the maximum delta. The actual adjustment will be supplied by INTELSAT when the location of the earth station is known.

SPACECRAFT CATEGORY	FEC CODE RATE	SATELLITE RECEIVE BEAM (UPLINK)	DELTA EIRP
(1) V (F1-F9) and VA (F10-F12)	3/4	East	+3 dB
(2) VB (F13-F15)	1/2	Between west outer and inner	Up to +5 dB
(3) VI (F1-F3)	3/4	East	+3 dB
(4) VI (F4-F5)	3/4	Between west outer and inner	Up to +3 dB

Source. Reference 16.

TABLE 4.19(C). Standard E Maximum EIRP
(INTELSAT VB Multibeam Connectivity Modes)

Mode	Link	Transponder	Reference FEC	Delta EIRP (see Note 3)
Full connectivity	Direct	3-4	1/2	3.5 dB
	Cross-strapped	3-4	1/2	3.5 dB
Spot-to-hemi connectivity	Cross-strapped	7-8	1/2	1.0 dB

Notes:

1. Full connectivity is a multibeam interconnection such that a transmission in either the hemispheric or spot beam, east or west, can be received simultaneously in both spot beams at K band, and in both hemi beams at C band. This mode is achieved when the satellite is placed in the extra high gain step (to make up for the loss of gain created by connection losses). It is assumed that all other C-and-K band transponders are operated in the high gain step whenever the full connectivity mode is used.

2. Spot-to-hemi connectivity is a multibeam interconnection such that a transmission in either one of the spot beams (east or west) at K band can be received simultaneously in both hemispheric beams at C band.

3. The delta EIRP values shown above assume the transmit station is located within the west inner contour or within the east beam. If the transmit station is located in the region between the beam edge contours of the inner and outer west spot beam, the delta EIRP values could be increased by up to an additional 5 dB. The actual increase will be supplied by INTELSAT when the location of the transmit station in the beam is known.

4. For the multibeam modes of operation on the INTELSAT VB, it is necessary to identify the highest EIRP required among all intended receive earth stations, as per Column 2 of Tables 4.19(a) and 4.19(b). To this highest EIRP, the value under "Delta EIRP" shown should be added to obtain the uplink EIRP for a particular application.

Source. Reference 16.

clear sky to disturbed weather related propagation conditions) can be factored to achieve power control at 14 GHz.

4.5.7 INTELSAT Standard F Earth Stations

The Standard F earth station operates in INTELSAT's International Business Service and is similar in most respects to the Standard E earth station, except that it operates in the 6/4 GHz frequency band pair. The expected BER performance is similar to the Class E earth station.

Three minimum Standard F G/T values are specified to allow suitable range of relatively small earth station sizes that can be located on or near business facilities. The selection of one of these values for G/T will depend on user requirements for performance and maximum transmission rate. The three

G/T values are stated below:

$$\text{Standard F-1: } \frac{G}{T} \geq 22.7 + 20\log_{10}(f/4) \quad \text{dB/K} \qquad (4.42a)$$

$$\text{Standard F-2: } \frac{G}{T} \geq 27.0 + 20\log_{10}(f/4) \quad \text{dB/K} \qquad (4.42b)$$

$$\text{Standard F-3: } \frac{G}{T} \geq 29.0 + 20\log_{10}(f/4) \quad \text{dB/K} \qquad (4.42c)$$

where G is the antenna gain measured at the input of the LNA relative to an isotropic radiator; T is the receiving system noise temperature referred to the input of the LNA, relative to 1 K; and f is the frequency in gigahertz. The G/T values are for clear sky conditions, in light wind, and for any frequency within at least one of the band segments defined in Table 4.20. The minimum elevation angle is $5°$.

The modulation type and access is similar to the Standard E earth station as is the envisioned data network types.

The minimum antenna transmit gain at 6 GHz measured at the antenna feed shall be no less than

$$\text{Standard F-1: } 47.7 - K \quad \text{(dBi)}$$
$$\text{Standard F-2: } 51.6 - K \quad \text{(dBi)}$$
$$\text{Standard F-3: } 53.0 - K \quad \text{(dBi)}$$

where K is equal to 0.02 $(\alpha - 10)$ dB for hemi and zone beams or 0.06

TABLE 4.20. Minimum Bandwidth Requirements for Standard F Earth Stations

Earth Station Mode	Band Segment	Frequency Band (MHz)	Bandwidth (MHz)	Available Transponders
Transmit (Uplink)	1	5925–6256	331	1-2[a], 3-4, 5-6, 7-8
	or			
	2	6094–6425	331	5-6, 7-8, 9, 10, 11, 12
Receive (Downlink)	1	3700–4031	331	1-2[a], 3-4, 5-6, 7-8,
	or			
	2	3869–4200	331	5-6, 7-8, 9, 10, 11, 12

[a]Earth station owners should consider in their design the possibility of extending their usable bandwidth to include transponders (1'-2') of INTELSAT VI to 3.625 GHz for receive and 5.845 GHz for transmit.

Source. Reference 21.

TABLE 4.21(A). Standard F Maximum Clear Sky EIRP[a] (6 to 4 GHz Direct Links)

Receiving Station	EIRP (dBW)																								
	F-1					F-2					F-3					B					A				
S/C Category	(1)	(2)	(3)	(4)	(5)	(1)	(2)	(3)	(4)	(5)	(1)	(2)	(3)	(4)	(5)	(1)	(2)	(3)	(4)	(5)	(1)	(2)	(3)	(4)	(5)
Information Rate (bit/sec)																									
64 k	70.1	70.1	67.5	65.9	62.8	66.1	66.1	63.5	61.9	60.0	64.9	64.9	62.3	60.7	58.8	63.2	63.2	60.6	59.0	57.4	56.2	56.2	53.6	52.0	50.4
128 k	73.1	73.1	70.5	68.9	65.8	69.1	69.1	66.5	64.9	63.0	67.9	67.9	65.3	67.7	61.8	66.2	66.2	63.6	62.0	60.4	59.2	59.2	56.6	55.0	53.4
256 k	76.1	76.1	73.5	71.9	68.8	72.1	72.1	69.5	67.9	66.0	70.9	70.9	68.3	66.7	64.8	69.2	69.2	66.6	65.0	63.4	62.2	62.2	59.6	58.0	56.4
384 k	77.9	77.9	75.3	73.7	70.6	73.9	73.9	71.3	69.7	67.8	72.7	72.7	70.1	68.5	66.6	71.0	71.0	68.4	66.8	65.2	64.0	64.0	61.4	59.8	58.2
512 k	79.1	79.1	76.5	74.9	71.8	75.1	75.1	72.5	70.9	69.0	73.9	73.9	71.3	69.7	67.8	72.2	72.2	69.6	68.0	66.4	65.2	65.2	62.6	61.0	59.4
768 k	80.9	80.9	78.3	76.7	73.6	76.9	76.9	74.3	72.7	70.8	75.7	75.7	73.1	71.5	69.6	74.0	74.0	71.4	69.8	68.2	67.0	67.0	64.4	62.8	61.2
1.544 M	84.0	84.0	81.4	79.8	76.7	80.0	80.0	77.4	75.8	73.9	78.8	78.8	76.2	74.6	72.7	77.1	77.1	74.5	72.9	71.3	70.1	70.1	67.5	65.9	64.3
2.048 M	85.2	85.2	82.6	81.0	77.9	81.2	81.2	78.6	77.0	75.1	80.0	80.0	77.4	75.8	73.9	78.3	78.3	75.7	74.1	72.5	71.3	71.3	68.7	67.1	65.5
4.096 M	88.2	88.2	85.6	84.0	80.9	84.2	84.2	81.6	80.0	78.1	83.0	83.0	80.4	78.8	76.9	81.3	81.3	78.7	77.1	75.5	74.3	74.3	71.7	70.1	68.5
6.312 M	90.1	90.1	87.5	85.9	82.8	86.1	86.1	83.5	81.9	80.0	84.9	84.9	82.3	80.7	78.8	83.2	83.2	80.6	79.0	77.4	76.2	76.2	73.6	72.0	70.4
8.448 M	91.3	91.3	88.7	87.1	84.0	87.3	87.3	84.7	83.1	81.2	86.1	86.1	83.5	81.9	80.0	84.4	84.4	81.8	80.2	78.6	77.4	77.4	74.8	73.2	71.6

[a]The EIRP values listed in this table are for 10° elevation angle. Correction factors for other elevation angles are listed in the set of notes given below.

Notes for Tables 4.21(a) and 4.21(b) (Earth Station EIRP):

1. All EIRP values shown in Tables 4.21(a) and 4.21(b) apply to the hemi and zone beams. For operation with global beams the maximum EIRP values shown in Table 4.21(a) are increased by 2.5 dB. All EIRP are subject to CCIR Rec 524-1 earth station off-beam emission criteria.

2. The EIRP values shown in Tables 4.21(a) and 4.21(b) were derived for the high-gain mode and include 1 dB mandatory margin and 2 dB allowance for satellite flux density variation. In some situations it may be possible to operate transponders in the extra high gain mode, however, this will be determined at a later date after operational experience has been gained with this service. If the extra high gain mode is employed, the maximum up-link EIRP requirement will be lowered by several decibels.

3. The EIRP values shown in Tables 4.21(a) and 4.21(b) may be corrected by the formula $0.02 (\alpha - 10)$ dB for hemi and zone beams and $0.06 (\alpha - 10)$ dB for global beams; where α is the earth station elevation angle, between 5° and 90°.

4. Earth station owners should note that their normal EIRP is likely to be 2 dB less than the maximum values shown in these tables.

SPACECRAFT CATEGORY	FEC CODE RATE
(1) IVA, F1 to F6 (6/4 GHz only)	3/4
(2) V, F1 to F4	3/4
(3) V, F5 to F9; VA, F10 to F12; and VB, F13 to F15 (11 GHz)	3/4
(4) VB, F13 to F15 (12 GHz)	1/2
(5) VI, F1 to F5	3/4

Source. Reference 21.

TABLE 4.21(B). Standard F Maximum Clear-Sky EIRP[a] (6 to 11 or 12 GHz Cross-Strapped Links)

Receiving Station	EIRP																			
S/C Category	E-1					E-2					E-3					C				
	(1)	(2)	(3)	(4)	(5)	(1)	(2)	(3)	(4)	(5)	(1)	(2)	(3)	(4)	(5)	(1)	(2)	(3)	(4)	(5)
Information Rate (bit/sec)																				
64 k	—	66.5	63.9	65.4	61.3	—	64.0	61.4	62.9	59.1	—	58.6	56.0	57.5	57.6	—	57.4	54.8	—	55.1
128 k	—	69.5	66.9	68.4	64.3	—	67.0	64.4	65.9	62.1	—	61.6	59.0	60.5	60.6	—	60.4	57.8	—	58.1
256 k	—	72.5	69.9	71.4	67.3	—	70.0	67.4	68.9	65.1	—	64.6	62.0	63.5	63.6	—	63.4	60.8	—	61.1
384 k	—	74.3	71.7	73.2	69.1	—	71.8	69.2	70.7	66.9	—	66.4	63.8	65.3	65.4	—	65.2	62.6	—	63.1
512 k	—	75.5	72.9	74.4	70.3	—	73.0	70.4	71.9	68.1	—	67.6	65.0	66.5	66.6	—	66.4	63.8	—	64.1
768 k	—	77.3	74.7	76.2	72.1	—	74.8	72.2	73.7	69.9	—	69.4	66.8	68.3	68.4	—	68.2	65.6	—	66.1
1.544 M	—	80.4	77.8	79.3	75.2	—	77.9	75.3	76.8	73.0	—	72.5	69.9	71.4	71.5	—	71.3	68.7	—	69.1
2.048 M	—	81.6	79.0	80.5	76.4	—	79.1	76.5	78.0	74.2	—	73.7	71.1	72.6	72.7	—	72.5	69.9	—	70.2
4.096 M	—	84.6	82.0	83.5	79.4	—	82.1	79.5	81.0	77.2	—	76.1	74.1	75.6	75.7	—	75.5	72.9	—	73.3
6.312 M	—	86.5	83.9	85.4	81.3	—	84.0	81.4	82.9	79.1	—	78.6	76.0	77.5	77.6	—	77.4	74.8	—	75.2
8.448 M	—	87.7	85.1	86.6	82.5	—	85.2	82.6	84.1	80.3	—	79.2	77.2	78.7	78.8	—	78.6	76.0	—	76.4

[a] The EIRP values listed in this table are for 10° elevation angle. Correction factors for other elevation angles are listed in the set of notes given at the bottom of Table 4.21(a).

SPACECRAFT CATEGORY	FEC CODE RATE
(1) IVA, F1 to F6 (6/4 GHz only)	3/4
(2) V, F1 to F4	3/4
(3) V, F5 to F9; VA, F10 to F12; and	3/4
VB, F13 to F15 (11 GHz)	
(4) VB, F13 to F15 (12 GHz)	1/2
(5) VI, F1 to F5	3/4

Source. Reference 21.

TABLE 4.21(C). Standard F Maximum Clear-Sky EIRP (INTELSAT VB Full Multibeam Connectivity Mode)

	Transponder	Reference FEC	Delta EIRP (See Note 2)
Cross-strapped connections only of full connectivity mode	3-4	Rate 1/2	+1.0 dB
All other connections of full connectivity mode	3-4	Rate 1/2	+2.0 dB

Notes:

1. Full connectivity is a multibeam interconnection such that a transmission in one hemispheric or spot beam, east or west, can be received simultaneously in both spot beams at K band, and in both hemi beams at C band. This mode is achieved when the satellite is placed in the extra high gain (to make up for the loss of gain created by connection losses). It is assumed that all other C-and-K band transponders are operated in the high gain step whenever the full connectivity mode is used.

2. In the case of the "full connectivity" mode of multibeam operation on the INTELSAT VB, it is necessary to identify the highest EIRP required among all intended receive earth stations, as per Column 4 of Tables 4.21(a) and 4.21(b). To this highest EIRP, the value under "Delta EIRP" shown above should be added to obtain the uplink EIRP for a particular application.

Source. Reference 21.

$(\alpha - 10)$ dB for global beams, where α is the earth station elevation angle in degrees.

The objective of this requirement is to place a limit on transmission power by defining the minimum contribution of on-axis antenna gain to the EIRP levels that are to be generated. This limitation is necessary to ensure that the off-beam emission criterion stated in CCIR Rec. 524-1 is not exceeded.

The required EIRP per carrier for the Standard F earth stations are shown in Tables 4.21(a) and 4.21(b) for normal transponder operation and Table 4.21(c) for the multibeam connectivity modes of operation. These values are valid for clear sky conditions. The values stated in the tables are maximums, and the operational values are established by INTELSAT, which are a function of transponder G/T, outage margins, and optimized loading of the transponder in question. The EIRP levels in the direction of the satellite shall, except under adverse weather conditions, be maintained to within $+1, -1.5$ dB for the Standard F-1 and F-2 earth stations and to within ± 0.5 dB for Standard F-3 earth stations. This tolerance includes all earth station factors contributing to EIRP variation such as HPA output power level instability, antenna beam pointing and/or tracking error, added on a root sum square basis.

Digital carrier reference parameters are given in Tables 4.22(a) and 4.22(b). The values are based on coherent QPSK modulation, maximum EIRP, specified FEC code rate, transmission rate, and bandwidth unit.

INTELSAT in BG-58-95E (Ref. 21) shows considerable flexibility. FEC schemes other than those specified may be selected by the user, or the user may elect not to use FEC. Such changes are acceptable provided that the transmission rate and bandwidth units given in Tables 4.22(a) and 4.22(b) are

TABLE 4.22(A). IBS Reference Transmission Parameters for Direct 6 / 4 GHz and Cross-Strapped Links[a] (INTELSAT IVA, V, VA, and VI; Rate 3 / 4 FEC)

Maximum Information Rate (bit/sec)	Transmission Rate (bit/sec)	Occupied Bandwidth Unit (Hertz)	Allocated Bandwidth Unit (Hertz)	C/T (dBw/K) 10^{-8}	C/N_0 (dB/Hz) 10^{-8}	C/N (dB) (10^{-8})
64 k	94 k	56 k	67.5 k	-171.0	57.6	
128 k	188 k	112 k	135.0 k	-168.8	60.6	
256 k	376 k	224 k	270.0 k	-165.0	63.6	
384 k	564 k	336 k	405.0 k	-163.2	65.3	10.1
512 k	752 k	448 k	540.0 k	-162.0	66.6	
768 k	1128 k	672 k	810.0 k	-160.2	68.4	
1.544 M	2.3 M	1.38 M	1.625 M	-157.1	71.5	
2.048 M	3.0 M	1.80 M	2.125 M	-155.9	72.7	
4.096 M	6.0 M	3.60 M	4.250 M	-152.9	75.7	
6.312 M	9.3 M	5.58 M	6.500 M	-151.1	77.5	
8.448 M	12.4 M	7.44 M	8.750 M	-149.8	78.9	
Notes 3 and 6	Note 4		Note 5			

[a] For cross-strapped K to C band or C to K band links the BER will be better than or equal to 10^{-6} for all but 1% of the year. This means that under clear sky conditions these carriers will be operated with enough margin to ensure that the 1% criterion is met. (Clear sky condition for the cross-strapped links is expected to be 10^{-8} or better.) The C/N, C/T, and C/N$_0$ values for 10^{-6} are 1.1 dB less than those shown for 10^{-8}.

Notes:

1. Carrier sizes not conforming to the reference units are acceptable under the conditions stated in paragraph 2.1 of INTELSAT document BG-58-95E W/3/84.

2. C/T, C/N$_0$, and C/N values have been calculated with Rate 3/4 FEC as a baseline. The assumed information rate E_b/N_0 is 9.2 dB for a BER of 10^{-8}.

3. The maximum information rate applies to the data rate before 10% overhead and encoding. It is assumed that earth station owners may add up to 10% overhead bits to their information rates for such functions as signaling, alarms, earth station control, encryption key control, or voice channels accompanying teleconferencing transmissions. In some cases the information rate may already include inband overhead channels.

4. Transmission rate = information rate plus 10% overhead $\times \frac{4}{3}$.

5. For carrier sizes from 64 kbit/sec to 768 bit/sec the bandwidth allocated to the carrier in the satellite transponder is a multiple of 22.5 kHz. For carrier sizes of 1.544 Mbit/sec and above the bandwidth allocated to the carrier is a multiple of 125 kHz.

6. The technical parameters for the 4.096, 6.312, and 8.448 Mbit/sec carrier sizes are shown here since they may find future application. Initial service is expected with carrier sizes from 64 kbit/sec to 2.048 Mbit/sec.

Source. Reference 21.

TABLE 4.22(B). IBS Reference Transmission Parameters for Direct 6 / 4 GHz and Cross-Strapped Links[a] (INTELSAT VB, Rate 1 / 2 FEC)

Maximum Information Rate (bit/sec)	Transmission Rate (bit/sec)	Occupied Bandwidth Unit (Hertz)	Allocated Bandwidth Unit (Hertz)	C/T (dBW/K) (10^{-8})	C/N_0 (dB/Hz) (10^{-8})	C/N (dB) (10^{-8})
64 k	141 k	85 k	112.5 k	−172.4	56.2	
128 k	282 k	170 k	202.5 k	−169.4	59.2	
256 k	564 k	340 k	405.0 k	−166.4	62.2	
284 k	846 k	510 k	607.5 k	−164.7	63.9	6.8
512 k	1128 k	680 k	900.0 k	−163.4	65.2	
768 k	1692 k	1020 k	1350.0 k	−161.6	67.0	
1.544 M	3.4 M	2.0 M	2.500 M	−158.7	69.9	
2.048 M	4.5 M	2.7 M	3.250 M	−157.4	71.2	
4.096 M	9.0 M	5.4 M	6.375 M	−154.4	74.2	
6.312 M	13.9 M	8.34 M	9.875 M	−152.5	76.1	
8.448 M	18.6 M	11.16 M	13.00 M	−151.3	77.3	
Notes 3 and 6	Note 3		Note 5			

[a] For cross-strapped K to C band or C to K band links the BER will be better than or equal to 10^{-6} for all but 1% of the year. This means that under clear sky conditions these carriers will be operated with enough margin to ensure that the 1% criterion is met. (Clear sky conditions for the cross-strapped links is expected to be 10^{-8} or better.) The C/N, C/T, and C/N_0 values for 10^{-6} are 1.6 dB less than those shown for 10^{-8}.

Notes:

1. Carrier sizes not conforming to the reference units are acceptable under the conditions stated in paragraph 2.1 of INTELSAT document BG-58-95E W/3/84.

2. C/T, C/N_0, and C/N values have been calculated with Rate 1/2 FEC as a baseline. The assumed information rate E_b/N_0 is 7.6 dB for a BER of 10^{-8}.

3. The maximum information rate applies to the data rate before 10% overhead and encoding. It is assumed that earth station owners may add up to 10% overhead bits to their information rates for such functions as signaling, alarms, earth station control, encryption key control, or voice channels accompanying teleconferencing transmissions. In some cases the information rate may already include inband overhead channels.

4. Transmission rate = information rate plus 10% overhead × 2.

5. For carrier sizes from 64 kbit/sec to 768 bit/sec the bandwidth allocated to the carrier in the satellite transponder is a multiple of 22.5 kHz. For carrier sizes of 1.544 Mbit/sec and above the bandwidth allocated to the carrier is a multiple of 125 kHz.

6. The technical parameters for the 4.096, 6.312, and 8.448 Mbit/sec carrier sizes are shown here since they may find future application. Initial service is expected with carrier sizes from 64 kbit/sec to 2.048 Mbit/sec.

Source. Reference 21.

not exceeded. If the user desires more powerful FEC schemes, the information throughput data rate may be correspondingly reduced for a given transmission rate.

4.5.8 Revised INTELSAT Standards

In March of 1986, INTELSAT released revised standards for Standard A and C earth stations. These revisions are briefly described in the following text.

4.5.8.1 Standard A*

Gain-To-Noise Temperature Ratio. Approval of an earth station in the Standard A category will only be obtained if the following minimum condition is met in the direction of the satellite for the required polarizations for each satellite series under clear sky conditions, in light wind and for any frequency in the band 3.7 to 4.2 MHz.

$$G/T = 35.0 + 20\log(f/4) \tag{4-43}$$

where G is the antenna gain measured at the input to the LNA relative to an isotropic radiator; T is the system noise temperature also referred to the input of the LNA, relative to 1 Hz; and f is the receive frequency in GHz.

Existing Antennas, Transmit Sidelobes. At angles greater than 1° away from the main beam axis, no more than 10 percent of the sidelobe peaks in the copolarized and cross-polarized senses is required to exceed an envelope described by the following two expressions:

$$G = 32 - 25\log\theta \text{ dBi}, \quad 1° \leq \theta \leq 48° \tag{4-44a}$$

$$G = 10 \text{ dBi}\theta > 48° \tag{4-44b}$$

where G is the gain of the sidelobe envelope relative to an isotropic antenna in the direction of the geostationary orbit in dBi; θ is the angle in degrees between the main beam axis and the direction considered; and 10 percent is taken to apply to the total number of peaks with the orbital boundaries defined in CCIR Rec. 580 and at any frequency within the transmit feed system bandwidth, as defined in the following text.

Existing Antennas, Receive Sidelobes (Recommended, not mandatory). The following copolarized and cross-polarized sidelobe characteristics apply to the receive band. Interference protection will be afforded only to the following

---

*Section 4.5.8.1 has been extracted from INTELSAT IESS 201 (Rev. 1) of March 1986.

envelope described by the envelope peaks (CCIR Rec. 465-1):

$$G = 32 - 25 \log \theta \text{ dBi}, \quad 1° \leq \theta \leq 48° \tag{4-45a}$$

$$G = -10 \text{ dBi} \, \theta > 48° \tag{4-45b}$$

New Antennas, Transmit Sidelobes (After March 1986). At angles greater than 1° away from the main beam axis, no more than 10 percent of the sidelobe peaks in the copolarized and cross-polarized senses can exceed an envelope described by the following four expressions:

$$G = 29 - 25 \log \theta \text{ dBi} \quad 1° \leq \theta \leq 20° \tag{4-46a}$$

$$G = -3.5 \text{ dBi} \quad 20° < \theta \leq 26.3° \tag{4-46b}$$

$$G = 32 - 25 \log \theta \text{ dBi} \quad 26.3° < \theta \leq 48° \tag{4-46c}$$

$$G = -10 \text{ dBi} \quad \theta > 48° \tag{4-46d}$$

where G, θ and the 10 percent factor are defined previously.

New Antennas, Receive Sidelobes (Recommended, not mandatory). Interference protection will be afforded only to the following envelope described by the sidelobe peaks:

$$G = 29 - 25 \log \theta \text{ dBi} \quad 1° \leq \theta \leq 20° \tag{4-47a}$$

$$G = -3.5 \text{ dBi} \quad 20° < \theta \leq 26.3° \tag{4-47b}$$

$$G = 32 - 25 \log \theta \text{ dBi} \quad 26.3° < \theta \leq 48° \tag{4-47c}$$

$$G = -10 \text{ dBi} \quad \theta > 48°$$

where G and θ are described previously.

Polarization. INTELSAT IVA satellites require left-hand circular polarization for transmission from the earth station and right-hand circular polarization for reception by the earth station. (Note: Senses of polarization are defined in ITU Radio Regulation 1-18, No. 148 and 149).

Earth station polarization requirements for operation with INTELSAT V, VA, VA (IBS) and VI satellites in the 6/4 GHz frequency bands are shown in Table 4.23.

Axial Ratio For INTELSAT IVA. The voltage axial ratio of transmission in the direction of the satellite shall not exceed 1.4. It is recommended that this ratio not be exceeded for reception.

TABLE 4.23. Earth Station Polarization Requirements to Operate with INTELSAT V, VA, VA(IBS), and VI Satellites (6 / 4 GHz)

Coverage	INTELSAT V		INTELSAT VA/VA(IBS)		INTELSAT VI	
	Earth Station Transmit	Earth Station Receive	Earth Station Transmit	Earth Station Receive	Earth Station Transmit	Earth Station Receive
1. Global A	LHCP	RHCP	LHCP	RHCP	LHCP	RHCP
2. Global B	–	–	RHCP	LHCP	RHCP	LHCP
3. West Hemisphere	LHCP	RHCP	LHCP	RHCP	LHCP	RHCP
4. East Hemisphere	LHCP	RHCP	LHCP	RHCP	LHCP	RHCP
5. NW Zone (Z1)[a]	RHCP	LHCP	RHCP	LHCP	RHCP	LHCP
6. NE Zone (Z3)[a]	RHCP	LHCP	RHCP	LHCP	RHCP	LHCP
7. SW Zone (Z2)[a]	–	–	–	–	RHCP	LHCP
8. SE Zone (Z4)[a]	–	–	–	–	RHCP	LHCP

[a] Z1, Z2, Z3, Z4 nomenclature applies to INTELSAT VI only.

Note: LHCP = Left-Hand Circularly Polarized; RHCP = Right-Hand Circularly Polarized.

For new antennas it is highly recommended that the axial ratio not exceed 1.06 to accommodate dual polarizations for satellite generations succeeding INTELSAT IVA.

For those earth stations that must be placed in global beam transponders of INTELSAT V because their geographical location restricts placement in a frequency reuse beam, the voltage axial ratio shall not exceed 1.4. It is recommended that this ratio not be exceeded for reception.

For INTELSAT VA, VA(IBS) and VI satellites, the voltage axial ratio for transmission shall not exceed 1.06, except for retrofitted antennas in frequency reuse operation which may have an axial ratio as great as 1.09. All new antennas shall have an axial ratio no higher than 1.06.

Satellite Station-Keeping Limits. For satellites IVA, V, VA and VA(IBS): North–South: $\pm 0.1°$; East–West: $\pm 0.1°$; satellite VI (provisional): North–South: $\pm 0.02°$; East–West: $\pm 0.06°$.

Satellite Beacon Frequencies. For satellites IVA and V: 3947.5 and 3952.5 MHz; for satellites VA, VA(IBS), and VI: 3947.5, 3948.0, 3952.0, and 3952.5 MHz.

Only two of the four beacons on each satellite INTELSAT VA, VA(IBS), and VI can be operated simultaneously, one of the lower frequency pair and one of the higher frequency pair.

Feed System Bandwidth. Receiving feed system (required) 3.700 to 4.200 GHz, design goal: 3.625 to 4.200 GHz; transmitting feed system (required) 5.825 to 6.425 GHz, design goal: 5.850 to 6.425 GHz.

Total Receive Power Flux Density. The maximum expected total receive power flux densities for earth stations are shown in the following list. These represent the worst case configurations of any combination of global, zone, hemispheric, and 4 GHz spot beams as they would apply to a given spacecraft series.

Satellite	A Pol Beams	B Pol Beams
	(dBW/m^2)	(dBW/m^2)
INTELSAT IVA	− 124.3	—
INTELSAT V	− 130.2	− 132.0
INTELSAT VA	− 122.4	− 122.4
INTELSAT VI	− 123.5	− 124.0

where A-pol = A polarization (RHCP), which includes global, hemispheric, and 4 GHz spot beams

B-pol = B polarization (LHCP), which includes global, zone, and 4 GHz spot beams

4.5.8.2 Standard C *

Gain-To-Noise Temperature Ratio. Approval of Standard C earth station C will only be obtained if the following conditions are met for operation in the direction of the satellite.

Condition 1. This condition applies to clear sky operation, in light wind and for the frequency bands 10.95 to 11.2 GHz and 11.45 to 11.7 GHz and has the following requirement:

$$G/T = 37.0 + 20\log(f/11.2) \text{ dB/K} \tag{4-48}$$

where G is the receiving antenna gain referred to the input of the LNA relative to an isotropic radiator; T is the receiving system noise temperature referred to the input of the LNA relative to 1 Kelvin; and f is the frequency in GHz. Clear sky is taken to be the condition of intrinsic atmospheric attenuation due to gases and water vapor without excess attenuation due to rain and snow.

Condition 2. This condition applies to operation under degraded weather. After reviewing local rain statistics and comparing them against the reference downlink degradation margins given in Table 4.24, it can be determined whether or not Condition 2 is satisfied. If the local weather statistics show that the downlink degradation will exceed these values, then a G/T value greater than that given in Condition 1 is needed, according to this equation:

$$G/T = 37.0 + 20\log(f/11.2) + X\,\text{dB/K} \tag{4-49}$$

* Section 4.5.8.2 has been extracted from INTELSAT document IESS-203 (Rev. 1), March 1986.

TABLE 4.24. Reference Downlink Degradation Margin for G/T Determination[a]

Satellite Orbital Location	Earth Station Location in Satellite Beam	Reference Downlink Degradation Margin (dB)	Percentage of Year Margin can be Exceeded (%)
325.5° to 341.5° and	West Spot	13	0.03
174° to 180°E	East Spot	11	0.01
307.0° to 310.0° E	West Spot	13	0.02
	East Spot	11	0.02

[a] The margins shown are those available under clear sky conditions.

where G, T, and f are defined for Condition 1. X is the excess of the downlink degradation predicted by local rain statistics over the reference downlink degradation margin from Table 4.24 for the same percentage of time. Downlink degradation is defined as the sum of the precipitation attenuation (in dB) and the increase in the receiving system noise temperature (in dB) for the given percentage of time.

Consideration should be given to the fact that an increase in antenna gain translates dB-per-dB into G/T required for Condition 2 (degraded case), while a decrease in the low noise temperature amplifier is less effective in compensating for the downlink degradation (i.e., 1 dB decrease in the LNA temperature produces less than 1 dB increase in the G/T required by Condition 2.) These factors should be considered in the trade-off between antenna size and LNA temperature in relation to downlink degradation.

Existing Antennas, Transmit Sidelobes. At angles greater than 1° away from the main-beam axis, no more than 10 percent of the sidelobe peaks in the copolarized and cross-polarized senses can exceed an envelope described by the following two expressions:

$$G = 32 - 25 \log \theta \text{ dBi}, \qquad 1° \leq \theta \leq 48° \qquad (4\text{-}48a)$$

$$G = -10 \text{ dBi} \qquad\qquad \theta > 48° \qquad (4\text{-}48b)$$

where G and θ are described in Section 4.5.8.1 (equation 4-44). Ten percent is taken to apply to the total number of peaks within the orbital boundaries defined by CCIR Rec. 580 and at any frequency within the transmit frequency band of 14.0 to 14.5 GHz.

Existing Antennas, Receive Sidelobes. Are described in Section 4.5.8.1.

New Antennas, Transmit Sidelobes. At angles greater than 1° away from the main-beam axis, no more than 10 percent of the sidelobe peaks in the copolarized and cross-polarized sense can exceed an envelope described by the following two expressions:

$$G = 32 - 25 \log \theta \text{ dBi}, 1° \leq \theta \leq 48° \qquad (4\text{-}49a)$$

$$G = -10 \text{ dBi} \quad \theta > 48° \qquad (4\text{-}49b)$$

where G, θ, and the 10 percent factor are defined in Section 4.5.8.1. At angles between 1° and 6° from the main beam axis, it is recommended that no peak exceed the envelope described by $G = 32 - 25 \log \theta$ dBi.

New Antennas, Receive Sidelobes. Interference protection will be afforded only to the following envelope described by the sidelobe peaks (CCIR Rec. 465-1):

$$G = 32 - 25 \log \theta \text{ dBi} \quad 1° \leq \theta \leq 48° \qquad (4\text{-}50a)$$

$$G = -10 \text{ dBi} \quad \theta > 48° \qquad (4\text{-}50b)$$

where G and θ are described in Section 4.5.8.1. This is not a mandatory requirement imposed by INTELSAT.

Polarization. In the 14/11 GHz frequency band the INTELSAT, V, VA, and VI, satellites, polarization is linear where the beam generating the east coverage is orthogonal to the beam generating the west coverage. The transmission and reception polarizations for a given coverage will be orthogonal. It is possible to match the spacecraft polarization angle within 1 degree under clear weather conditions. It is assumed that earth stations will have an elliptical polarization pattern that will meet the axial ratio specified as follows.

Although the INTELSAT V, VA, and VI spacecrafts do not employ dual polarization in the 14/11 GHz frequency bands, consideration should be given to the possible future use of dual linear polarization.

Axial Ratio. The voltage axial ratio of transmission in the direction of the satellite shall exceed 31.6 (30 dB polarization discrimination everywhere within a cone centered on the main beam axis with a cone angle defined by the pointing error.) INTELSAT recommends that this ratio be exceeded for reception.

Satellite Beacon Characteristics. Beacon transmit frequencies for IN-TELSAT V, VA, and VI satellites are 11.198 and 11.452 MHz. The beacons are unmodulated and the long-term frequency stability is 3 parts in 10^6. Beacon carriers are right-hand circularly polarized with a voltage axial ratio not exceeding 1.12 for INTELSAT V and VA and 1.03 for INTELSAT VI. The nominal beacon EIRP is +6 dBW minimum. However, earth stations should be designed to operate satisfactorily with a beacon EIRP of −10 dBW as may occur under certain contingency circumstances.

Feed System Bandwidth, Receive. 10.95 to 11.20 GHz and 11.45 to 11.70 GHz. Transmit Feed System: 14.00 to 14.50 GHz

Receiving System Bandwidth. Low noise receiving equipment shall be designed to receive any carrier in the bands 10.95 to 11.2 GHz and 11.45 to 11.7 GHz.

Transmitting System Bandwidth. The transmitting system shall be capable of transmitting one or more carriers simultaneously anywhere within the frequency band of 14.0 to 14.5 GHz.

Total Receive Power Flux Density. The maximum receive power flux density that may be expected at an earth station is:

Satellite	11 GHz Spot Beam (Linear)
INTELSAT V	−116.3 dBW/m^2
INTELSAT VA	−112.3 dBW/m^2
INTELSAT VI	−110.4 dBW/m^2

4.6 DOMESTIC AND REGIONAL SATELLITE SYSTEMS

4.6.1 Introduction

Domestic or regional satellite communication systems are attractive to regions of the world with a large community of interest. These regions may be just one country with a comparatively large geographical expanse such as the United States, India, Indonesia, Mexico, or the Soviet Union, or a group of countries with a common interest such as Europe or a common culture such as the Arab countries, or a common language such as Hispanic America. Such systems also serve areas that are sparsely populated to bring in quality telephone service and TV programming. Canada's TeleSat is a good example. Still another family of systems serves the business community, providing basically a digital offering. There are two approaches to the latter business systems:

1. DTU or direct-to-user (sometimes called CPE or customer premise equipment).

2. Trunking systems, where a city may have a "teleport" or central earth station, which is connected to business premises by the standard local telephone connection, usually by wire pair.

Businesses are also establishing private networks by leasing all or a portion of a satellite transponder.

4.6.2 Rationale

The basic, underlying guideline for any communication system is cost effectiveness. As a general rule to achieve a cost-effective system is that for a low-population terminal segment, less investment is placed in the space segment and more in the terminal segment. On the other hand, for a high-population terminal segment, more investment is placed in the space segment, allowing us to reduce the cost of each terminal in the terminal segment.

Generally speaking, domestic and regional satellite systems fall into the latter category, where the terminal segment is fairly highly populated to very highly populated. An example of this latter is Direct Broadcast Satellite (television), where it would be hoped that every home would have a terminal.

4.6.3 Approaches to Cost Reduction

Terminal cost reduction can be achieved by:

☐ Reducing performance
☐ Eliminating connecting links
☐ Reduce/optimize bandwidth
☐ Augmenting the space segment
☐ Mass production of terminals

There are a number of measures of performance. There are those that seriously impact customer satisfaction and, of course, should not be reduced or remain unmodified. There are others that have less impact such as link availability, small reductions in S/N and possibly BER, small increases in postdial delay, and increases in blockage probability.

Link performance can be reduced without impacting subscribers when connecting links can be eliminated. This is certainly the case for DTU systems, where the local connecting links to the nearest earth station are eliminated.

As an example, INTELSAT links (i.e., ground–satellite–ground) are built to a 10,000 pWp specification for analog voice channel service. This noise value is based on the CCIR 2500-km hypothetical reference circuit (HRC). The HRC value is based on the CCITT reference connection of no more than 12 links in tandem. Here there is one, or possibly two, links in tandem; 50,000 pWp may be a more appropriate value. An analogous reasoning can be shown for BER.

If we consider the INTELSAT Standard A earth station as a model, then the major cost item is the antenna system with its autotracking feature. Reduction of antenna aperture can reduce cost by a power law relationship. As the aperture size reduces, beamwidth increases. Depending on station keeping of a satellite in question, there is some point where autotracking can be eliminated entirely, and the station operator need only trim up the antenna pointing periodically—daily, weekly, monthly, or even annually. At this point, dramatic savings on earth station first cost can be achieved.

Without compromising performance, antenna aperture reductions can be realized in several ways. The Standard A earth station must accommodate the entire 500-MHz bandwidth of the satellite. Suppose that a particular domestic earth station were only required to accommodate a 36-MHz transponder rather than the entire 500-MHz bandwidth encompassing all transponders. This can permit a direct decibel for decibel reduction. A rule of thumb is that if we reduce gain requirements of an antenna by 6 dB, reflector diameter can be cut in half; by 3 dB, 25%; etc. In this case the gain reduction is $10 \log_{10}(500/36)$ or 11.4 dB. The 30-m dish can be reduced to about 8 m.

A similar rationale can be used by beefing up the space segment. For every decibel of improvement in satellite G/T, earth station EIRP can be reduced by the same amount for a given performance. Likewise, for every decibel improvement of satellite EIRP, earth station G/T can be reduced by the same amount. Of course, satellite EIRP has a limit if we are to follow CCIR recommendations, as outlined in Section 4.3.5.

4.6.4 Minimum Aperture

4.6.4.1 Baseline Model

We use the standard approach to determine minimum aperture antenna for TV service, various capacity FDM implementations, and SCPC. This approach involves a baseline model where system parameters can be varied to meet specific requirements such as SNR, noise in the voice channel, link availability, and so forth. Only the downlink will be considered. It should be kept in mind that most systems are downlink driven. In other words, the antenna size will be determined more by the downlink than the uplink.

For the baseline model the satellite is of the "bent pipe" variety with an EIRP of $+30$ dBW. The downlink is in the 4-GHz band. The earth terminal receiving system noise temperature (T_{sys}) is 150 K, and the range to the geostationary satellite is 22,208 nautical miles or 25,573 statute miles. The free space loss is [from equation (1.7a)]

$$\mathrm{FSL}_{dB} = 36.58 + 20 \log 4000 + 20 \log 25{,}573$$

$$= 196.78 \text{ dB}$$

Let us now set up an abbreviated link budget for the downlink and let the G/T of the earth terminal equal zero:

Satellite EIRP	+30 dBW
Free space loss	−196.78 dB
Miscellaneous losses*	−2.0 dB

Isotropic receive level	−168.78 dBW
Terminal G/T	0.0

Sum	−168.78 dBW
Boltzmann's constant	−(−228.6) dBW

C/N$_0$	59.82 dB

The gain of a uniformly illuminated parabolic dish antenna, which has a diameter of D feet operating at a frequency of F MHz can be expressed by the formula

$$G_{dB} = 20 \log F_{MHz} + 20 \log D_{ft} + 10 \log \eta - 49.92 \text{ dB} \qquad (4.51)$$

where η is the aperture efficiency expressed as a percentage decimal. The value of η is usually in the range of 0.50–0.85. The value of G will be taken from the term G/T. Once G, the antenna gain in decibels is derived, we can then calculate the antenna diameter in feet from equation (4.51).

In the preceding initial link budget, the G/T value was set at 0 dB/K. A reasonable value for T_{sys} at 4 GHz would be 150 K. In this case we use an economic LNA with a noise temperature on the order of 100–110 K. We now derive a value for G in decibels; from equation (4.27)

$$\frac{G}{T} = G_{dB} - 10 \log 150$$

$$0 = G_{dB} - 21.76 \text{ dB}$$

$$G_{dB} = 21.76 \text{ dB}$$

*Miscellaneous losses: 0.5 dB pointing loss, 1.0 dB gaseous absorption losses, and 0.5 dB polarization losses.

This value is now substituted into the left-hand side of equation (4.51)

$$21.76 \text{ dB} = 20 \log(4000) + 20 \log D_{ft} + 10 \log \eta - 49.92 \text{ dB}$$

Let η have the value 55% or 0.55. This is the conventional value for antenna aperture efficiency for run-of-the-mill parabolic dish antennas. Now solve for D:

$$20 \log D_{ft} = 21.76 \text{ dB} + 49.92 \text{ dB} - 72.04 \text{ dB} - 10 \log 0.55$$

$$= 2.236$$

$$D = 1.29 \text{ ft}$$

This tells us that a 1.29-ft antenna will provide a 59.82 dB C/N_0 at 4000 MHz at a 5° elevation angle.

4.6.4.2 Video Systems

First we examine the case for video downlink at 4 GHz. Using equation (4.36) with a receiver bandwidth of 16.8 MHz (sometimes referred to as half-transponder TV), a $C/N = 10$ dB will derive a signal-to-noise ratio in the video baseband of about 42 dB. This is sufficient S/N for home-quality TV. The $C/N = 10$ dB value assumes FM improvement threshold.

If $C/N = 10$ dB, what is C/N_0? The C/N value is taken in a 16.8-MHz bandwidth and the C/N_0 bandwidth, by definition, is 1 Hz. To calculate the C/N_0 value we add $10 \log(16.8 \times 10^6)$ to the C/N value or $C/N_0 = C/N + 10 \log(16.8 \times 10^6)$. Then $C/N_0 = 10 + 72.25$ or 82.25 dB.

The objective is to determine the antenna aperture diameter. To do this, compare this C/N_0 value with that derived in the model downlink budget. In that link budget we saw that a $G/T = 0$ produces a C/N_0 of 59.82 dB with no link margin. To achieve the 82.25 dB value, the G/T must be $82.25 - 59.82 = +22.43$ dB/K.

T_{sys} is held at 150 K, then

$$\frac{G}{T} = G - 10 \log 150$$

and substituting the preceding G/T value,

$$22.43 = G - 21.76$$

$$G = 44.19 \text{ dB}$$

Calculate the antenna diameter using equation (4.51) with a conventional antenna efficiency value (η) of 0.55 or 55%:

$$44.19 \text{ dB} = 20 \log 4000 + 20 \log D_{ft} + 10 \log 0.55 - 49.92 \text{ dB}$$

$$20 \log D_{ft} = 24.67$$

$$D = 17.1 \text{ ft}$$

Now the question arises of how we can reduce the diameter still further. Examine the model. There are several steps that can be taken.

On a first critique of the preceding analysis, three areas of improvement can be suggested:

- ☐ Improve antenna aperture efficiency
- ☐ Increase satellite EIRP
- ☐ Increase minimum elevation angle

We keep the receiving system noise temperature at 150 K and improve the antenna efficiency to 0.65, which can be done today at very small additional cost. There will then be a decibel for decibel impact on the G/T value as shown below. We still require a G/T of 22.43 dB/K, and the resulting antenna system gain requirement is 44.19 dB. We carry out the same exercise as the preceding one to calculate D, the antenna diameter, but use $10 \log 0.65$ rather than $10 \log 0.55$:

$$44.19 = 20 \log 4000 + 20 \log D_{ft} + 10 \log 0.65 - 49.92$$

$$20 \log D_{ft} = 23.94$$

$$D = 15.74 \text{ ft}$$

Any further increase in antenna efficiency would probably increase antenna cost considerably. The objective is low cost.

If we increase the EIRP to $+34$ dBW, using the improved antenna improves the C/N_0 value by 4 dB ($+34$ dBW $- 30$ dBW $= 4$ dB). This can be shown to be true by examination of the model link budget. With a $G/T = 0$ and no link margin, the new $C/N_0 = 63.82$ dB. We still require 82.25 dB C/N_0 for the video link. The new G/T value is then 82.25 dB $- 63.82$ dB $= 18.43$ dB/K. Calculate G:

$$\frac{G}{T} = G - 10 \log 150$$

$$18.43 \text{ dB/K} = G - 10 \log 150$$

$$G = 40.19 \text{ dB}$$

Calculate the antenna aperture diameter from equation (4.51):

$$40.19 \text{ dB} = 20 \log 4000 + 20 \log D_{ft} + 10 \log 0.65 - 49.92$$

$$20 \log D_{ft} = 19.95$$

$$D = 9.9 \text{ ft} \quad (3.02 \text{ m})$$

By increasing the minimum elevation angle to say 25°, three improvements occur: range to the satellite is reduced, gaseous absorption is reduced to about 0.3 dB, and antenna noise is reduced such that T_{sys} is about 125 K rather than 150 K, as before.

The range improvement (range now equals 24,295 statute miles) is about 0.45 dB; the gaseous absorption improvement is about 0.7 dB; and the resultant noise improvement is 0.79 dB. The total improvement is about 1.94 dB, and the antenna gain can be reduced by this amount. The new antenna gain requirement is 38.25 dB. Turning again the equation (4.51):

$$38.25 = 20 \log 4000 + 20 \log D_{ft} + 10 \log 0.65 - 49.92$$

$$20 \log D_{ft} = 18.0$$

$$D = 7.94 \text{ ft} \quad (2.42 \text{ m})$$

This is the derived value for minimum aperture for TV service for half-transponder with a weighted $S/N = 42$ dB in the baseband.

It will be observed that we are analyzing only the downlink. What about the uplink? Equation (4.33) shows that with a bent pipe satellite the downlink does not operate alone. The received signal-to-noise ratio at an earth terminal is also affected by the uplink S/N and the IM products generated in the satellite transponder by multicarrier operation. In the preceding argument and those that follow, it is assumed that uplink S/N is at least 10 dB better than the downlink S/N. IM products in the satellite transponder have been neglected.

4.6.4.3 Telephony

The objective in this section is to determine minimum aperture for telephony service using a 4-GHz downlink. Voice channel noise at the receiving earth terminal is relaxed to 50,000 pWp from the 10,000 pWp value that is the basis of the CCIR 2500-km hypothetical reference circuit and is also specified for INTELSAT service.

Consider an SCPC system on a 36-MHz transponder where half (18 MHz) is devoted to the downlink and the other half to the uplink. Channel spacing is 45 kHz. The EIRP per channel is +6 dBW; the peak channel deviation is 12.2 kHz; and the predetection noise bandwidth based on Carson's rule is 31.2 kHz with a 3.4-kHz voice channel. Assume a 25° elevation angle. From these parameters we derive a model link budget shown below:

Satellite EIRP	+6 dBW
Free space loss	−196.33 dB
Miscellaneous losses*	−1.5 dB
Isotropic receive level	−191.83 dBW
Terminal G/T	0.0 dB
Sum	−191.83 dBW
Boltzmann's constant	−(−228.6) dBW
C/N_0	36.77 dB/Hz

The next step is to find the required C/N_0 for an SCPC channel. With that value, using the preceding model, the required G/T can be established. We will assume T_{sys} is again 125 K, and now G can be calculated easily. Once we have a value of G, the antenna aperture diameter is calculated using equation (4.51). Reference 4 relates S/N to C/N_0 for an SCPC FM channel by

$$\frac{S}{N} = \frac{C}{N_0} + 10\log\left[3\left(\frac{\Delta F}{f_m}\right)^2\right] - 10\log(2B_a) + P + C \qquad (4.52)$$

where ΔF = peak deviation (12.2 kHz)

$\quad f_m$ = highest baseband frequency (3.4) kHz

$\quad B_a$ = baseband noise bandwidth (3.4 kHz)

$\quad P$ = pre-emphasis advantage (6.3 dB)

$\quad C$ = companding advantage (17.0 dB)

The pre-emphasis advantage and companding advantage values are suggested in Ref. 4. Also see equation (4.39) relating S/N to C/N, not to C/N_0.

---

*Miscellaneous losses: 0.5 dB pointing loss, 0.5 dB polarization loss, and 0.5 dB atmospheric gaseous loss.

For 50,000 pWp noise value in the voice channel, a S/N of approximately 43 dB is required. Substitute the 43-dB value for S/N in equation (4.52):

$$43 \text{ dB} = \frac{C}{N_0} + 10 \log\left[3\left(12.2 \times 10^3/3.4 \times 10^3\right)^2\right] - 10 \log 6800 + 6.3 + 17.0$$

$$\frac{C}{N_0} = 42.17 \text{ dB}$$

If C/N_0 is 42.17 dB, what is C/N? Test this value for FM improvement threshold. The predetection noise bandwidth is 31.2 kHz. Take 10 log of this value ($10 \log 32{,}100 = 45.06$ dB). Then $C/N = C/N_0 - 45.06$ dB or -2.89 dB. To reach FM improvement threshold the C/N value must be equal to or greater than 10 dB. To reach FM improvement threshold, we are short $10 + 2.89$ dB or 12.89 dB. The new C/N_0 value is 42.17 dB + 12.89 dB or 55.07 dB.

Subtract the value in the model link budget for C/N_0 where $G/T = 0$ dB/K to derive the required minimum G/T:

$$55.07 - 36.77 = 18.3 \text{ dB/K}$$

and $G/T = 18.3$ dB/K.

Calculate the value G from G/T assuming again that T_{sys} is 125 K:

$$\frac{G}{T} = G - 10 \log 125$$

$$18.3 \text{ dB/K} = G - 20.97$$

$$G = 39.27 \text{ dB}$$

Again assume a 65% efficiency antenna and calculate the aperture diameter from equation (4.51).

$$39.27 = 20 \log 4000 + 20 \log D_{ft} + 10 \log 0.65 - 49.92$$

$$20 \log D_{ft} = 19.01$$

$$D = 8.93 \text{ ft} \quad (2.72 \text{ m})$$

Note that the antenna aperture is driven by the low satellite EIRP.

Let us consider the case of FDM/FM/FDMA operation where the earth terminal receives an entire 36-MHz transponder. Operation is again in the 4-GHz band, and its operation will be trunking. This means that the facility is a relay point of a larger network. Thus, we will allow only 10,000 pWp of noise in the derived receive voice channel. Again we assume that the uplink C/N is at least 10 dB greater than the downlink. This permits us to look at the downlink in isolation because the uplink S/N will have little effect on its companion downlink. The downlink will have the full transponder EIRP of $+34$ dBW. As before, the initial terminal G/T is set at 0 dB/K. Establish a baseline link budget:

Satellite EIRP	$+34$ dBW	
Free space loss	-196.33 dB	(25° elevation angle)
Miscellaneous losses	-1.5 dB	(as above)

Isotropic receive level	-163.83 dBW	
G/T	0.0 dB/K	
Boltzmann's constant	$-(-228.6$ dBW)	

C/N_0	$+64.77$ dB

We now calculate the required C/N_0 to achieve the noise power in the derived voice channel and will adjust the G/T accordingly. Given 10,000 pWp of noise power, calculate the required S/N to just achieve this noise power using the following formula:

$$\text{Noise power}_{\text{pWp}} = \frac{10^9 \times 0.56}{\log^{-1}(\text{S/N})} \tag{4.53}$$

$$\frac{S}{N} = 47.48 \text{ dB}$$

Turning to equation (4.34) and assume a 972 channel FDM configuration;

$$\frac{S}{N} = \frac{C}{N} + 20\log\left(\frac{\Delta F_{tt}}{f_{ch}}\right) + 10\log\left(\frac{B_{if}}{B_{ch}}\right) + P + W$$

where ΔF_{tt} = rms test tone deviation

f_{ch} = highest voice channel frequency (e.g., highest baseband frequency)

B_{if} = IF bandwidth (e.g., occupied bandwidth)

B_{ch} = voice channel bandwidth (3100 Hz)

P = top channel emphasis in decibels (4 dB)

W = psophometric weighting factor (2.5 dB)

The value for B_{if} will be 36 MHz in this example because the entire 36-MHz transponder is occupied. The top baseband frequency is 4028 kHz (f_{ch}). ΔF_{tt} is 802 kHz. The above values are taken from Table 4.6.

Substitute the S/N value of 47.48 dB into equation (4.34):

$$47.48 \text{ dB} = \frac{C}{N} + 20\log(802 \times 10^3 / 4028 \times 10^3)$$

$$+ 10\log(36 \times 10^6 / 3.1 \times 10^3) + 4 \text{ dB} + 2.5 \text{ dB}$$

$$47.48 = \frac{C}{N} + 33.12$$

$$\frac{C}{N} = 14.35 \text{ dB}$$

Calculate C/N_0 by adding $10\log(36 \times 10^6)$ to the value above:

$$C/N_0 = 14.35 + 10\log(36 \times 10^6)$$

$$C/N_0 = 89.91 \text{ dB}$$

To calculate G/T use the value of C/N_0 obtained in the minilink budget, where $G/T = 0$ dB/K, or $89.91 - 64.77 = 25.14$ dB/K.

Again assume that $T_{sys} = 125$ K and calculate G (antenna system gain) in the G/T identity:

$$\frac{G}{T} = G - 10\log(125 \text{ K})$$

$$25.14 = G - 20.97$$

$$G = 46.11 \text{ dB}$$

Calculate the antenna aperture from equation (4.51) assuming an antenna efficiency of 65%:

$$46.11 \text{ dB} = 20\log 4000 + 20\log D_{ft} = 10\log 0.65 - 49.92$$

$$D = 19.63 \text{ ft} \quad (5.98 \text{ m or approximately 6 m})$$

In the preceding discussions dealing with minimum aperture antennas, we have let the link margins be zero. This is not good practice for an operating installation. Of course, every decibel of margin costs money and may be directly translatable to antenna aperture, improved antenna efficiency, or reduced system noise. We have also left out any margin for rainfall; 1 dB for rainfall at 4 GHz is probably sufficient. If the satellite in question uses a spot beam, EIRPs may be greater if flux density limits on the earth's surface remain in CCIR recommended limits, but then we will have a satellite pointing loss and a footprint loss. This latter, of course, assumes that the earth terminal in question is not located on beam center. Many footprint contours are 3 dB, thus 2 dB may be a good value for this latter loss.

How much will an improved system noise value buy us? Keep G/T constant at 25.14 dB/K and reduce T_{sys} to 100 K. Now the value of $10 \log T$ is 20 rather than the 20.97 dB value used previously. We picked up only 0.97 dB. If we reduced T_{sys} to 80 K, we would pick up 1.94 dB. Much of the noise improvement would be achieved by using a better LNA. This costs more. Of course, there is a cost tradeoff here.

4.7 FREQUENCY PLANNING AND INTERFERENCE CONTROL

4.7.1 Introduction

It was pointed out in Section 4.3.5 that the assigned satellite communication frequency bands are shared with other services, most often a complementary wideband terrestrial radio relay service. Because of this sharing between (among) services, there is the possibility that one service interferes with another. We must then consider two situations:

☐ Terrestrial service interferes with satellite service.
☐ Satellite service interferes with terrestrial service.

In this section we will only consider the analog situation.

Taking guidance from CCIR, the measure of interference is the resulting noise in pW0p (or dBm0p) in the derived 4-kHz voice channel.

We first consider the interference in a satellite system (or systems) from a radio relay (e.g., radiolink LOS) system or systems. CCIR Rec. 356-4 (Ref. 22) recommends the following:

That systems in the Fixed Satellite Service and radio-relay systems sharing the same frequency bands be designed in such a manner that the interference noise power, at a fixed zero relative level in any telephone channel of a hypothetical reference circuit of a system in the Fixed Satellite Service, caused by the aggregate of the transmitters of radio-relay stations, conforming to Recommendation 406-4 (Ref. 23), should not exceed:

☐ 1000 pW0p psophometrically weighted one-minute mean power for 20% of any month
☐ 50,000 pW0p psophometrically weighted one-minute mean power for more than 0.03% of any month.

Next we consider interference from a satellite service (or services) into a radio-relay system (LOS radiolink). Quoting from CCIR Rec. 357-3 (Ref. 24):

> That systems in the Fixed Satellite Service and line-of-sight analog angle-modulated radio-relay systems which share the same frequency bands, should be designed in such a manner, that in any telephone channel of a 2500 km hypothetical reference circuit for frequency division multiplex, analog angle-modulated radio-relay systems, the interference noise power at a point of zero relative level, caused by the aggregate of the emission of earth stations and space stations of the systems of the fixed satellite service, including associated telemetering, telecommand and tracking transmitters, should not exceed:
>
> ☐ 1000 pW0p psophometrically weighted one-minute mean power for more than 20% of any month;
> ☐ 50,000 pW0p psophometrically weighted one-minute mean power for more than 0.01% of any month.

4.7.2 Conceptual Approach to Interference Determination

A new earth station is to be installed. We know its location, its antenna elevation angle and azimuth, and its radiation patterns for its passband. We will carry out a survey of nearby radio-relay facilities that are within LOS of the new earth station facility operating in the same band.

These facilities have been licensed by a national regulatory agency and the characteristics of each emitter are known. For a sample emitter, we need to know its location, height above sea level, azimuth and elevation angle of its main beam, EIRP of main beam and antenna sidelobe characteristics, and the azimuth and elevation angle relative to the new earth station. From the radio-relay emitter we calculate a new EIRP of its antenna sidelobe in the direction of the earth station antenna. The ray beam formed by this sidelobe will probably strike the antenna of the new earth station installation on one of its sidelobes, which displays a gain (or loss) to the receiving system of the new installation. We are now faced with a traditional radiolink path analysis problem. Of course, the potential offender can be eliminated if its derived RSL into the new facility is below the noise threshold of the new facility.

Let us run an example. A potential offender is 4 miles away from a new earth station. The offender transmits a carrier at 4 GHz, and the new earth station receives a carrier from the satellite just at 4-GHz center frequency. The offender has an EIRP of +30 dBW and an antenna sidelobe in the direction of the new earth station antenna that is down 42 dB striking a sidelobe on the earth station antenna that is down 39 dB from main beam which has a gain of 40 dB. Line losses to the earth station LNA are 1.5 dB. The system noise temperature of the earth station is 100 K and its passband is 36 MHz.

First calculate the earth station noise threshold. Equation (2.12), with bandwidth expansion, applies:

$$\text{Noise threshold}_{\text{dBW}} = -228.6 \text{ dBW} + 10\log 100 + 10\log(BW_{\text{Hz}})$$

where BW is the bandwidth of the receiving system in Hz.

$$\text{Noise threshold}_{dBW} = -228.6 \text{ dBW} + 20 \text{ dB} + 10\log(36 \times 10^6)$$

$$= -228.6 + 20 + 75.56$$

$$= -133.04 \text{ dBW}$$

Calculate the isotropic receive level (IRL) of the potential offender on the new earth station antenna (see Section 4.3.4):

$$\text{IRL}_{dBW} = \text{EIRP}_{dBW(0)} - 42 \text{ dB} - \text{FSL}_{dB}$$

where $\text{EIRP}_{(0)}$ is the EIRP in dBW of the potential offender and FSL is the free space loss at 4 GHz between the potential offender and the new earth station.

$$\text{IRL} = +30 \text{ dBW} - 42 \text{ dB} - \text{FSL}$$

Calculate FSL using equation (1.7a):

$$\text{FSL}_{dB} = 36.58 + 20\log 4000 + 20\log 4$$

$$= 120.66 \text{ dB}$$

$$\text{IRL} = +30 \text{ dBW} - 42 \text{ dB} - 120.66 \text{ dB}$$

$$= -132.66 \text{ dBW}$$

Calculate RSL from equation (2.8):

$$\text{RSL} = -132.66 \text{ dBW} - 39 \text{ dB} - 1.5 \text{ dB} + 40 \text{ dB}$$

$$\text{RSL} = -133.16 \text{ dBW}$$

For this example, the RSL is just below noise threshold and the only effect will be to raise the noise threshold ≈ 3 dB. In fact, for FM-operating systems with sufficient deviation to attain the FM capture effect, for offenders with a C/N in the offended system of 10 dB or less can be treated as noise sources.

For a rigorous analysis of interference from one radio system into another, EIA "Telecommunications Systems Bulletin No. 10D" (Ref. 25) is recommended.

4.7.3 Coordination Distance

Coordination distance is defined by the FCC in Part 25 of FCC Rules and Regulations (Ref. 26) as "the distance from an earth station, within which there is a possibility of the use of a given transmitting frequency at this earth station causing harmful interference to stations in the fixed or mobile service, sharing the same band, or the use of a given frequency for reception at this earth station receiving harmful interference from such stations in the fixed or mobile service."

Both the FCC and CCIR treat interference issues. The FCC in Rules and Regulations (Ref. 26) treats "maximum permissible interference power" in part 25.252 and "determination of coordination distance for near great circle propagation mechanisms" in part 25.253.

The methodology used by the FCC (and in essence by CCIR) is to calculate the maximum interference power in dBW in the reference bandwidth of the potentially interfered-with station. And then with that value calculate the normalized basic transmission loss. This loss is then applied to a family of curves for three weather zones (climatic zones) to derive coordination distance. A family of coordination distances around an earth station define a coordination contour.

Returning to our previous section (4.7.2), the FCC (part 25.252) (Ref. 26) offers the following formula to calculate the maximum permissible interference power (P_{max}) (20%) in dBW in the reference bandwidth of the potentially interfered-with station, not to be exceeded for all but 20% of the time from each source of interference:

$$P_{max}(20\%) = 10\log(kT_rB) + J - W - 10\log(n_{20}) \qquad (4.54)$$

where n_{20} = number of assumed simultaneous interference entries of equal power level

T_r = thermal noise temperature of the receiving system in kelvins

$k = -228.6$ dBW

B = reference bandwidth in hertz

J = ratio in decibels of the maximum permissible long-term interfering power to the long-term thermal noise power in the receiving system, where long term refers to 20% of the time

W = equivalence factor in decibels relating to the effect of interference to that of thermal noise of equal power in the reference bandwidth. In the direction terrestrial station to earth station, the FCC recommends that $W = 4$ dB, and in the reverse direction, terrestrial to earth station, the FCC recommends 0 dB. (Reference Table 1, FCC Section 25.252).

The FCC makes the following recommendations for the value of J:

The factor J in dB is defined as the ratio of total permissible long-term (20% of the time) interference power in the system, to the long-term thermal noise power in a single receiver. For example, in a 50-hop terrestrial hypothetical reference circuit, the total allowable additive interference power is 1000 pW0p (CCIR Rec. 357-1) and the mean thermal noise power in a single hop may (then) be assumed to be 25 pW0p. Therefore, since in a FDM/FM system the ratio of interference noise power to thermal noise power in a 4-kHz band is same before and after demodulation, $J = 16$ dB. In a fixed-service satellite system, the total allowable interference power is also 1000 pW0p (CCIR Rec. 356-2) (Ref. 22), but the thermal noise contribution of the down path (downlink) is not likely to exceed 7000 pW0p, hence J equals or is less than 8.5 dB. In digital systems it may be necessary to protect each communication path individually, and in that case, long-term interference power may be of the same order of magnitude as long-term thermal noise, hence $J = 0$ dB.

4.7.4 Some General Measures for Interference Reduction

4.7.4.1 Earth Station Antenna Sidelobes

Comparing potential interference offenders, the earth station has the potential of being the greater offender when compared to its terrestrial (radiolink) counterpart. The rationale is simple. The earth station generates a considerably greater EIRP than a terrestrial radiolink to access a satellite adequately. Obviously, the earth station must overcome free space loss in excess of 196 dB, whereas the terrestrial LOS radiolink must overcome free space losses on the order of 120–145 dB. To reduce earth station interference levels, sidelobe characteristics of earth station antennas have more stringent requirements.

Part 25.209 of the FCC Rules and Regulations (Ref. 26) states:

(a) Any antenna to be employed in transmission at an earth station in the Communication-Satellite Service shall conform to the following standard:

Outside the main beam, the gain of the antenna shall lie below the envelope defined by

$$\begin{array}{ll} 32 - 25 \log_{10}(\theta) \text{ dBi} & 1° \leq \theta \leq 48° \\ -10 \text{ dBi} & 48° \theta \leq 180° \end{array} \qquad (4.55)$$

where θ is the angle in degrees from the axis of the main lobe, and dBi refers to dB relative to an isotropic radiator. For the purposes of this section, the peak gain of an individual sidelobe may be reduced by averaging its peak level with the peaks of the nearest sidelobes on either side, or with the peaks of two nearest sidelobes on either side, provided that the level of no individual sidelobe exceeds the gain envelope given above by more than 6 dB.

4.7.4.2 Earth Station Power Limits

Its a fair generality to say that an earth station has the potential of causing the greatest interference to nearby terrestrial facilities along the horizontal relative to its antenna. FCC Rules and Regulations Section 25.204 (Ref. 26) states that (a) within the band 5925–6425 MHz the mean effective radiated power transmitted in any direction in the horizontal plane by an earth station shall not exceed $+45$ dBW in any 4-kHz band; and (b) within the band 7900–8400

MHz, the mean effective radiated power transmitted in any direction in the horizontal plane by an earth station shall not exceed $+55$ dBW in any 4-kHz band, except upon showing a need for greater power, in which case the maximum of $+65$ dBW may be authorized.

4.7.4.3 Maximum Transmitter Output and Maximum EIRP for Terrestrial Radiolink Facilities

Now we turn in the other direction—interference from terrestrial LOS facilities. The FCC in Section 21.107 (Ref. 27) of the Rules and Regulations limits transmitter rated output power to 20 W in the frequency range 512–10,000 MHz and to 10 W above 10,000 MHz. It further stipulates in a footnote that in the band 5925–6425 MHz that the maximum EIRP of a terrestrial station in the fixed service shall not exceed $+55$ dBW.

CCIR Rec. 406-4 (Ref. 23) provides the following guidance on EIRP maximums for terrestrial LOS radio relay transmitters. For the band 1–10 GHz, the power delivered to the antenna shall not exceed $+13$ dBW and the maximum EIRP shall not exceed $+55$ dBW. For proper system coordination, a review of CCIR Rec. 406-4 is recommended.

PROBLEMS AND EXERCISES

1. Draw a simplified block diagram of a "bent pipe" satellite. Show the local oscillator mixing frequency to derive the current 4-GHz band from the 6-GHz uplink band.

2. List four advantages and four disadvantages of geostationary satellites used for telecommunications relay.

3. What is the range (distance) to a geostationary satellite if the elevation angle for a particular earth station is 27°?

4. Calculate the free space (spreading) loss to a geostationary satellite at 6105 MHz when the elevation angle is 23°.

5. Calculate the isotropic receive level at an earth station if a satellite transponder radiates $+31$ dBW on 7.305 GHz and the elevation angle is 15°. Consider only free space loss.

6. Calculate the flux density in dBW per meter impinging on a satellite antenna if the EIRP of an earth station were $+65$ dBW and the earth station was directly under the satellite (i.e., at the subsatellite point).

7. Downlink signals are generally of low level at the earth's surface.
   a. Give at least two reasons why.
   b. Give at least two general ways in which we can achieve sufficient "sensitivity" at an earth station to satisfactorily utilize these weak signals.

8. What is the thermal noise level in dBW in 1-Hz bandwidth of the theoretically perfect receiver operating at absolute zero.

9. What is the receiving system noise temperature in kelvins when the antenna noise is 105 K and the receiver noise is 163 K.

10. As one lowers the elevation angle of an earth station antenna toward the horizon, what happens to T_{sys}? Give at least two reasons why.

11. The noise figure of a certain LNA is 1.25 dB. What is the effective noise temperature of the LNA in kelvins?

12. A section of waveguide has an ohmic loss of 0.3 dB. What is its (approximate) equivalent noise temperature when inserted in the transmission line system?

13. What is the spectral noise density in 1-Hz bandwidth (N_0) of an earth station receiving system where T_{sys} is 97 K?

14. If the receive signal level (RSL) at the input of a certain LNA is -131 dBW where T_{sys} is 141 K at the same reference point, what, then, is the value of C/N_0?

15. G/T for a particular earth station is given as 40.7 dB/K where G at the reference plane is 61.5 dB. What is T_{sys}?

16. Name at least three components of sky noise.

17. Calculate the antenna noise of an earth station where the elevation angle is $10°$, the operating frequency is 7300 MHz, feed loss is 0.1 dB, waveguide losses are 1.9 dB, and the bandpass filter has an insertion loss of 0.5 dB at 7300 MHz. The reference plane is at the input of the LNA.

18. Calculate T_{sys} from question 17 where the LNA has a noise figure of 1.19 dB. Disregard noise contributions of subsequent receiver stages after the LNA.

19. Calculate receiver noise temperature where an LNA has a noise figure of 1.05 dB and a gain of 30 dB, and the noise temperature of a subsequent postamplifier is 450 K.

20. Calculate the value of G/T of a satellite earth terminal given the following parameters: sky noise is 53 K, total ohmic losses at the reference plane of the antenna subsystem is 1.94 dB, receiver noise figure is 1.65 dB, gross antenna gain is 47 dB. Disregard nonohmic losses.

21. Calculate downlink C/N_0 for an earth coverage antenna of a satellite where the EIRP of that beam is $+34$ dBW, the earth station G/T is 21.5 dB/K, and the total downlink losses are 202 dB.

22. Calculate the required G/T of a satellite where the required uplink C/N_0 at the satellite transponder is 85 dB/Hz, the terminal EIRP is $+62$ dBW, the operating frequency is 6250 MHz, the elevation angle is $21°$; the satellite uses a spot beam and the earth terminal is somewhere inside the 3-dB contour footprint. Use reasonable loss values.

23. What is the system C/N_0 where the uplink C/N_0 is 85 dB; the downlink C/N_0 is 71 dB; and the transponder IM C/N_0 is 96 dB?

24. A satellite FDM/FM/FDMA system with 972 VF channels on one carrier is designed for a noise power value on the downlink VF channel derived output of 10,000 pW0p. Calculate S/N in the derived voice channel where the emphasis improvement factor is 4 dB. Use INTELSAT values.

25. Compare FDMA with TDMA.

26. Using the rule of thumb, how much backoff is required to achieve a C/IM of 24 dB when initially the C/IM value is 16 dB?

27. Where would SCPC DAMA have application?

28. Name the three methods used to control DAMA operation.

29. Why does INTELSAT Standard A earth stations require such a large G/T?

30. On an INTELSAT Standard A earth station system, 80% of the noise in a derived voice channel is allocated to the space segment. Give at least two reasons why explaining the sources of the noise.

31. We are to design a domestic satellite system from scratch. Using INTELSAT Standard A as a model for starting point, name six ways to reduce system cost.

32. What is the minimum aperture antenna assuming 60% antenna efficiency for an SCPC system to receive just one 45-kHz channel and allow 20,000 pWp noise in the derived voice channel. The system operates in the 6/4 GHz band pair. Assume satellite EIRP of +10 dBW and an elevation angle of 25°. Use reasonable loss values.

33. Define a coordination contour. Why go to all this trouble in the first place?

REFERENCES

1. R. L. Freeman, *Reference Manual for Telecommunications Engineering*, Wiley, New York, 1985.
2. *Reference Data for Radio Engineers*, 5th ed., ITT-Howard W. Sams, Indianapolis, IN, 1968.
3. *World Administrative Radio Congress–1979* (WARC-79), ITU, Geneva, 1980.
4. *Satellite Communications Symposium '82*, Scientific-Atlanta, Atlanta, GA, 1982.
5. CCIR Recommendation 358-2, "Recommendations and Reports of the CCIR 1978," XIVth Plenary Assembly, Kyoto, 1978, ITU, Geneva, 1978.
6. *Transmission Systems for Communications*, 5th ed., Bell Telephone Laboratories, Holmdel, NJ 1982.
7. CCIR Report 720, "Recommendations and Reports of the CCIR 1978," XIVth Plenary Assembly, Kyoto, 1978, ITU, Geneva, 1978.
8. R. Pettai, *Noise in Receiving Systems*, Wiley, New York, 1984.

9. INTELSAT (Specification) BG-23-10E W/9/76, INTELSAT, Washington, DC, 1976.

10. CCIR Recommendation 421-3, "Recommendations and Reports of the CCIR 1978," XIVth Plenary Assembly, Kyoto, 1978, ITU, Geneva, 1978.

11. H. L. Van Trees, *Satellite Communications*, IEEE Press, New York, 1979, Section 3.6, "Multiple Access."

12. CCIR Report 384-3, "Recommendations and Reports of the CCIR 1978," XIVth Plenary Assembly, Kyoto, 1978, ITU, Geneva, 1978.

13. R. L. Freeman, *Telecommunication System Engineering*, Wiley, New York, 1980.

14. H. L. Van Trees, *Satellite Communications*, IEEE Press, New York, 1979, Section 3.6.5, "Demand Assignment."

15. H. L. Van Trees, *Satellite Communications*, IEEE Press, New York, 1979, Section 3.6.3.1, "Time-Division Multiple Access."

16. INTELSAT (Specification) BG-53-86 (Rev. 1), INTELSAT, Washington, DC, 1983.

17. INTELSAT (Specification) BG-28-74E M/6/77, INTELSAT, Washington, DC, 1977.

18. INTELSAT (Specification) BG-28-73E M/6/77, INTELSAT, Washington, DC, 1977.

19. INTELSAT (Specification) BG-56-72E W/9/83, INTELSAT, Washington, DC, 1983.

20. CCITT Recommendation G-162, "CCITT Reports and Recommendations," Vol. III, VIIth Plenary Assembly, ITU, Geneva, 1980.

21. INTELSAT (Specification) BG-58-95E W/3/84, INTELSAT, Washington, DC, 1984.

22. CCIR Recommendation 356-4, "Recommendations and Reports of the CCIR 1978," XIVth Plenary Assembly, Kyoto, 1978, ITU, Geneva, 1978.

23. CCIR Recommendation 406-4, "Recommendations and Reports of the CCIR 1978," XIVth Plenary Assembly, Kyoto, 1978, ITU, Geneva, 1978.

24. CCIR Recommendation 357-3, "Recommendations and Reports of the CCIR 1978," XIVth Plenary Assembly, Kyoto, 1978, ITU, Geneva, 1978.

25. *EIA Telecommunication System Bulletin No. 10D*, Electronic Industries Association, Washington, DC, August, 1983.

26. "Rules and Regulations," U.S. Federal Communications Commission, Washington, DC, September, 1982, Part 25.

27. "Rules and Regulations," U.S. Federal Communications Commission, Washington, DC, September, 1982, Part 21.

28. R. C. Dixon, *Spread Spectrum Systems*, 2nd ed., Wiley, New York, 1984.

29. Warren L. Flock, "Propagation Effects on Satellite Systems at Frequencies, Below 10 GHz," Univ. Colorado 1983. Prepared for NASA. NASA Ref. Pub. 1108.

30. INTELSAT (Specification) BG-28-72E M/6/77, INTELSAT, Washington, DC, August, 1977.

31. INTELSAT Earth Station Standard (IESS)-203, INTELSAT, Washington, DC, March, 1986.

32. INTELSAT Earth Station Standard (IESS)-201, INTELSAT, Washington, DC, March, 1986.

DIGITAL COMMUNICATIONS BY SATELLITE

5.1 INTRODUCTION

The world's telecommunication network is evolving to an all-digital network. The evolution process is estimated by some to be completed in under 15 years and by others in over 30 years. Many private/industrial networks are already digital. A number of satellite telecommunication systems now in service are all-digital, such as SBS (Satellite Business Systems) and the U.S. military systems: DSCS and FltSat. INTELSAT has implemented hybrid systems where one or more transponders operate in a digital mode.

In this chapter we will discuss two basic classes of digital satellites. The first is the familiar bent pipe satellite where "what comes down is a reasonable replica of what goes up," but with some degradation. The second class of digital satellite is the processing satellite, where "what comes down is not necessarily a replica of what went up." In this latter case the satellite demodulates and regenerates the uplink signal and may carry out decoding/recoding and various levels of switching. The trend today for advanced satellites is the latter class such as NASA's 30/20 GHz program.

TDMA will be discussed in detail and DSI (digital speech interpolation) will be described. Channel coding with interleaving is often used to augment performance. Several coding approaches are discussed and compared. Digital DAMA systems are reviewed for both basic speech operation and low-data-rate computer networks.

Digital transmission offers many advantages (see Ref. 7, Chapter 11). Regeneration at all digital network nodes and repeaters prevent noise accumulation, a primary concern on analog systems. The principal disadvantages are error accumulation and critical timing requirements.

5.2 DIGITAL OPERATION OF A BENT PIPE
SATELLITE SYSTEM

5.2.1 General

There are two approaches to digital bent pipe satellite operation:

□ FDMA mode
□ TDMA mode

In either case the satellite does not regenerate the digital signal.

5.2.2 Digital FDMA Operation

An example of digital bent pipe operation is DSCS (Defense Satellite Communication System) on main transponders, which operate in an FDMA mode (except transponder 1, which operates in a CDMA mode using direct sequence spread spectrum modulation). Rather than place an FDM/FM signal in a preassigned transponder frequency slot as discussed in Chapter 4, a digital waveform is transmitted in that segment using modulation techniques covered in Section 2.11.3. Because there is no signal regeneration in the satellite transponder, the downlink signal suffers a certain amount of distortion due to the satellite. For one thing, there is additive thermal noise, and when there is multichannel activity on the same transponder, IM noise can be a major impairment. A primary source of IM noise is the transponder TWT final amplifier. Solid-state amplifiers, which are now being implemented in new satellite designs, tend to reduce IM products because of improved linearity over their TWT counterparts. Both types of amplifiers remain notoriously inefficient. Noise and distortion result in degraded error performance.

5.2.3 TDMA Operation on a Bent Pipe Satellite

5.2.3.1 General

With a time division multiple access (TDMA) arrangement on a bent pipe satellite, each earth station accessing a transponder is assigned a time slot for its transmission, and all uplinks use the same carrier frequency on a particular transponder. (See Section 4.4.3.) We recall that a major limitation of FDMA systems is the required backoff of drive in a transponder to reduce IM products developed in the TWT final amplifier owing to simultaneous multicarrier operation. With TDMA, on the other hand, only one carrier appears at the transponder input at any one time, and, as a result, the TWT can be run to saturation minus a small fixed backoff to reduce waveform spreading. This results in more efficient use of a transponder and permits greater system capacity. In some cases capacity can be doubled when compared to an equivalent FDMA counterpart. Another advantage of a TDMA system is that the traffic capacity of each access can be modified on a nearly instantaneous basis. The loading of a long haul system can be varied as the busy hour moves across it, assuming that accesses are located in different time zones. This is difficult to achieve on a conventional FDMA system.

5.2.3.2 Simplified Description of TDMA Operation

An important requirement of TDMA is that transmission bursts do not overlap. To ensure nonoverlap, bursts are separated by a guard time, which is analogous to a guard band in FDMA. The longer we make the guard time, the greater assurance we have of nonoverlap. However, this guard time reduces the efficiency of the system. Decreasing guard time improves system efficiency. The amount of guard time, of course, is a function of system timing. The better the timing system is, the shorter we can make guard times. Typical guard times for operating systems are on the order of 100–200 nsec.

Figure 5.1 shows a typical TDMA frame. A frame is a complete cycle of bursts, usually with one burst per access. The burst length per access need not be of uniform duration; in fact, it is usually not. Burst length can be made a function of the traffic load of a particular access at a particular time. This nonuniformity is shown in the figure. The frame period is the time required to sequence the bursts through a frame.

The number of accesses on a transponder can vary from 3 to over 100. Obviously, the number of accesses is a function of the traffic intensity of each access, assuming a full-capacity system. It is also a function of the transponder

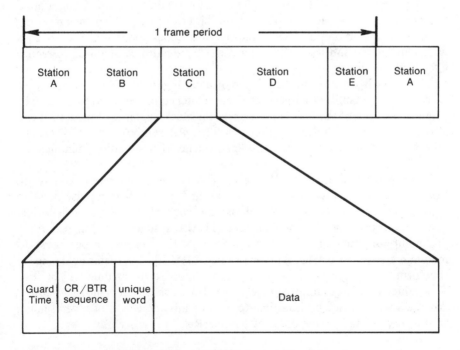

FIGURE 5.1 Typical TDMA frame and burst formats.

bandwidth and the digital modulation employed (e.g., the packing ratio or the number of bits per hertz). For example, more bits can be packed per unit bandwidth with an 8-ary PSK signal of fixed duration than a BPSK signal of equal duration. For high-capacity systems, frame periods vary from 100 μsec to over 2 msec. As an example, the INTELSAT TDMA system has a frame period of 2 msec.

Figure 5.1 also shows a typical burst format, which we call an access subframe. The first segment of the subframe is called the CR/BTR, which stands for carrier recovery (CR), bit timing recovery (BTR). This symbol sequence is particularly necessary on a coherent PSK system, where the CR is used by the PSK demodulator in each receiver to recover local carrier and the BTR to synchronize (sync) the local clock. In the INTELSAT system CR/BTR is 176 symbols long. Other systems may use as few as 30 symbols. We must remember that CR/BTR is overhead, and thus it would be desirable to shorten its duration as much as possible.

Generally, the minimum number of bits (or symbols—in QPSK systems one symbol or baud carries two bits of information) required in a CR/BTR sequence is only roughly a function of system bit or symbol rate. Carrier recovery and bit timing recovery at the receiver must be accomplished by realizable phase-lock loops and/or filters that have a sufficiently narrow bandwidth to provide a satisfactory output signal-to-noise ratio. There is a tradeoff in system design between acquisition time (implying a wider bandwidth) and SNR (implying a narrower bandwidth). Reference 1 suggests CR/BTR bandwidths of 0.5–2% of the bit or symbol rate providing a good compromise between acquisition time and bit error rate performance resulting from a finite-loop-output SNR. Adaptive phase-lock loops that acquire in a wide band mode and track with a narrower bandwidth can be used to reduce CR/BTR overhead.

The next bit sequence in the burst subframe (Figure 5.1) is the unique word (UW), which establishes an accurate time reference in the received burst. The primary purpose of the UW is to perform the clock alignment function. It can also be used as a transmit station identifier. Alternatively, the UW can be followed by a transmit station identifier sequence (SIC—station identification code).

The loss of either the BR/CTR or the UW is fatal to the receipt of a burst. For voice traffic, a lost burst causes clicks, and sounds like impulse noise to the listener. In the case of a data bit stream, large blocks of data can be lost due to a "skew" or slip of alignment. TDMA systems are designed for a probability of miss or false detection of 1×10^{-8} or better per burst to maintain a required threshold bit error rate of 1×10^{-4}. A major guideline we should not lose sight of is the point where supervisory signaling will be lost. This value is a BER of approximately 1×10^{-3}. The design value of 1×10^{-8} threshold will provide a mean time to miss or false detection of several hours with a frame length in the order of 1 msec (Ref. 1).

TDMA system design usually allows for some errors in the UW without loss of alignment. One approach to reduce chances of misalignment suggests a change in waveform during the UW interval. For instance, on an 8-ary PSK system, we might transmit BPSK during this interval, providing more energy per bit, ensuring an improved error performance during this important interval, thus improving the threshold performance.

Other overhead or housekeeping functions are inserted in the burst subframe between the UW and the data text. These may include voice and teleprinter orderwires, BERT (bit error rate test) and other sequences, alarms, and a control and delay channel. Preambles or burst overhead usually require between 100 and 600 bits. INTELSAT uses 288 symbols or 576 bits (with QPSK, 1 symbol is equivalent to two information bits).

The efficiency of a TDMA system depends largely on how well we can amortize system overhead and reduce guard times, in both length and number. Frame length affects efficiency. As the length increases, the number of overhead bits per unit time decrease. Also, as the length increases, the receiving system buffer memory size must increase, increasing the complexity and cost of terminals.

5.2.3.3 TDMA Channel Capacity

Satellite communication systems may be bandwidth limited or power limited. For the bandwidth-limited case, the nominal capacity of a satellite transponder using TDMA may be approximated by the following expression (Ref. 1):

$$R_b = W + B - C_w \qquad (5.1)$$

where R_b = link bit rate expressed in dB (i.e., 100 bps is equivalent to 20 dB, 1000 bps to 30 dB, 1 Mbps to 60 dB, 2 Mbps to 63 dB, and so forth)

W = bandwidth of the satellite transponder expressed in decibels

B = bit rate to symbol rate ratio expressed in decibels

C_w = ratio of the transponder bandwidth to the possible band-limited symbol rate through the transponder (if no other value available, use 0.8 dB)

Example 1. A typical transponder has a bandwidth of 36 MHz and QPSK modulation is employed. What is the satellite link transmission bit rate?

$$R_b = 75.6 \text{ dB} + 3 \text{ dB} - 0.8 \text{ dB}$$

$$R_b = 77.8 \text{ dB}$$

The bit rate is

$$R = \log^{-1}(77.8/10) = 60.26 \text{ Mbps}$$

If a satellite channel is power-limited on the downlink, the following expression may be used to determine R_p, (Ref. 1):

$$R_p = \text{EIRP}_{\text{dBW}} - P_L + \frac{G}{T} - K - \frac{E_b}{N_0} - M \qquad (5.2)$$

where R_p = satellite transmission link bit rate expressed in decibels for the power-limited case

EIRP = effective isotropically radiated power of the transponder in dBW

P_L = path loss of the downlink in decibels (for the 4-GHz case, use 197 dB)

K = Boltzmann's constant (-228.6 dBW/Hz/K)

E_b/N_0 = value for the required BER

M = total system link margin in dB

G/T = the earth station in question G/T.

Example 2. Given an EIRP from a satellite transponder of $+22.5$ dBW, G/T of the earth terminal of 40.7 dB/K, coherent QPSK modulation, an 8-dB margin, which includes modulation implementation loss, and an E_b/N_0 of 9.6 dB for a BER of 1×10^{-5}, then what is the bit rate for the power-limited case?

$$R_p = 22.5 \text{ dBW} - 197 \text{ dB} + 40.7 \text{ dB/K} + 228.6 \text{ dBW} - 9.6 \text{ dB} - 8 \text{ dB}$$

$$R_p = 77.2 \text{ dB}$$

which is equivalent to 52.48 Mbps.

We note that if the satellite transponder EIRP is increased 10 dB, the bit rate would increase by an equivalent 10 dB and forces us into the band-limited regime, where we will then use equation (5.1). Obviously, we are not going to get 524 Mbps through a 36-MHz transponder utilizing QPSK or even 16-ary PSK.

5.2.3.4 TDMA System Clocking, Timing, and Synchronization

It was previously stressed that an efficient TDMA system must have no burst overlap, on the one hand, and as short a guard time as possible between bursts, on the other hand. We are looking at guard times in the nanosecond regime. The satellites under discussion here are geostationary. For a particular TDMA system the range to a satellite can vary from 23,000 to 26,000 statute miles. We can express these range values in time equivalents by dividing by the velocity of propagation in free space or 186,000 miles/sec. These values are

23,000/186,000 and 26,000/186,000 or 123.469 and 139.573 msec. The time difference for a signal to reach a geostationary satellite from a very-low-elevation-angle earth station and a very-high-elevation-angle earth station is 16.104 msec (e.g., 139.573 − 123.469) or 16,104 μsec or 16,104,335 nsec. Of course, this is a worst case, but still feasible. The TDMA system must be capable of handling these orders of time differences among the accessing earth stations. How do we do it and meet the guidelines set out previously? We must also keep in mind that geostationary satellites actually are in motion in a suborbit causing an additional time difference and doppler shift, both varying dynamically with time.

There are two generic methods used to handle the problem: "open loop" and "closed loop." Open-loop methods are characterized by the property that an earth station's transmitted burst is not received by that station. We mean here that an earth station does not monitor the downlink of its own signal for sync and timing purposes. By not using its delayed receiving signal for timing, the loop is not closed, hence, it is open.

Closed-loop covers those synchronization techniques in which the transmitted signals are returned through the bent pipe transponder repeater to the transmitting station. This permits nearly perfect synchronization and high-precision ranging. The term "closed loop" derives from the looping back through the satellite of the transmitted signal permitting the transmitting TDMA station to compare the time of the transmitted-burst leading edge to that of the same burst after passing through the satellite repeater. The TDMA transmitter is then controlled by the result, an early or late arrival relative to the transmitting station's time base. (*Note*: These definitions of open loop and closed loop should be taken in context and not confused with open-loop and closed-loop tracking discussed subsequently in Chapter 7.)

One open-loop method uses no active form of synchronization. It is possible to achieve accuracies from 5 μsec to 1 msec (Ref. 2) through what can be termed "coarse sync." The system is based on very stable free-running clocks, and an approximation is made of the orbit parameters where burst positioning can be done to better than 200-μsec accuracy. The method was used on some early TDMA trial systems and on some military systems and will probably be employed on satellite-based data networks, particularly with long frames.

One of the most common methods used to synchronize a family of TDMA accesses is by a reference burst. A reference burst is a special preamble only, and its purpose is to mark the start of frame with a burst codeword. The station transmitting the reference burst is called a reference station. The reference bursts are received by each member of the family of TDMA accesses, and all transmissions of the family are locked to the time base of the reference station. This, of course, is a form of open-loop operation. Generally, a reference burst is inserted at the beginning of frame. Since reference bursts pass through the bent pipe repeater and usually occupy the same bandwidth as traffic bursts, they provide each station in the family with information on doppler shift, time delay variations due to satellite motion, and channel characteristics. However, the reference burst technique has some drawbacks. It can only serve one repeater and a specific pair of uplinks and downlinks. There are difficulties with this technique in transferring such results accurately to other repeaters, other beams, or stations in different locations. The reference bursts also add to system overhead by using a bandwidth/time product not strictly devoted to the transfer of useful, revenue-bearing data/information. However, we must accept that some satellite capacity must be devoted to achieve synchronization.

5.2.4 Some Techniques for Calculating TDMA System Efficiency (Ref. 2)

We will discuss frame efficiency, burst transmission efficiency and burst system efficiency. Frame efficiency is defined here as the ratio of the portion of a frame available for useful data text interchange to the total frame length. For a large class of systems it has the general form

$$\text{eff}(\%) = 1 - \frac{[S + \sum_{i=1}^{n}(G_i + P_i + Q_i)]T_s}{F} \qquad (5.3)$$

where F = frame length in microseconds

S = number of symbols in synchronization bursts

G = guard time in microseconds

P = number of symbols in the preamble

Q = number of symbols in postamble [in some applications a postamble is used for decoder quenching (initialization) for the next burst]

T_s = symbol length in microseconds
$T_s = (\log_2 A)/R$
where A = number of symbols in the alphabet (code set)
R = the transmission rate in Mbps

n = number of accesses

Transmission efficiency of a burst, which we will call Te, is the ratio of useful message or text data information to the total bits transmitted:

$$Te = rM/(P + M + Q) \qquad (5.4)$$

where r = coding rate of the codec

M = message symbols per burst

P = number of symbols in the preamble

Q = number of symbols in the postamble

System efficiency is defined as the ratio of the useful capacity (traffic that bears revenue) measured in bps to the available capacity. Available capacity is the theoretical capacity of the channel limited only by thermal noise:

$$Ca = B \log_2\left(1 + \frac{E_s T_s}{N_0 B}\right) \qquad (5.5)$$

where Ca = available capacity

B = channel bandwidth

E_s = energy per symbol

T_s = symbol duration

N_0 = noise spectral density (thermal noise)

System efficiency also depends on traffic patterns, networking, method of multiple access, modulation, demodulation approach, coding, and type of decoder. The desired end product is net user throughput as in any other data transmission system.

5.3 DIGITAL SPEECH INTERPOLATION (DSI)

DSI is designed for speech operation to increase system capacity. It is based on the fact that there is active speech on a full duplex voice circuit only a fraction of the time. For one thing, there is the talk–listen effect. In normal operation, while one end of a speech circuit talks, the other end listens. For this effect alone, there is only 50% usage. Also, there are many pauses in normal speech. DSI exploits these periods of nonusage and speech pauses.

A similar system was implemented on undersea cables operating in the analog mode. It was called TASI (time-assigned speech interpolation). With a significantly large number of voice channels, TASI could enhance transmission capacity by a factor of 2.

When describing TASI or DSI, we talk about the time occupied by a caller's speech as a speech spurt. With 100% of a TASI terminal connected to active circuits, speech is actively present on a busy channel only about 40% of the time. If, on the other hand, all circuits are not busy, the average speech activity on a TASI terminal is further decreased. The percentage of busy circuits is called the incoming channel activity, and the percentage of time that speech spurts occupy a channel is the speech spurt activity, or simply speech activity.

5.3.1 Freeze-Out and Clipping

The operation of TASI and DSI exploits low speech spurt activity by assigning transmission channels only when a speech spurt is present. As the number of channels increase, the process becomes more efficient. If two speech users were to use one channel through interpolation, a large portion of the speech will be lost owing to competition for simultaneous occupancy. The spurt of one user occupying the channel will "freeze-out" any other attempt to use the channel by another user, and that freeze-out will continue until the spurt is terminated.

When a larger portion of users use a comparatively smaller number of available channels, there is always a finite probability that the number of conversations requiring service will exceed the number of channels providing that service. This competition causes an impairment called "competitive clipping," where the initial portion of a speech spurt is clipped. The percentage of time that speech is lost due to such competition is called percentage of freeze-out or freeze-out fraction. In the design of a TASI or DSI system, the fraction of speech lost must be acceptably small. Reference 3 gives a freeze-out fraction of 0.5% for TASI systems.

The most common freeze-outs clip initial portions of speech spurts from near zero to several hundred milliseconds. Clips longer than 50 msec cause perceptible mutilation, and the percentage of clips longer than 50 msec should be kept to less than 2%.

Another form of clipping is "connect clipping." This type of clipping is caused by the channel-assignment process. The presence of speech on an incoming telephone channel on the transmit side of a TASI terminal is sensed by a speech detector, which initiates a request for a channel. A processor assigns an idle transmission channel to the incoming channel in response to that request and also informs the distant end specifying the outgoing channel to which the transmission channel is to be connected. During the time required to make the total connection, speech is clipped. This type of clipping only occurs when the demand for channels exceeds the operational transmit channels available. As the demand increases for service, connect clipping becomes more prevalent and, on a fully loaded TASI system, connect clipping may occur on every speech spurt. Thus, a system design objective is to minimize transmit channel connect time.

5.3.2 TASI-Based DSI

One type of DSI is based on the TASI concept. It operates with 8-bit PCM words, 8000 samples per second. The incoming PCM signals in TDM format are processed by a transmit assignment processor. When speech activity is detected by the processor on an incoming PCM time slot, it is assigned an available transmit slot. The distant-end processor is alerted, via a control channel, of the slot assignment, and makes the corresponding connection on its terrestrial side. The control channel is carried on the same PCM TDM frame. Figure 5.2 shows a TASI type DSI system.

Digital TASI has a number of advantages over its analog counterpart. The fact that it is all-digital lends itself more to digital processor control. Connect clipping is reduced because switching and control is faster. When the system

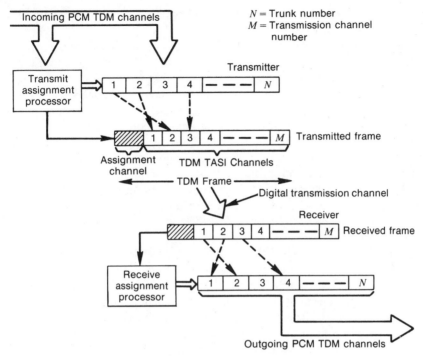

FIGURE 5.2 TASI-type DSI system.
From Ref. 3. Reprinted with permission.

becomes more loaded, competitive clipping can be reduced or eliminated by dropping the least significant bit on each PCM word. This increases quantizing distortion somewhat, but is a lesser impairment than competition clipping. However, the time required when bit reduction is invoked is very low and, thus, the impairment is hardly noticeable.

5.3.3 Speech Predictive Encoding DSI

The basic idea of this type of DSI is that it eliminates PCM frame-to-frame redundancy that exists in ordinary PCM transmission. This redundancy includes that due to pauses and redundancy in the speech spurts themselves. It permits a two or more times increase in speech channel capacity compared to conventional, unprocessed PCM.

Redundancy is reduced by storing a frame and transmitting one frame behind. The two consecutive frames are compared in the transmit processor and only the nonredundant information is transmitted to the distant end. The redundant information is called predictable and the nonredundant information is called unpredictable. We need some way to tell the distant end how much of the information is predictable and where it is. This is done with a "sample assignment word" (SAW).

One system described in Ref. 3 can transmit up to 64 active speech channels in the frequency spectrum allotted to a 32-channel conventional PCM system corresponding to a bit rate of 2.048 Mbps. The PCM sample derived during each sample period from the 64 incoming channels is compared with samples previously sent to the receiver and stored in memory at the transmitter. Any samples that differ by an amount equal to or less than some given number of quantizing steps, called the aperture, are discarded and not sent to the receiver. These are the predictable samples. The remaining "unpredictable" samples are transmitted to the receiver and replace the values formerly stored in memories at both the transmitter and receiver. The aperture is adjusted automatically as a function of activity observed over the 64 incoming channels on each frame so that the number of samples transmitted is nearly constant.

The transmission frame of the predictive system is composed of an initial SAW slot followed by a number of 8-bit time slots that carry the individual PCM samples that are unpredictable. The SAW contains one bit for each of the incoming telephone channels. Thus, for a 64-channel terrestrial system, the SAW contains 64 bits. The bit corresponding to a given channel is a "1" if the frame contains a sample for that particular channel and a "0" if it does not. Thus, the SAW contains all of the information needed to distribute the samples among the 64 outgoing channels at the distant receive end.

At the receiver the unpredictable samples received in the transmission channel frame replace previously stored samples in the receiver's 64-channel memory as directed by the SAW. The samples in memory, in the form of a conventional PCM frame, are reslotted into the outgoing terrestrial channels at the appropriate rate. The most recent frame thus contains new samples on the channels that have been updated by the most recent transmission channel frame and repetitions of the samples that have not been updated.

The term DSI advantage is the ratio of incoming terrestrial channels to the transmitter to the number of required transmit channels of the DSI system. If there were 120 terrestrial channels occupying only 54 satellite transmission channels, the DSI advantage is 120/54 or 2.22.

Both the TASI and predictive methods of DSI offer significant enhancements in the capacity of digital transmission of speech. The two methods achieve interpolation advantages greater than two, with the predictive methods achieving a slightly higher value than TASI with systems carrying a smaller number of channels (on the order of 120 PCM channels or less).

The TASI method degrades transmission by initial clips of speech spurts when approaching full-capacity loading. The frequency of occurrence of destructive initial clips can be kept low enough to produce little degradation by properly adjusting the DSI advantage. The probability of initial clips longer than 50 msec should be no greater than 2% to meet the degradation acceptability criteria. The technique of bit reduction during periods of heavy traffic loading (e.g., from 8-bit samples to 7-bit samples) can reduce the probability of initial clips. Just the 1-bit reduction reduces the probability of occurrence by more than an order of magnitude.

Prediction distortion is the major cause of degradation of the predictive methods of DSI. The amount of this distortion varies as the fraction of samples predicted varies in response to changes in the ensemble average of voice spurt activity. Again the DSI advantage is the controlling figure. The prediction noise produced is controlled by the DSI advantage. Reference 3 suggests designing the DSI advantage such that the probability that more than 25% of the samples are predicted during speech spurts is 0.25. This results, on the average, in a 0.5-dB degradation in the subjectively assessed speech-power-to-quantization-noise-power ratio. The predictive method is adaptive and yields to occasions of higher than average voice spurt activity by automatically increasing the fraction of samples predicted. It is this feature that eliminates the possibility of damaging initial clips.

5.4 THE INTELSAT TDMA / DSI SYSTEM *

The terminal segment of the INTELSAT TDMA/DSI comprises four reference stations per satellite and a number of traffic terminals. The system has been designed to operate with satellites having four coverage areas compatible with INTELSAT V and VI operation. These are east hemispheric beam, west hemispheric beam, east zone beam, and west zone beam. (See Section 4.5.) Normally, zone beam coverage areas will also be contained within hemispheric beam coverage areas. Zone and hemispheric beams use opposite senses of polarization. Figure 5.3 shows a satellite with typical east-to-west and west-to-east connectivities of both the zone and hemispheric beams. Two dual-polarized reference stations located in each zone coverage area are thus able to monitor and control both zone and hemispheric beam transponders. Each reference station generates one reference burst per transponder, and each transponder will be served by two reference stations. This provides redundancy by enabling traffic terminals to operate with either reference burst. The two pairs of reference stations provide network timing and control the operation of traffic terminals and other reference stations.

Reference stations include a TDMA system monitor (TSM), which is used to monitor system performance and diagnose system faults. In addition, the TSM is used to assist users in carrying out their traffic terminal lineups.

The traffic terminals operate under control of a reference station and transmit and receive bursts containing traffic and system management information. Traffic terminals include interfaces that are used to connect terminals to the terrestrial networks. The INTELSAT TDMA/DSI system uses two

* This subsection is extracted from Pantan, Dicks, et al., "The INTELSAT TDMA/DSI System" as set forth in Ref. 4, Section 13-4.12. Reprinted with permission.

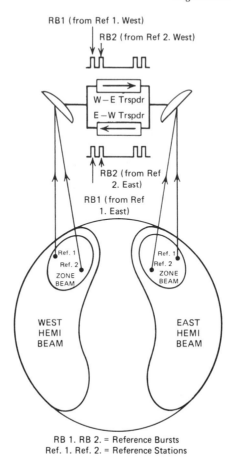

RB 1. RB 2. = Reference Bursts
Ref. 1. Ref. 2. = Reference Stations

FIGURE 5.3 An overview of the INTELSAT TDMA system.

types of interfaces: DSI for voice traffic (together with a limited amount of nonvoice traffic) and digital noninterpolated (DNI) for data and noninterpolated voice traffic.

The principal features of the system are summarized in Table 5.1, and the major functions of the reference stations and traffic terminals are summarized in Table 5.2.

5.4.1 Burst and Frame Format

The INTELSAT TDMA frame contains traffic bursts and reference bursts RB1 and RB2 as shown in Figure 5.4. Each reference burst is transmitted by a separate reference station and, under normal conditions, both reference sta-

TABLE 5.1. INTELSAT TDMA / DSI System Features

The frame length is 2 msec

Four-phase coherent phase-shift keying (CPSK) modulation without
differential encoding is employed at a nominal bit rate of
120.832 Mbps

Forward error correction (FEC) is applied to the traffic portion
of selected traffic bursts

Open-loop acquisition and feedback closed-loop synchronization
for traffic and reference terminals

Burst time plan rearrangement can be made automatically
without loss of traffic

Each pair of reference stations can control up to 32 terminals
including other reference stations

Each traffic terminal can transmit up to 16 bursts and receive up to 32 bursts
per frame

Each traffic terminal can transponder hop across up to
four transponders

Digital speech interpolation interface can be applied in
channel groupings of up to 240 terrestrial channels

The digital noninterpolated interface can accommodate channel
groupings of up to 128 terrestrial channels

Plesiochronous interconnection to the terrestrial network can be provided

tions are active. One reference station is designated a primary reference
station, and the other is designated a secondary (backup) reference station.
The traffic terminals respond to the secondary reference station only when the
primary reference station fails. The nominal guard time between bursts is 64
symbols.

Figure 5.5 shows the reference and traffic burst formats. The reference burst
consists of a preamble and a control and delay channel (CDC). The traffic
burst consists of the preamble and a traffic section consisting of one or more
DSI and/or DNI (digital noninterpolated) subbursts. The preamble includes
the carrier and bit timing recovery sequence, the unique word, teleprinter
order wire channels, the service channel, and the voice order wire (VOW)
channels.

The CR/BTR sequence enables the modem to acquire and synchronize
received bursts. The 24-symbol unique word is used to differentiate reference
bursts and traffic bursts, resolve the four-fold phase ambiguity inherent in
QPSK modulation, and mark the beginning of a multiframe. Eight teleprinter
order wires and two voice order wires are allocated 8 and 64 symbols,
respectively, in each reference burst and traffic burst. Eight symbols form a
service channel that is used to exchange control and housekeeping information
throughout the TDMA network. Finally, in the reference burst, eight symbols

TABLE 5.2. Major Functions of Reference Stations and Traffic Terminals

Reference Station Functions	Traffic Terminal Functions
Perform satellite position determination needed for acquisition	Perform acquisition and synchronization under control of a reference station
Provide open-loop acquisition information to traffic terminals and other reference stations	Generate and receive bursts containing traffic and housekeeping information
Provide synchronization information to to traffic terminals and controlled reference stations	Perform transponder hopping where necessary
Provide TDMA system monitoring	Under the coordination of the reference station, carry out synchronous burst time plan changes
Provide network management by transmitting the appropriate messages or codes to traffic terminals and other reference stations	Provide voice and teletype order wires
Provide common synchronization across multiple satellite transponders (permitting transponder hopping)	
Provide voice and teletype order wires	
Provide access to the INTELSAT Operations Center (IOC) for network voice and teletype order wires and for transmittal of status information	

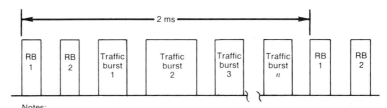

Notes:
 1) RB 1 is the Reference Burst from Reference Station 1.
 2) RB 2 is the Reference Burst from Reference Station 2.

FIGURE 5.4 Structure of INTELSAT TDMA frame.

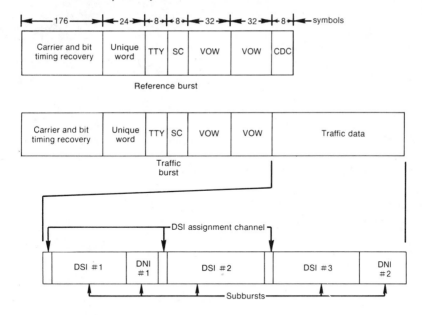

FIGURE 5.5 TDMA burst format.

are allocated for the CDC, which is used to control the traffic terminals' acquisition and synchronization.

5.4.1.1 Unique Words

The 24-symbol unique word contains two consecutive 12-symbol patterns that serve three purposes. The last symbol of the word marks the position of a burst relative to the start of frame. Next, the pattern of the first 12 symbols is used to resolve the fourfold phase ambiguity of QPSK. Finally, the pattern of the second 12 symbols relative to the first is an identifier used to distinguish between reference bursts and traffic bursts. For 15 consecutive frames, all bursts use the same identifier (UW0) but on the 16th burst, which is called a multiframe marker, the identifier changes to identify the burst as RB1, RB2, or a traffic burst. A multiframe is defined for the INTELSAT TDMA system as the 16 frames starting with the multiframe marker. Figure 5.6 shows the multiframe format.

5.4.1.2 Service Channel (SC)

A service channel message consists of an 8-bit function code, a 22-bit parameter, and a 2-bit parity check. This word is transmitted over one multiframe (e.g., 16 TDMA frames) at a rate of 2 bits per frame. For redundancy, each bit is repeated eight times in every burst.

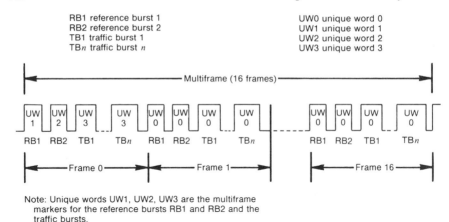

Note: Unique words UW1, UW2, UW3 are the multiframe
markers for the reference bursts RB1 and RB2 and the
traffic bursts.

FIGURE 5.6 INTELSAT TDMA multiframe format.

5.4.1.3 Control and Delay Channel

The CDC cyclically addresses each controlled traffic terminal and reference station in successive multiframes using a 32 multiframe cycle referred to as a control frame. Transmission of a 32-bit CDC message is accomplished in a manner similar to the service channel message. The structure of the 32-bit CDC message is shown in Figure 5.7. Except for terminal number 0, which is used for reference station status codes and the burst time plan number, each message is directed to a particular terminal identified by the terminal number. The destination-directed messages control terminal operation by means of a 2-bit control code and a 22-bit word providing the transmit delay.

5.4.2 Acquisition and Synchronization

In this context acquisition is the process by which a TDMA terminal initially places its burst into the assigned position within the TDMA frame. This process must be executed without interference to other bursts in the frame. A terminal may enter the acquisition phase if it receives, via the CDC, an acquisition control code from the reference station together with a value of transmit delay. The transmit delay is the time between the reception of a reference burst and the transmission of the acquiring terminal's own burst. The reference station calculates the value of the transmit delay based on knowledge of the satellite position. This method is referred to as "open-loop" acquisition. (See Section 5.2.3.4.)

During the acquisition process, the terminal transmits only its preamble (short burst). The reference station measures the position of the short burst within the frame and transmits a new value of delay that causes the short burst to move to its nominal position. When the reference station has verified that

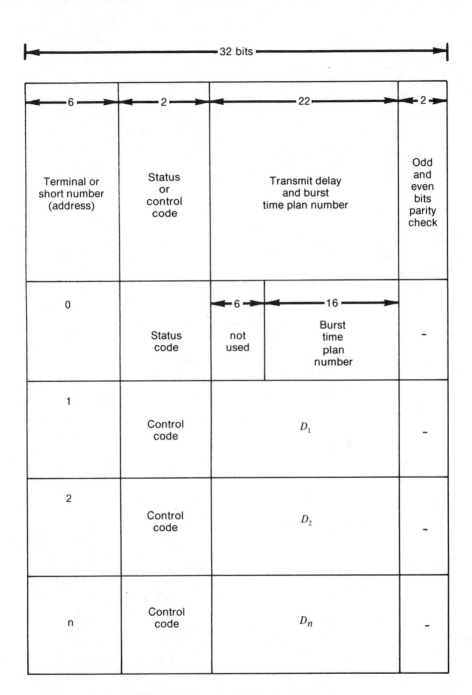

Terminal or short number (address)	Status or control code	Transmit delay and burst time plan number	Odd and even bits parity check
6	2	22	2
0	Status code	not used (6) / Burst time plan number (16)	-
1	Control code	D_1	-
2	Control code	D_2	-
n	Control code	D_n	-

FIGURE 5.7 Control and delay channel structure.

343

the short burst is in its nominal assigned position, it transmits the "synchronization" code to the terminal. The reception of the synchronization notifies the terminal that the acquisition phase is complete and that traffic subbursts can be added to the preamble.

Synchronization is the process by which bursts are maintained in their assigned positions within the frame. The reference station monitors the position of the burst in the frame and continuously modifies the transmit delay values to maintain their bursts in their proper positions. This process is referred to as "feedback-closed-loop" synchronization.

5.4.3 Transponder Hopping

The traffic terminals are designed to hop across a maximum of four transponders, which can be separated in frequency and/or polarization. Since the TDMA/DSI system employs mutually synchronized reference bursts in all transponders, traffic bursts transmitted into different transponders will be separated by fixed time intervals. This allows reference stations to control the position of only one of the terminal's bursts, since the others are synchronized by fixed time offsets.

5.4.4 TDMA Reference Station

Each reference station is equipped with sufficient redundancy to provide a high degree of reliability. However, in order to ensure a high degree of continuity of service, each reference station operating in a given coverage area will have one backup reference station. This leads to the concept of two reference bursts per frame (primary and secondary). Although one of the stations has a standby role, both stations are simultaneously active. In the event of a failure in the primary reference station, system control passes to the secondary reference station without disturbance to the network.

A simplified block diagram of the reference station is shown in Figure 5.8. The reference station equipment consists of the antenna, RF/IF equipment, the reference terminal equipment, and a TDMA system monitor (TSM). Reference stations are located at existing earth stations with antennas meeting INTELSAT Standard A requirements. The RF/IF equipment includes high-power amplifiers (HPA), low-noise amplifiers (LNA), interfacility links, and up and down converters. In addition, at INTELSAT's Operations Center (IOC) in Washington, DC, display and storage facilities are provided. These facilities interface with all reference stations operating in the network.

Figure 5.9 is a simplified block diagram of the reference terminal equipment (RTE). This equipment comprises two on-line signal processing equipment (SPE) units, a local timing source, and peripheral equipment. An SPE will accommodate up to four transmit and receive chains. Transponder hopping at the IF or RF level is not employed at reference stations and, as a consequence, each up and down chain is equipped with separate modems. Each signal

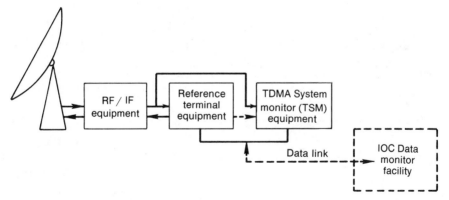

FIGURE 5.8 Simplified block diagram of a reference station.

processing unit generates reference bursts, but only the output of one of the units is selected for transmission. The other unit serves as an active backup.

To detect abnormalities in its operation, the RTE is equipped with a monitor and diagnostics (MAD) unit, which monitors a number of SPE conditions. Upon detection of a failure condition, the standby SPE is switched on line.

The switchover to redundant RF/IF equipment is controlled by local switchover logic, which is independent of the MAD unit. The switchover time of the failure-sensing logic and the switchover time of the RTE-RF/IF equipment are chosen such that no interaction occurs between the RF/IF equipment and the RTE.

The TDMA System Monitor (TSM) is shown in Figure 5.10. It performs measurements to monitor performance of the TDMA system. The measurement equipment is switched between the reference station downlink chains. The TSM is equipped to measure relative burst power, burst carrier frequency, burst position, pseudo-bit-error rate, and transponder backoff. Burst measurements are gated by the TSM controller. The measurement data are processed within the TSM prior to display and storage. TSM data are available to the IOC via packet-switched data lines used between the RTE and the IOC.

Each reference station is provided with control equipment and displays of local equipment status and TDMA system status. The local equipment displays include mimic diagrams of equipment, protocols, and a facility for examining the events that led to a failure condition. This latter display continuously stores the last four control frames of data to give an indication of the state of the protocols within the terminal. The system status displays include a list of stations controlled by the RTE and their current status.

The IOC monitors and controls the operation of all INTELSAT satellite systems. For the TDMA/DSI system, the IOC is linked by packet-switched

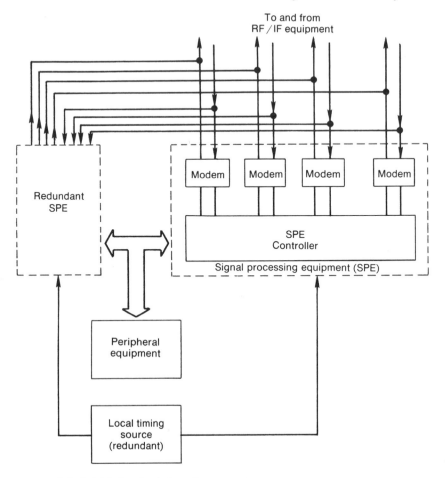

FIGURE 5.9 Simplified block diagram of the reference terminal equipment.

data lines to each reference station, and is equipped with display equipment similar to that at reference stations. This enables the IOC to use the majority of the displays available to the local operator and can perform remote system startup and burst time-plan rearrangements using commands sent over the packet-switched data links.

5.4.5 Traffic Terminals

Traffic terminals are used by INTELSAT terminals to interface voice and data traffic with the INTELSAT TDMA system. These terminals acquire and synchronize to the TDMA frame under the control of a reference station. Once synchronized, a TDMA traffic terminal transmits bursts and subbursts in

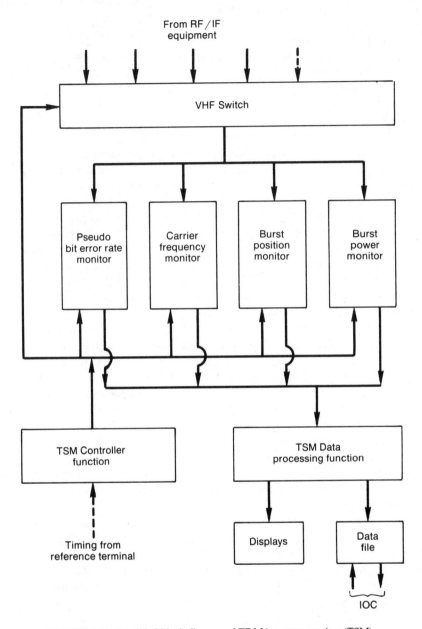

FIGURE 5.10 Simplified block diagram of TDMA system monitor (TSM).

accordance with the INTELSAT burst time plan. The subbursts, which contain the traffic from that terminal, originate from either DSI or DNI units.

To make the most use of satellite capacity, most traffic will use DSI. Each DSI unit can accommodate up to 240 terrestrial channels. The usable capacity of the DSI unit may vary from one terrestrial channel to its maximum capacity in increments of one terrestrial channel. The corresponding capacity of a DSI unit is 127 satellite channels with its actual utilization varying from one satellite channel to its maximum capacity. Individual DSI units may be implemented for either multidestination or single-destination operation. Units operating in the multidestination mode are capable of transmitting to eight destinations in one subburst and receiving eight subbursts.

Each DSI subburst contains an assignment channel (DSI-AC) located at the beginning of each DSI subburst and occupying 128 bits (equivalent to one satellite channel). The assignment channel carries assignment messages and DSI alarm messages. The assignment messages are used to inform the distant end receive DSI unit of the terrestrial channel to satellite associations made at the transmitting end. The DSI-AC consists of 128 bits, which carry three 16-bit assignment messages. Rate 1/2 Golay coding (see Section 5.6.4.3) is used to correct all one-, two-, and three-bit errors.

Each DSI unit contains a special assignment channel check procedure, which is used to automatically check the end-to-end DSI channel assignments. It consists of a special channel on the transmit side of the DSI that generates a request for assignment once every 10 sec. At the receive side of the DSI a special receive channel expects to receive the assignment every 10 sec. When it is not received, an alarm is activated and sent back to the originating DSI unit.

The ratio of the number of terrestrial channels to the number of satellite channels to be employed by each DSI unit in the system is set such that competitive clipping lasting more than 50 msec will occur on less than 2% of the voice spurts. In order to meet this requirement while maximizing satellite capacity, the DSI units employ bit reduction. When the number of simultaneously active terrestrial channels exceeds the number of satellite channels allocated for the DSI unit, additional satellite channels can be derived by "stealing" the least significant bit of the 8-bit PCM word of each channel. These derived overload channels are used to prevent "freeze-out" when no satellite channel is available for an active terrestrial channel. This process reduces the affected satellite channels from 8 bits to 7 bits during periods of overload. (See Section 5.3.3.)

The DSI unit may be used to carry noninterpolated traffic by preassigning terrestrial channels to satellite channels. The noninterpolated channels are transmitted at the end of the subbursts, and the number of channels is expandable in one-channel steps to a maximum of 127 satellite channels.

A DNI unit can accommodate small capacity links and preassigned data. The capacity of a DNI unit may range from 1 to 128 satellite channels in single-channel increments. The DNI unit may also be used for forming higher bit rate channels in increments of 64 kbps up to a maximum of 8.192 Mbps.

The terrestrial interface at an INTELSAT TDMA terminal uses the plesiochronous method for the interconnection of national digital networks. This requires that national digital networks be interconnected by means of buffers sized to accommodate the surplus or deficiency of bits arising from the difference in bit rates between the two networks. This is acomplished by repeating a block of PCM bits if a buffer is approaching exhaustion, or deleting a block if the buffer is full. These blocks are chosen to be PCM frames, since deleting or repeating PCM frames (slip) will not cause a significant disturbance to the network carrying speech traffic.

To further limit the disturbance, slip is only allowed every 72 days in accordance with CCITT Rec. G.811. This in turn dictates that the national networks and the TDMA frame rate must be held to within 1 part in 10^{11} of their design frequencies over 72 days. This necessitates that one reference station derives its frame timing from a high-stability clock, which provides the necessary time reference for the TDMA network.

Plesiochronous interfacing of the TDMA terminals also requires doppler buffers to remove path-length variations caused by satellite movement. The TDMA/DSI system buffers are able to accommodate up to 1.1 msec of peak-to-peak path length variation.

5.4.6 System Coordination

System coordination functions are concerned with facilitating and maintaining the correct operation and interworking of the TDMA network. These include three function groups: voice and teleprinter (TTY) communications (order-wires), interterminal alarms, and automatic burst time rearrangements. Consistent with its responsibility for overall coordination and maintenance of each TDMA network, the IOC has access to all system coordination functions.

The INTELSAT TDMA system provides for two independent voice orderwires and up to eight independent telegraph (TTY) orderwires on each traffic and reference burst. These circuits are used by the terminal operators to coordinate commissioning, operation, and faultfinding of the equipment. Automatic signaling and call routing is provided by switching computers in each earth station. The IOC gains access to each network by using the reference stations' switching computers as "gateway exchanges." This gives the IOC full voice and teleprinter communications with all reference and traffic terminals in the TDMA system.

All TDMA terminals are capable of exchanging two types of alarms corresponding to the loss of unique word and the occurrence of high BER on a received burst. The alarms are selectively addressed to the originating terminal of the burst in question, using the service channel.

Unique word alarms are generated when a particular burst's unique word is declared lost for more than 500 msec. The alarm message is sent every second until the unique word is declared present. High BER alarms are generated whenever a burst is perceived to have an error rate worse than 1×10^{-3}. This alarm message is sent every 4 sec until the BER improves.

The reference stations have the capability within the display equipment to log all alarm conditions and can assist traffic terminal operators in diagnosing faults. The reference station displays are made available to the IOC, which can also log the alarm incident and assist in fault finding, especially when a larger-scale network problem exists.

A burst time plan for the INTELSAT system contains all the operational parameters for all the terminals in the network. Each burst time plan (BTP) is represented by a unique number, and the current time plan number verification is part of the terminal's acquisition procedure. Thus, whenever a change in burst time plan occurs, all stations are updated.

To ensure that critical elements of the BTP have been correctly loaded into the terminals, critical elements are grouped into a special format and transmitted over a secure data circuit using the CCITT X.25 data protocol which provides error correction using ARQ. When successfully received, the data are translated into a form suitable for use by the terminal and stored in the background memory, which will be used to control the terminal. These data will then be transmitted back to the IOC to enable operators to perform bit-by-bit comparison with the original transmission.

The INTELSAT TDMA/DSI system provides fully synchronous burst time plan rearrangements. This involves changing the position and/or length of some or all bursts within the frame of any TDMA transponder.

Immediately prior to implementation of a new burst time plan, the controlling primary reference station will send "start of plan change" to the primary in the other coverage area. After an appropriate delay, which synchronizes the messages, both reference stations will send a "request for ready to change" message to the terminals over the service channel (see Figure 5.11a). This message will activate a facility that permits the terminal to react to a countdown signal. When this facility is activated, the terminal transmits a "ready to change" signal over the service channel. When both reference stations have confirmed that all terminals involved in the time plan change are enabled and the new time plan has been correctly placed in the background memories of the terminals concerned, the reference stations declare "ready to change." To inform the controlling reference station that it is ready to change, the reference station sends a message over the service channel for 1 sec. When the controlling primary station receives the message, it declares "ready to initiate countdown." Figure 5.11b shows the burst time plan rearrangements sequence. The controlling reference station sends "initiate countdown" to the other reference station, and, after delays necessary to synchronize the countdown, both stations send "notification of time plan change" to all terminals using a countdown sequence over the service channel. When terminals receive

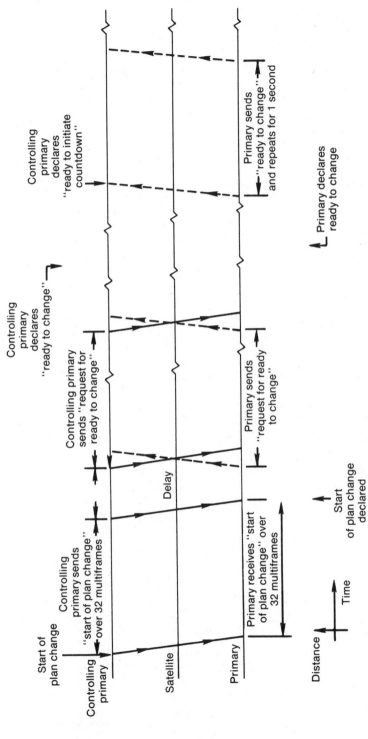

FIGURE 5.11 (*a*) Preparation for BTP change.

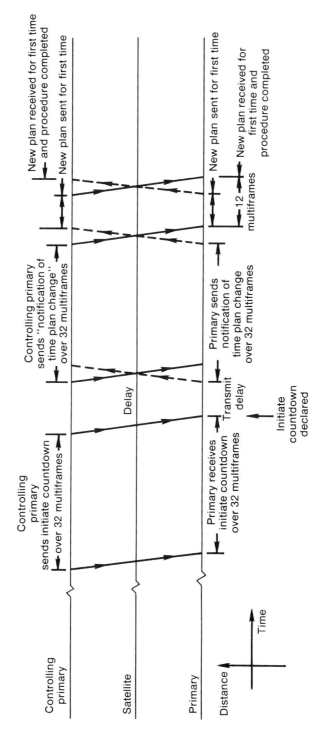

FIGURE 5.11 (*b*) BTP change timing diagram.

the final countdown message, they adopt the new transmit time plan on the next transmitted frame, while on the receive side, the plan is adopted 12 multiframes after the final message. This results in a synchronized system-wide change without interruption to the network.

In the INTELSAT VI SS/TDMA (switched satellite TDMA) system, burst time plan rearrangements may require synchronous changes in satellite switch connectivities.

5.5 PROCESSING SATELLITES

Processing satellites, as distinguished from "bent pipe" satellites, operate in the digital mode, and as a minimum, demodulate the uplink signal to baseband for regeneration. As a maximum, at least as we envision today, they operate as digital switches in the sky. In this section we will discuss satellite systems that demodulate and decode in the transponder and then present some ideas on switching schemes suggested in the NASA 30/20 GHz system. This will be followed by a section on coding gain and a section on link analysis for processing satellites.

5.5.1 Primitive Processing Satellite

The most primitive form of satellite processing is the implementation of on-board regenerative repeaters. This only requires that the uplink signal be demodulated and passed through a hard limiter or a decision circuit. The implementation of regenerative repeaters accrues the following advantages:

☐ Isolation of the uplink and downlink by on-board regeneration prevents the accumulation of thermal noise and interference. Cochannel interference is a predominant factor of signal degradation because of the measures taken to augment communication capacity by such means as frequency reuse.

☐ Isolating the uplink and downlink makes the optimization of each link possible. For example, the modulation format of the downlink need not be the same as that for the uplink.

☐ Regeneration on the satellite makes it possible to implement various kinds of signal processing on board the satellite. This can add to the communication capacity of the satellite and provide a more versatile set of conveniences for the user network.

Reference 5 points out that it can be shown on a PSK system that a regenerative repeater on board can save 6 dB on the uplink budget and 3 dB on the downlink budget over its bent pipe counterpart, assuming the same BER on both systems.

Applying this technique to a digital TDMA system requires carrier recovery and bit timing recovery. Although the carrier frequencies and clocking are quite close among all bursts, coherency of the carrier and the clocking recovery may not be anticipated between bursts. To regenerate baseband signals on TDMA systems effectively, carrier and bit timing recovery are done in the preamble of each burst and must be done very rapidly to maintain a high communication efficiency. There are two methods that can be implemented to resolve the correct phase of the recovered carrier. One method uses reference code words in the transmitted bit stream at regular intervals. The other solution is to use differential encoding on the transmitted bit stream. Although this latter method is simpler, it does degrade BER considering equal C/N_0 of each approach.

An ideal regenerative repeater for a satellite transponder is shown in Figure 5.12 for PSK operation. It will carry out the following functions:

☐ Carrier generation
☐ Carrier recovery
☐ Clock recovery
☐ Coherent detection
☐ Differential decoding
☐ Differential encoding
☐ Modulation
☐ Signal processing
☐ Symbol/bit decision

The addition of FEC coding/decoding on the uplink and on the downlink carries on-board processing one step further. FEC coding and decoding is discussed subsequently. In a fading environment such as one might expect with satellite communication systems operating above 10 GHz, during periods of heavy rainfall, an interleaver would be added after the coder and a deinterleaver before the decoder. Fading causes burst errors, and conventional FEC schemes handle random errors. Interleavers break up a digital bit stream by shuffling coded symbols such that symbols in error due to the burst appear

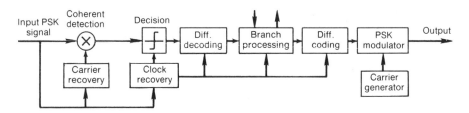

FIGURE 5.12 Configuration of ideal regenerative repeater.

to the decoder as random errors. Of course, interleaving intervals or spans should be significantly longer than the fade period expected.

5.5.2. Switched-Satellite TDMA (SS / TDMA)

We now carry satellite processing one step further by employing antenna beam switching in conjunction with TDMA. This technique provides bulk trunk routing increasing satellite capacity by frequency reuse. Figure 5.13 depicts this concept. TDMA signals from a geographical zone are cyclically interconnected to other beams or zones so that a set of transponders appears to have beam-hopping capability. A sync window or reference window is usually required to synchronize the TDMA signals from earth terminals to the on-board switch sequence.

Sync window is a generic method to allow earth stations to synchronize to a switching sequence being followed in the satellite. A switching satellite, as described here, consists of a number of transmitters cross-connected to receivers by a high-speed time-division switch matrix. The switch matrix connections are changed throughout the TDMA frame to produce the required interconnections of earth terminals. A special connection at the beginning of the frame is the sync window, during which signals from each spot-beam zone are looped back to their originating spot-beam zone, thus forming the timing reference for all zones. This establishes closed-loop synchronization.

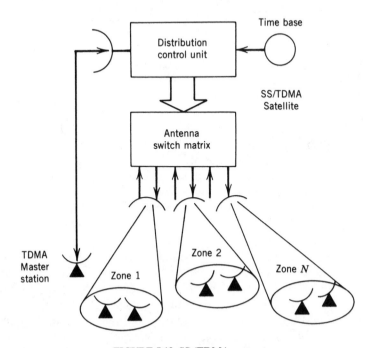

FIGURE 5.13 SS/TDMA concept.

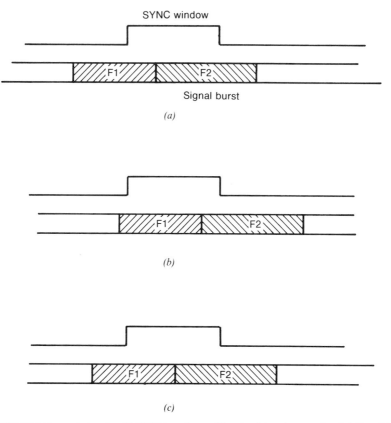

FIGURE 5.14 Sync window for SS-TDMA and use of bursts of two frequencies. (*a*) Signal burst too early; (*b*) signal burst too late; (*c*) synchronized to SYNC window. From Ref. 2. Reprinted with permission.

Figure 5.14 shows a scheme for locking and tracking a sync window in a satellite switching sequence. A burst of two tones, F_1 and F_2, is transmitted by a single access station. Only the portion that passes through the sync window is received back at that access station.

The basic concept is to measure and compare the received subbursts F_1 and F_2 as shown in the figure. Although a very narrow bandwidth and full RF power are used, digital averaging over many frames still is required. The difference is used to control the F_1/F_2 burst to a resolution of one symbol, and the process is continually repeated in closed loop. The sync bursts to the TDMA network are slaved, and the network is thus synchronized to the sync window and the switching sequence on the satellite.

With SS/TDMA the network connectivity and the traffic volume between zones can be adapted to changing needs by reprogramming the processor antennas. Also, of course, the narrow beams (e.g., higher-gain antennas) increase the uplink C/N and the downlink EIRP for a given transponder HPA power output. However, the applicability of the system must be carefully analyzed. Since there is a limit to the speed at which the antenna beam can be switched, as the data symbol rate increases, guard times become an increasing fraction of the message frame, resulting in a loss of efficiency. This can be offset somewhat by the use of longer frames, but this will require increased buffering and will increase the transmission time.

The scheme becomes very inefficient if a large proportion of the traffic is to be broadcast to many zones. SS/TDMA is not applicable to channelized satellites where multiple transponders share common transmit and receive antennas and the signals in each transponder cannot be synchronized for transmission between common terminal zones.

Figure 5.15 is a block diagram of a beam-switching processor in which narrow scanning beams are implemented by the use of a processor-controlled MBA (multiple-beam antenna) for both uplinks and downlinks. The output of the address selector is modified by the memory to provide control signals to the beam-switching network, which, in turn, controls the antenna-weighting networks. This process steers the uplink and downlink antennas to the selected zones for a time interval that matches the time interval of a data frame to be exchanged between the selected zones. Initial synchronization of the system is obtained from a synchronizing signal transmitted to the processor from the TDMA master station. The memory can be updated at any time by signals from the master terminal via a control channel. Often, in practice, two memories are provided so that one can be updated while the other is on-line, thus preserving traffic continuity.

5.5.3. IF Switching

When the principal requirement for switching is for traffic routing rather than just for antenna selection, one method is to convert all uplink signals to a common IF, followed by a switching matrix and upconverting translators. This provides greater switching capability and flexibility. It allows for a wider choice of switching components in the design, since many more device types show good performance at the lower IF frequencies than at the higher uplink and downlink RF frequencies. Table 5.3 lists four switch matrix architectures and their advantages and disadvantages. Of these, Ref. 6 states that the coupler crossbar offers the best performance for a large switching matrix (e.g., on the order of 20×20—20 inlets, 20 outlets). Figure 5.16 is a block diagram of such a crossbar matrix. In this case the crosspoints could be PIN diodes or dual-gate FETs.

5.5.4 Intersatellite Links

Carrying the on-board processing concept still further, we now consider intersatellite links that can greatly extend the coverage area of a network. An

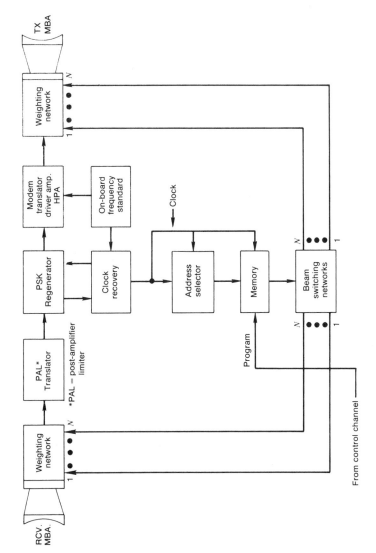

FIGURE 5.15 Block diagram, beam-switching processor.

TABLE 5.3. Comparison of Switch Matrix Architectures

Description	Advantages	Disadvantages
Fan-out/ fan-in	Broadcast mode capability	High input VSWR High insertion loss Redundancy difficult to implement
Single pole/ multiple throw	Low insertion loss	Poor reliability
Rearrangeable switch	Low insertion loss	Poor reliability Random interruptions Control algorithm complicated
Coupler crossbar	Planar construction (minimum size, weight and volume) Broadcast mode capability Redundancy easy to implement Good input/output VSWR* Signal output level independent of the path Enhanced reliability	Difficult feedthroughs Difficult broad- banding Isolation hard to maintain

*VSWR = Voltage Standing Wave Ratio
Source. Reference 6.

intersatellite link (ISL) is a full duplex link between two satellites, and other similar links can be added providing intersatellite connectivity among satellites in a large constellation. A number of military satellite constellations (e.g., MILSTAR) have cross-linking capability. The 58–62-GHz band has been assigned for this purpose by the ITU. Laser cross-links are actively under consideration by the U.S. Department of Defense.

An ISL capability can have a significant impact on system design. The following three system design characteristics are most directly affected:

1. *Connectivity.* The ISL can be used to provide connectivity among users served by regional satellites still retaining the required level of interconnectivity within a community of users. It is assumed that each satellite in the system will serve a region with a high intraregion community of interest. The community of interest among regions will be lower. Thus, each satellite will handle the high-traffic-intensity intraregion and cross-links will serve as tandem relays for the lower intensity traffic between regions.

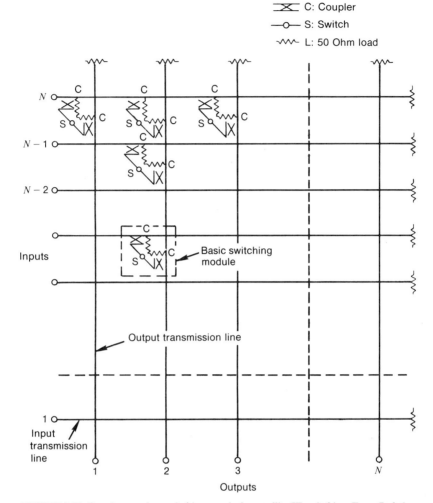

FIGURE 5.16 Coupler crossbar switching matrix for satellite IF switching. From Ref. 6.

2. *Capacity.* As described previously, uplinks and downlinks will have a high fill factor, and with cross-links implemented, low-traffic-intensity uplinks and downlinks will not be required to serve other regions with low interregional community of interest.

3. *Coverage.* Many satellite systems are designed for worldwide coverage where users of one satellite footprint require connectivity to users not in view of that satellite. The use of cross-links eliminates the need of earth relay at a dual-antenna earth station. The cross-link saves money and reduces propagation delay. Generally, an ISL is more economic than adding the additional uplink and downlink to the next satellite.

The cross-link concept incurs other advantages:

☐ Not affected by climatic conditions (e.g., the link does not pass through the atmosphere, assuming geostationary satellites).
☐ Low antenna noise temperature.
☐ Low probability of earth-based intercept.
☐ For military systems using 60-GHz band, there is a low probability of ground-based jamming owing to satellite antenna discrimination and the high atmospheric absorption in the 60-GHz band.

An intersatellite link or cross-link system consists of four subsystems: receiver, transmitter, antenna or lens subsystem, and acquisition and tracking subsystem. The subsystems perform the following functions (Ref. 6):

1. *Receiver.* The receiver, in the generic case, detects, demodulates, and decodes the received signal. It interfaces the uplink/downlink system by cross-strapping at baseband. In some implementations this may also require format/protocol conversion. In a primitive cross-link system there may be no decoding and minimal format conversion.

2. *Transmitter.* In the general case, the transmitter selects the traffic for cross-linking. This may be accomplished by detecting a unique header word for routing. The signal for cross-linking is then encoded, which, in turn, modulates a carrier (or a laser); the signal is then upconverted and amplified. In a primitive system, there may not be any coding step.

3. *Antenna / Lens Subsystem.* For conventional RF transmission a suitable dish or lens antenna is required to radiate the transmitted signal and receive the incoming signal. For an optical system, a suitable lens or mirror, or combination, would be required to direct the laser signal.

4. *Tracking and Acquisition.* The two satellites involved, whether in a geo- or nongeostationary orbit, require an antenna system that acquires and tracks to an appropriate accuracy, usually specified to some fraction of a beamwidth. Such techniques as raster scan and monopulse can be used for tracking. Initial pointing may be ground controlled from the TT & C* or master station/ stations. It should be noted that it is not necessary to orient the entire spacecraft to these accuracies, but only the ISL antenna reflector or feed.

5.6 CODING AND CODING GAIN

5.6.1 Introduction

Error rate is a principal design factor for digital transmission systems. PCM speech systems can tolerate a BER as degraded as 10^{-2} for intelligibility and

---

* TT & C = Telemetry, Tracking and Command.

CVSD (continuous variable slope delta modulation) remains intelligible with BERs of nearly 10^{-1}. For conventional speech telephony the gating error rate value is determined by supervisory signaling. This value of BER should be better than 1×10^{-3}. CCITT gives a value of 1×10^{-4} for telegraph/telex traffic end-to-end and 1×10^{-5} for computer data. Designers of transmission systems for computer networks will differ with this latter value, and BER values of 10^{-6} end-to-end or better may be the performance values required for the transmission of computer data.

To design a transmission system to meet a specified error performance, we must first consider the cause of errors. Let us assume that intersymbol interference is negligible. Then errors derive from insufficient signal-to-noise ratio, which results in bit mutilation by thermal noise peaks (additive white Gaussian noise or AWGN). Error rate is a function of signal-to-noise ratio or E_b/N_0 (see Section 2.11.1.1). Obviously, we can achieve a desired BER on a link by specifying a very high E_b/N_0. On many satellite and tropospheric scatter links, this may not be economically feasible.

Errors derive from insufficient E_b/N_0 because of inferior design, equipment deterioration, or fading. On unfaded links or during unfaded conditions, these errors are random in nature, and during fading errors are predominantly bursty in nature. The length of the burst can be related to the fade duration.

There are several tools available to the design engineer to achieve a desired link error rate. Obviously the first is to specify sufficient E_b/N_0 for the waveform selected, adding a margin for link deterioration due to equipment aging. We can use a similar approach for fading as described in Chapter 2.

Another approach is to specify a lower E_b/N_0 for a BER say in the range of 1×10^{-4} and implement ARQ (automatic repeat request). This is usually done on an end-to-end basis or a section-by-section basis when multilayer protocols are implemented. With ARQ, data messages are built up at the originating end on a block or packet basis. Each block or packet has appended a "block check count" or parity tail. This tail is generated at the originating end of the link by a processor that determines the parity characteristics of the message or uses a cyclic redundancy check (CRC), and the tail is the remainder of that check, often 16 bits in length. At the receiving end a similar processing technique is used, and the locally derived remainder is compared to the remainder received from the distant end. If they are the same, the message is said to be error free, and if not, the block or packet is in error.

There are several ARQ implementations. One is called stop-and-wait ARQ. In this case, if the block is error free, the receiver transmits an ACK (acknowledgment) to the transmitting end, which, in turn, sends the next block or packet. If the block is in error, the receiver sends a NACK (negative acknowledgment) to the transmitter, which now repeats the block just sent. It continues to repeat it until the ACK signal is received.

A second type of ARQ is variously called "continuous ARQ" or "go-back-N" ARQ. At the transmit end, in this case, sending of blocks is continuous. There is also accounting information or sequential block or packet numbering in each block header. Similar parity checking is carried out as before. When the receive end encounters a block in error, it identifies the errored block to the transmitter, which intersperses the repeated blocks (with proper identifier) with its regular, continuous transmissions. Obviously, continuous ARQ is a more complex implementation than stop-and-wait ARQ. Considerably more processing and buffer memory are required at each end. It also follows that the more circuit delay, the more memory is required.

On satellite circuits carrying data, delay is an important design consideration. The delay from earth to a geostationary satellite is 125 msec or somewhat more. On a stop-and-wait ARQ system, considering the satellite link only, the period of "wait" is 4×125 msec or 0.5 sec. This is wasted time and inefficient. Continuous or go-back-N ARQ does not waste this time, but is more expensive to implement.

Another approach is to use error-correction coding. It offers a number of advantages and at least one major disadvantage. It is up to the system designer to tradeoff system overbuild, ARQ implementations, and error correction for the optimum error-control technique to implement.

5.6.2 Basic Forward Error Correction

Forward error correction (FEC) is a method of error control that employs the adding of systematic redundancy at the transmit end of a link such that errors caused by the medium can be corrected at the receiver by means of a decoding algorithm.

Figure 5.17 shows a digital communication system with FEC. The binary data source generates information bits at R_s bits per second. These information bits are encoded for FEC at a code rate R. The output of the encoder is a binary sequence at R_c symbols per second. This output is related to the bit rate by the following expression:

$$R_c = \frac{R_s}{R} \tag{5.6}$$

R, the code rate, is the ratio of the number of information bits to the number of encoded symbols for binary transmission. For example, if the information bit rate were 2400 bps and code rate were $\frac{1}{2}$, then the symbol rate (R_c) would be 4800 symbols per second.

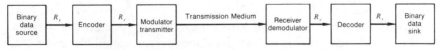

FIGURE 5.17 Simplified diagram of the FEC technique.

The encoder output sequence is then modulated and transmitted over the transmission medium or channel. Demodulation is then performed at the receive end. The output of the demodulator is R_c symbols per second, which is fed to the decoder. The decoder output to the data sink is the recovered 2400 bits per second (R_s).

The major advantages of an FEC system are:

☐ No feedback channel required as with the ARQ system.

☐ Constant information throughput (e.g., no stop-and-wait gaps).

☐ Decoding delay is generally small and constant.

☐ Significant coding gain for an AWGN channel.

There are two basic disadvantages to an FEC system. To effect FEC, for a fixed information bit rate, the bandwidth must be increased, because, by definition, the symbol rate on the transmission channel is greater than the information bit rate. There is also the cost of the added complexity of the coder and decoder.

5.6.3 Coding Gain

For a given information bit rate and modulation waveform (i.e., QPSK, 8-ary FSK, etc.), the required E_b/N_0 for a specified BER with FEC is less than the E_b/N_0 required without FEC. The coding gain is the difference in E_b/N_0 values.

The use of a selected FEC method in satellite communications can effect major savings simply by adding processors at each end of a link. There is no reason why FEC cannot be used on other digital systems where from 1 to 6 dB of additional gain is required under unfaded conditions. FEC with interleaving can show an even greater improvement under fading conditions.

Consider a satellite downlink where, by coding, the satellite EIRP can be reduced by half (e.g., 3 dB) without affecting performance. This could allow the use of a satellite transmitter with half the output power that would be required without FEC implemented. Transponder transmitters are a major weight factor in satellites. Reducing the output power in half could possibly reduce the transmitter weight by 75% (including its power supply). Battery weight may be reduced by perhaps 50% (batteries are used to power the transponder during satellite eclipse), with the concurrent reduction of solar cells. It is not only the savings in the direct cost of these items, but also the savings in lifting weight of the satellite to place it in orbit, whether by space shuttle or rocket booster.

With unfaded conditions, assuming random errors, coding gains from 1 to 6 dB or more can be realized. The amount of gain achievable under these conditions is a function of the modulation type (waveform), the code employed, the coding rate [equation (5.6)], the constraint length, the type of decoder, and the demodulation approach.

An FEC system can use one of two broad classes of codes: block and convolutional. These are briefly described below.

5.6.4 FEC Codes

5.6.4.1 Block Codes

With block-coding techniques each group of K consecutive information bits is encoded into a group of N symbols for transmission over the channel. Normally, the K information bits are located at the beginning of the N-symbol block code, and the last $N - K$ symbols correspond to the parity check bits formed by taking the modulo-2 sum of certain sets of K information bits. Block codes containing this property are referred to as systematic codes. The encoded symbols for the $(K + 1)$th information bit and beyond are completely independent of the symbols generated for the first K information bits and, hence, cannot be used to help decode the first group of K information bits at the far end decoder. This essentially says that blocks are independent entities, and one block has no enhancement capability on another.

Because N symbols are used to represent K bits, the code rate R of such a block code is K/N bits per symbol, or

$$R = \frac{K}{N} \tag{5.7}$$

Note that equation (5.7) is just a restatement of equation (5.6), but with different notation.

Block codes are often described with the notation such as $(7, 4)$, meaning $N = 7$ and $K = 4$. Here the information bits are stored in $K = 4$ storage devices and the device is made to shift $N = 7$ times. The first K symbols of the block output are the information symbols, and the last $N - K$ symbols are a set of check symbols that form the whole N-symbol word. A block code may also be identified with the notation (N, K, t), where t corresponds to the number of errors in a block of N symbols that the code will correct.

5.6.4.2 Bose–Chaudhuri–Hocquenghem (BCH) Codes

Binary BCH codes are a large class of block codes with a wide range of code parameters. The so-called primitive BCH codes, which are the most common, have codeword lengths of the form $2^m - 1$, $m \geq 3$, where m describes the "degree' of the generating polynomial (e.g., the highest exponent value of the polynomial). For BCH codes, there is no simple expression relating to the N, K, and t parameters. Table 5.4 gives these parameters for all binary BCH codes of length 255 and less. In general, for any m and t, there is a BCH code

TABLE 5.4. Primitive BCH Codes of Length ' 255 and Less

N	K	t	N	K	t	N	K	t
7	4	1	127	64	10	255	87	26
				57	11		79	
15	11	1		50	13		71	29
	7	2		43	14		63	30
				36	15		55	31
31	26	1		29	21		47	42
	21	2		22	23		45	43
	16	3		15	27		37	45
	11	5		8	31		29	47
	6	7	255	247	1		21	55
63	57	1		239	2		13	59
	51	2		231	3		9	63
	45	3		223	4			
	39	4		215	5			
	36	5		207	6			
	30	6		199	7			
	24	7		191	8			
	18	10		187	9			
	16	11		179	10			
	10	13		171	11			
	7	15		163	12			
				155	13			
127	120	1		147	14			
	113	2		139	15			
	106	3		131	18			
	99	4		123	19			
	92	5		115	21			
	85	6		107	22			
	78	7		99	23			
	71	9		99	25			

Source. Reference 6.

of length $2^m - 1$ that corrects any combination of t errors and requires no more than m/t parity check symbols.

BCH codes are cyclic and are characterized by a generating polynomial. A selected group of primitive generating polynomials is given in Table 5.5. The encoding of a BCH code can be performed with a feedback shift register of length K or $N - K$ (from Ref. 6).

5.6.4.3 Golay Code

The Golay (23, 12) code is a linear cyclic binary block code that is capable of correcting up to three errors in a block of 23 binary symbols. There are two

TABLE 5.5. Selected Primitive Binary Polynomials

m	$p(x)$
3	$1 + x + x^3$
4	$1 + x + x^4$
5	$1 + x^2 + x^5$
6	$1 + x + x^6$
7	$1 + x^3 + x^7$
8	$1 + x^2 + x^3 + x^4 + x^8$
9	$1 + x^4 + x^9$
10	$1 + x^3 + x^{10}$
11	$1 + x^2 + x^{11}$
12	$1 + x + x^4 + x^6 + x^{12}$
13	$1 + x + x^3 + x^4 + x^{13}$
14	$1 + x + x^6 + x^{10} + x^{14}$
15	$1 + x + x^{15}$

Source. Reference 6.

possible generator polynomials for the Golay code as follows:

$$g_1(x) = 1 + x^2 + x^4 + x^5 + x^6 + x^{10} + x^{11} \qquad (5.8)$$

$$g_2(x) = 1 + x + x^5 + x^6 + x^7 + x^9 + x^{11} \qquad (5.9)$$

The encoding of the Golay code can be performed with an 11-stage feedback shift register with feedback connections determined by the coefficients of either $g_1(x)$ or $g_2(x)$. There are efficient decoding techniques available that are implemented in hardware. It should be noted that often an overall parity check bit is appended to each codeword.

5.6.4.4 Convolutional codes

Viterbi (Refs. 7 and 8) defines a convolutional encoder as the following:

A linear finite state machine consisting of a K-stage shift register and n linear algebraic function generators. The input data which are usually, but not necessarily always, binary, are shifted along the register b bits at a time.

Figure 5.18 is an example of a convolutional encoder. If there were a five-stage shift register where the input data were shifted along 1 bit at a time and we had three modulo-2 adders (e.g., $n = 3$), using the Viterbi notation, the code would be described as a 5, 3, 1 convolutional code (e.g., $K = 5$, $n = 3$, $b = 1$).

In Figure 5.18 information bits are shifted to the right 1 bit at a time ($b = 1$) through the K-stage shift register as new information bits enter from

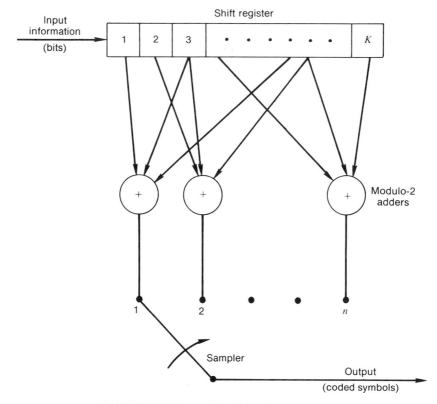

FIGURE 5.18 A convolutional encoder, $n = 3$, $b = 1$.

the left. Bits out of the last stage are discarded. The bits are shifted one position each T sec, where $1/T$ is the information rate in bits per second. The modulo-2 adders are used to form the output coded symbols, each of which is a binary function of a particular subset of the information bits in the shift register. The output coded symbols can be seen to depend on a sequence of K information bits, and thus we define the constraint length K. The constraint length K is defined as the total number of binary register stages in the encoder.

If we feed in 1 bit at a time to the encoder (e.g., $b = 1$), each coded symbol carries an average of $1/n$ information bits, and the code is said to have a rate R of $1/n$. For the more generalized case where $b \neq 1$, the rate R is expressed as

$$R = \frac{b}{n} \tag{5.10}$$

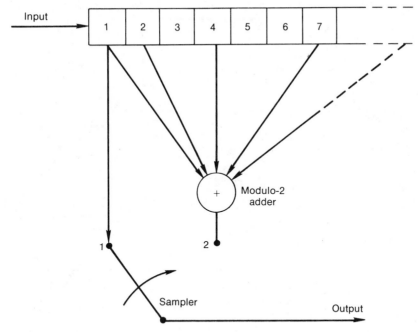

FIGURE 5.19 A systematic convolutional encoder.

where b = number of bits shifted into the register at a time. [See Equation (5.7), and the similarity.]

In Figure 5.18, when the first modulo-2 adder is replaced by a direct connection to the first stage of the shift register, the first symbol becomes a replica of the information bit. Such an encoder is called a systematic convolutional encoder, as shown in Figure 5.19.

Let us examine a rate $\frac{1}{2}$ constraint length $K = 3$ encoder shown in Figure 5.20. This figure indicates the outputs for a particular binary input sequence assuming the state (i.e., the previous two data bits in the register) were zero. Modulo-2 addition (e.g., $0 \oplus 0 = 0, 0 \oplus 1 = 1, 1 \oplus 0 = 1$, and $1 \oplus 1 = 0$) is used. With the input and output sequences defined from right-to-left, the first three input bits—0, 1, and 1—generate the code outputs 00, 11, and 01, respectively. The outputs are then multiplexed (commutated) into a single code sequence. For this rate $\frac{1}{2}$ case, of course, the output code sequence has twice the symbol rate as the incoming data sequence.

A convolutional code is often shown by means of a tree diagram, as shown in Figure 5.21. At each branch (node) of the tree the input information bit determines which direction (i.e., which branch) will be taken, following the convention "up" for zero and "down" for one. Restated, if the first input bit is a zero, the code symbols are those shown on the first upper branch, while if it

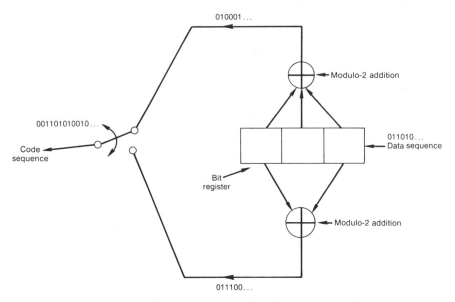

FIGURE 5.20 Rate $\frac{1}{2}$ convolutional encoder with constraint length $K = 3$.

is a one, then the output symbols are those shown on the first lower branch. Similarly, if the second input bit is a zero, we trace the tree diagram to the next upper branch, while if it is a one, we trace the diagram downward. In this manner all 32 possible outputs for the first five inputs may be traced.

From the tree diagram in Figure 5.21 it also becomes clear that after the first three branches the structure becomes repetitive. In fact, we readily recognize that beyond the third branch the code symbols on branches emanating from the two nodes labeled "a" are identical, and soon, for all the similarly labeled pairs of nodes. The reason for this is obvious from examination of the encoder. As the fourth bit enters the coder at the right, the first data bit falls off on the left and no longer influences the output code symbols. Consequently, the data sequences $100xy\ldots$ and $000xy\ldots$ generate the same code symbols after the third branch and, as is shown in the tree diagram, both nodes labeled "a" can be joined together.

This leads to redrawing the tree diagram as shown in Figure 5.22. This is called a trellis diagram, since a trellis is a treelike structure with remerging branches. A convention is adopted here where the code branches produced by a zero input bit are shown as solid lines and code branches produced by a one input bit are shown as dashed lines.

The completely repetitive structure of the trellis diagram suggests a further reduction in the representation of the code to the state diagram in Figure 5.23. The "states" of the state diagram are labeled according to the nodes of the trellis diagram. However, since the states correspond merely to the last two

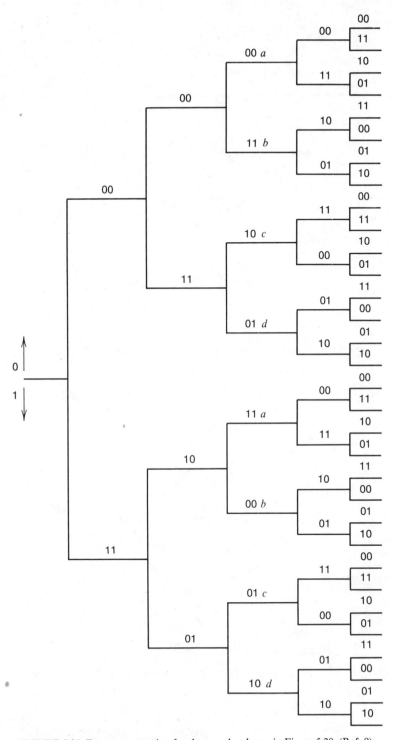

FIGURE 5.21 Tree representation for the encoder shown in Figure 5.20. (Ref. 9).

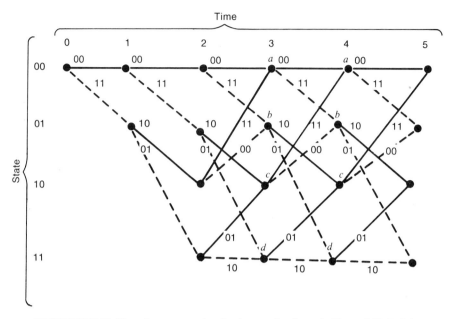

FIGURE 5.22 Trellis code representation for the encoder shown in Figure 5.20 (Ref. 9).

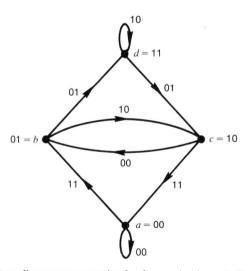

FIGURE 5.23 State-diagram representation for the encoder shown in Figure 5.20 (Ref. 9).

input bits to the coder, we may use these bits to denote the nodes or states of this diagram.

It can be observed that the state diagram can be drawn directly observing the finite-state machine properties of the encoder and, particularly, the fact that a four-state directed graph can be used to represent uniquely the input–output relation of the finite-state machine, since the nodes represent the previous two bits, while the present bit is indicated by the transition branch. For example, if the encoder (synonomous with finite-state machine) contains the sequence 011, this is represented in the diagram by the transition from state $b = 01$ to state $d = 11$ and the corresponding branch indicates the code symbol outputs 01.

This example will be used when we describe the Viterbi decoder.

5.6.4.5 Convolutional Decoding

Decoding algorithms for block and convolutional codes are quite different. Because a block code has a formal structure, advantage can be taken of the known structural properties of the words or the algebraic nature of the constraints among the symbols used to represent an information sequence. One class of powerful block codes with well-defined decoding algorithms is the BCH codes (Section 5.6.4.2).

The decoding of convolutional codes can be carried out by a number of different techniques. Among these are the simpler threshold and feedback decoders and those more complex decoders with improved performance (coding gain) such as the Viterbi decoder and sequential decoder. These techniques depend on the ability to home in on the correct sequence by designing efficient search procedures that discard unlikely sequences quickly. The sequential decoder differs from most other types of decoders in that when it finds itself on the wrong path in the tree, it has the ability to search back and forth, changing previously decoded information bits until it finds the correct tree path. The frequency with which the decoder has to search back and the depth of the backward searches is dependent on the value of the channel BER.

An important property of the sequential decoder is that, if the constraint length is large enough, the probability that the decoder will make an error approaches zero (i.e., a BER better than 1×10^{-9}). One cause of error is overflow, being defined as the situation in which the decoder is unable to perform the necessary number of computations in the performance of the tree search. If we define a computation as a complete examination of a path through the decoding tree, a decoder has a limit on the number of computations that it can make per unit time. The number of searches and computations is a function of the number of errors arriving at the decoder input, and the number of computations that must be made to decode one information bit is a random variable. An important parameter for a decoder is the average number of computations per decoded information bit. As long as the probability of bit error is not too high, the chances of decoder overflow will be low, and satisfactory performance will result.

For the previous discussion it has been assumed that the output of a demodulator has been a hard decision. By "hard" decision we mean a firm, irrevocable decision. If these were soft decisions instead of hard decisions, additional improvement in error performance (or coding gain) on the order of several decibels can be obtained. By a "soft" decision we mean that the output of a demodulator is quantized into four or eight levels (e.g., two- or three-bit quantization respectively), and then certain decoding algorithms can use this additional information to improve the output BER. Sequential and Viterbi decoding algorithms can use this soft decision information effectively, giving them an advantage over algebraic decoding techniques, which are not designed to handle the additional information provided by the soft decision (Ref. 9).

The soft decision level of quantization is indicated conventionallly by the letter Q, which indicates the number of bits in the quantized decision sample. If $Q = 1$, we are dealing with a hard decision demodulator output; $Q = 2$ indicates a quantization level of 4; $Q = 3$ a quantization level of 8; etc.

5.6.4.6 Viterbi Decoding

The Viterbi decoder is one of the more common decoders on links using convolutional codes. The Viterbi decoding algorithm is a path maximum-likelihood algorithm that takes advantage of the remerging path structure (see Figure 5.22) of convolutional codes. By path maximum-likelihood decoding we mean that all the possible paths through the trellis, a Viterbi decoder chooses the path, or one of the paths, most likely in the probabilistic sense to have been transmitted. A brief description of the operation of a Viterbi decoder using a demodulator giving hard decisions is now given.

For this description our model will be a binary symmetric channel (i.e., BPSK, BFSK). Errors that transform a channel code symbol 0 to 1 or 1 to 0 are assumed to occur independently from symbols with a probability of p. If all input (message) sequences are equally likely, the decoder that minimizes the overall path error probability for any code, block, or convolutional is one which examines the error-corrupted received sequence, which we may call y_1, $y_2, \ldots, y_j \ldots$ and chooses the data sequence corresponding to that sequence that was transmitted or $x_1, x_2, \ldots, x_j \ldots$ which is closest to the received sequence as measured by the Hamming distance. The Hamming distance can be defined as the transmitted sequence that differs from the received sequence by the minimum number of symbols.

Consider the tree diagram (typically Figure 5.21). The preceding statement tells us that the path to be selected in the tree is the one whose code sequence differs in the minimum number of symbols from the received sequence. In the derived trellis diagram (Figure 5.22) it was shown that the transmitted code branches remerge continually. Thus the choice of possible paths can be limited in the trellis diagram. It is also unnecessary to consider the entire received sequence at any one time to decide upon the most likely transmitted sequence or minimum distance. In particular, immediately after the third branch (Figure 5.22) we may determine which of the two paths leading to node or state "a" is more likely to have been sent. For example, if 010001 is received, it is clear that this is a Hamming distance 2 from 000000, while it is a distance 3 from 111011. As a consequence, we may exclude the lower path into node "a." For no matter what the subsequent received symbols will be, they will affect the Hamming distances only over subsequent branches after these two paths have remerged and, consequently, in exactly the same way. The same can be said for pairs of paths merging at the other three nodes after the third branch. Often, in the literature, the minimum distance path of the two paths merging at a given node is called the "survivor." Only two things have to be remembered: the minimum distance path from the received sequence (or survivor) at each node and the value of that minimum distance. This is necessary because at the next node level we must compare two branches merging at each node level, which were survivors at the previous level for different nodes. This can be seen in Figure 5.22 where the comparison at node "a" after the fourth branch is among the survivors of the comparison of nodes "a" and "c" after the third branch. For example, if received sequence over the first four branches is 01000111, the survivor at the third node level for node "a" is 000000 with distance 2 and at node "c" it is 110101, also with distance 2. In going from the third node level to the fourth, the received sequence agrees precisely with the survivor from "c" but has distance 2 from the survivor from "a." Hence, the survivor at node "a" of the fourth level is the data sequence 1100 that produced the code sequence 11010111, which is at minimum distance 2 from the received sequence.

In this way we may proceed through the received sequence and at each step preserve one surviving path and its distance from the received sequence, which is more generally called a "metric." The only difficulty which may arise is the possibility that, in a given comparison between merging paths, the distances or metrics are identical. In this case we may simply flip a coin, as is done for block code words at equal distances from the received sequence. For even if both equally valid contenders were preserved, further received symbols would affect both metrics in exactly the same way and thus not further influence our choice.

Another approach to the description of the algorithm can be obtained from the state diagram representation given in Figure 5.23. Suppose we sought that path around the directed state diagram arriving at node "a" after the kth transition, whose code symbols are at a minimum distance from the received sequence. But, clearly, this minimum distance path to node "a" at time k can only be one of two candidates: the minimum distance path to node "a" at time $k - 1$ and the minimum distance path to node "c" at time $k - 1$. The comparison is performed by adding the new distance accumulated in the kth transition by each of these paths to their minimum distances (metrics) at time $k - 1$.

Thus it appears that the state diagram also represents a system diagram for this decoder. With each node or state, we associate a storage register that remembers the minimum distance path into the state after each transition as well as a metric register that remembers its (minimum) distance from the received sequence. Furthermore, comparisons are made at each step between the two paths that lead into each node. Thus, four comparators must also be provided.

We will expand somewhat on the distance properties of convolutional codes following the example given in Figure 5.20. It should be noted that as with linear block codes, there is no loss in generality in computing the distance from the all-zeros code word to all other code words, for this set of distances is the same as the set of distances from any specific codeword to all the others.

For this purpose we may again use either the trellis diagram or state diagram. First of all we redraw the trellis diagram in Figure 5.22 labeling the branches according to their distances from the all-zeros path. Now consider all the paths that merge with the all-zeros path for the first time at some arbitrary node "j." From the redrawn trellis diagram (Figure 5.24), it can be seen that of these paths there will be just one path at distance 5 from the all-zeros path and this diverged from it three branches back. Similarly, there are two at distance 6 from it, one which diverged four branches back and the other which diverged five branches back, and so forth. It should be noted that the input bits for the distance 5 path are $00\ldots01000$ and thus differ in only one input

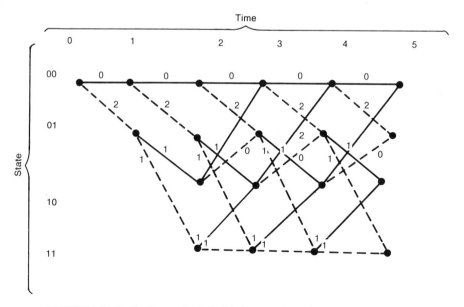

FIGURE 5.24 Trellis diagram labeled with distances from the all-zeros path (Ref. 9).

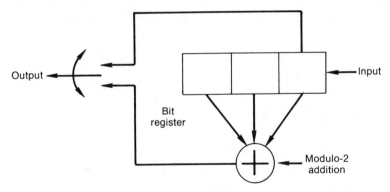

FIGURE 5.25 Systematic convolutional encoder, $K = 3$, $R = \frac{1}{2}$.

bit from the all-zero path. The minimum distance, sometimes called the "minimum free distance," among all paths is thus seen to be 5. This implies that any pair of channel errors can be corrected, for two errors will cause the received sequence to be at a distance 2 from the transmitted (correct) sequence, but it will be at least at distance 3 from any other possible code sequence. In this way the distances of all paths from all-zeros (or any arbitrary) path can be determined from the trellis diagram.

5.6.4.7 Systematic and Nonsystematic Convolutional Codes *

The term "systematic" convolutional code refers to a code on each of whose branches the uncoded information bits are included in the encoder output bits generated by that branch. Figure 5.25 shows an encoder for rate $\frac{1}{2}$ and $K = 2$ that is systematic.

For linear block codes, any nonsystematic code can be transformed into a systematic code with the same block distance properties. This is not the case for convolutional codes. The reason for this is that the performance of a code on any channel depends largely the relative distance between codewords and, particularly, on the minimum free distance. Making the code systematic, in general, reduces the maximum possible free distance for a given constraint length and code rate. For example, the maximum minimum-free-distance systematic code for $K = 3$ is that of Figure 5.24 and this has $d = 4$, while the nonsystematic $K = 3$ code of Figure 5.20 has a minimum free distance of $d = 5$. Table 5.6 shows the maximum free distance for $R = \frac{1}{2}$ systematic and nonsystematic codes for $K = 2$ through 5. It should be noted that for large constraint lengths the results are even more widely separated.

* Sections 5.6.4.5, 5.6.4.6, and 5.4.6.7 have been abstracted from Ref. 9.

TABLE 5.6. Comparison of Systematic and Nonsystematic $R = \frac{1}{2}$ Code Distances

	Maximum, Minimum Free Distance	
K	Systematic	Nonsystematic
2	3	3
3	4	5
4	4	6
5	5	7

Source. Reference 9.

5.6.5 Channel Performance of Uncoded and Coded Systems

For uncoded systems a number of modulation implementations are reviewed in the presence of additive white Gaussian noise (AWGN) and with Rayleigh fading. The AWGN performance of BPSK, QPSK, and 8-ary PSK is shown in Figure 5.26. AWGN is typified by thermal noise or wideband white noise jamming. The demodulator for this system requires a coherent phase reference.

Another similar implementation is differentially coherent phase-shift keying. This is a method of obtaining a phase reference by using the previously received channel symbol. The demodulator makes its decision based on the change in phase from the previous to the present received channel symbol. Figure 5.27 gives the performance of DBPSK and DQPSK with values of BER versus E_b/N_0.

Figure 5.28 gives performance for M-ary FSK.

Independent Rayleigh fading can be assumed during periods of heavy rainfall on satellite links operating above about 10 GHz (see Chapter 6). Such fading can severely degrade error rate performance. The performance with this type of channel can be greatly improved by providing some type of diversity. Here we mean providing several independent transmissions for each information symbol. In this case we will restrict the meaning to some form of time diversity that can be achieved by repeating each information symbol several times and using interleaving/deinterleaving for the channel symbols. Figure 5.29 gives binary bit error probability for several orders of diversity (L = order of diversity; $L = 1$, no diversity) for the mean bit energy-to-noise ratio ($\overline{E}_b/N_0$). This figure shows that for a particular error rate there is an optimum amount of diversity. The modulation is binary FSK.

Table 5.7 recaps error performance versus E_b/N_0 for the several modulation types considered. The reader should keep in mind that the values for E_b/N_0 are theoretical values. A certain modulation implementation loss should be added for each case to derive practical values. The modulation implementation loss value in each case is equipment driven. (See Table 2.16).

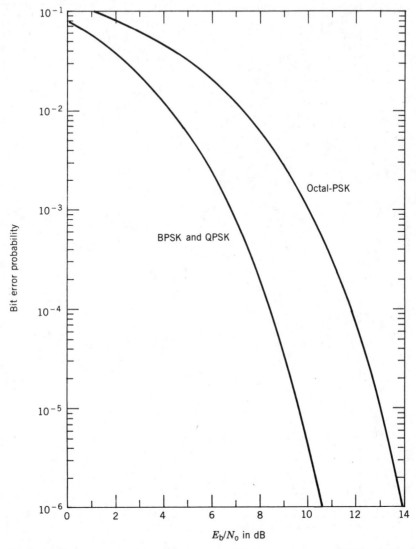

FIGURE 5.26 Bit error probability versus E_b/N_0 performance of coherent BPSK, QPSK, and 8-ary PSK. (Ref. 9.)

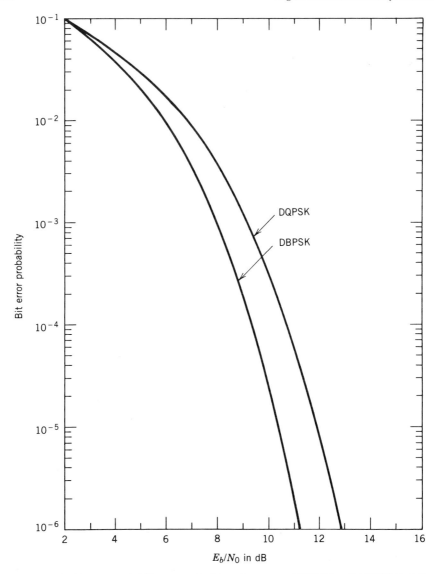

FIGURE 5.27 Bit error probability versus E_b/N_0 performance of DBPSK and DQPSK (Ref. 9).

5.6.5.1 Theoretical Performance for Coded Systems

Figure 5.30 gives the theoretical performance for extended Golay (block) coding with BPSK/QPSK modulation. Figure 5.31 provides uncoded channel error rate versus bit error rate for several BCH codes with block length of 127. k represents the largest number of information bits per block and E is the number of channel errors that each code is capable of correcting.

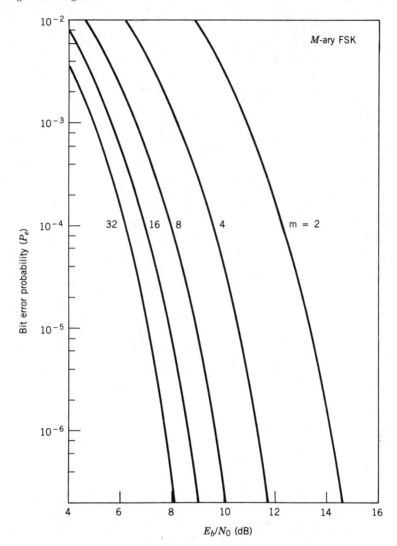

FIGURE 5.28 Bit error probability for M-ary FSK; $M = 2$ for BFSK (Ref. 6).

Figure 5.32 gives the coding gain for several convolutional codes where the demodulator provides three-bit quantizing. The modulation is BPSK. Figure 5.33 shows the performance of a rate $\frac{1}{2}$ convolutional code with $K = 7$ and the modulation is DBPSK. T is a quantization parameter, usually some value of the standard deviation of the unquantized demodulator outputs. In this case $T = 0.5$ or 0.5 times the standard deviation.

Table 5.8 is a summary of E_b/N_0 requirements and coding gains of $K = 7$, rate $\frac{1}{2}$ Viterbi-decoded convolutional coding systems with several modulation

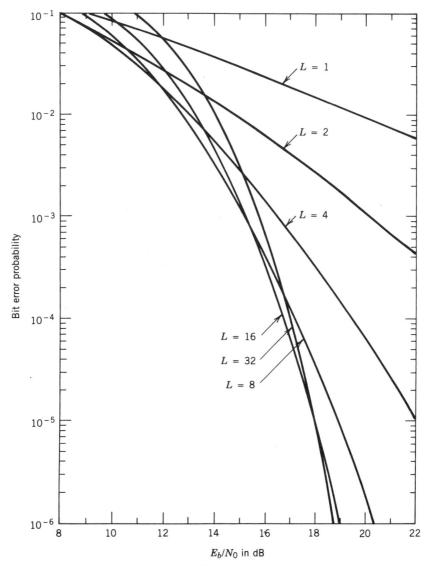

FIGURE 5.29 Bit error probability versus $\overline{E}_b/N_0$ performance of binary FSK on a Rayleigh fading channel for several orders of diversity. L = order of diversity. (Ref. 9.)

TABLE 5.7. Summary of Uncoded System Performance

Channel	Modulation/ Demodulation	E_b/N_0 (dB) Required for Given Bit Error Rate						
		10^{-1}	10^{-2}	10^{-3}	10^{-4}	10^{-5}	10^{-6}	10^{-7}
Additive white Gaussian noise	BPSK and QPSK	-0.8	4.3	6.8	8.4	9.6	10.5	11.3
	Octal-PSK	1.0	7.3	10.0	11.7	13.0	13.9	14.7
	DBPSK	2.1	5.9	7.9	9.3	10.3	11.2	11.9
	DQPSK	2.1	6.8	9.2	10.8	12.0	12.9	13.6
	Noncoherently demodulated binary FSK	5.1	8.9	10.9	12.3	13.4	14.2	14.9
	Noncoherently demodulated 8-ary MFSK	2.0	5.2	7.0	8.2	9.1	9.9	10.5
Independent Rayleigh fading	Binary FSK, $L = 1$	9.0	19.9	30.0	40.0	50.0	60.0	70.0
	Binary FSK, $L = 2$	7.9	14.8	20.2	25.3	30.4	35.4	40.4
	Binary FSK, $L = 4$	8.1	13.0	16.5	19.4	22.1	24.8	27.3
	Binary FSK, $L = 8$	8.7	12.8	15.3	17.2	18.9	20.5	22.0
	Binary FSK, $L = 16$	9.7	13.2	15.3	16.7	18.0	19.1	20.0
	Binary FSK, $L = 32$	10.9	14.1	15.8	17.1	18.1	18.9	19.7

Source. Reference 9.

types at a bit error rate of 10^{-5}. Table 5.9 summarizes a number of FEC schemes with their respective coding gains.

5.7 LINK BUDGETS FOR DIGITAL LINKS ON PROCESSING SATELLITES

The link budget is a tool used to dimension or size a satellite communication system. In Section 4.3.9 link budgets for analog systems were described. There an uplink with its associated downlink were considered jointly [equation (4.28)] to calculate system S/N. With digital systems utilizing a processing satellite, uplinks and downlinks are treated separately, calculating the required E_b/N_0 on each for a specified BER. Otherwise the approach is quite similar as that described in Section 4.3.9.

The uplink and downlink are just two more links of a larger system unless the user is colocated at the satellite terminal on each end. For the larger system, link BERs are usually specified in such a way that the user BER meets a need requirement.

Example 4. A specific uplink at 6 GHz working into a processing satellite is to have a BER of 1×10^{-5}. The modulation is QPSK and the data rate is

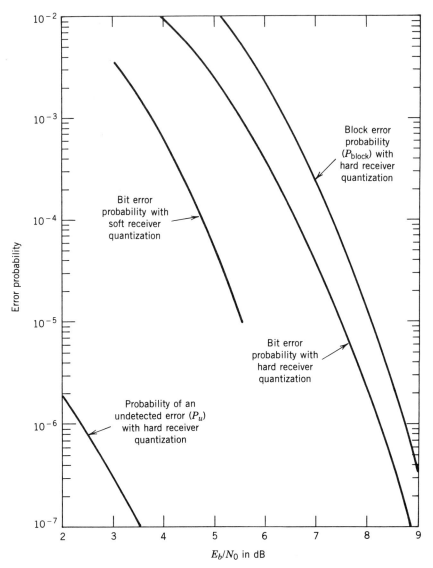

FIGURE 5.30 Block, bit, and undetected error probabilities versus E_b/N_0 for BPSK and QPSK modulation using extended Golay coding. AWGN channel assumed. (Ref. 9.)

10 Mbps. The terminal EIRP is $+65$ dBW. The free space loss to the satellite is 199.2 dB. What satellite G/T will be required without coding? What G/T will be required when FEC is employed? The satellite will use an earth coverage antenna. See Table 5.10.

Discussion. Select the required E_b/N_0 first for no coding and then with convolutional coding rate $\frac{1}{2}$ and $K = 7$, Viterbi-decoded. Use a modulation

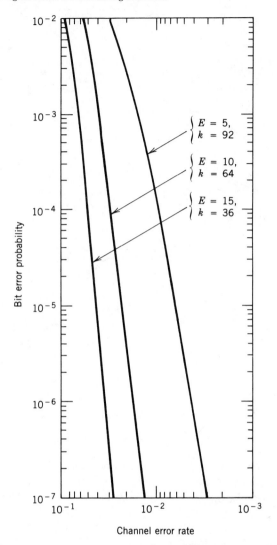

FIGURE 5.31 Decoded bit error probability versus channel (uncoded) error rate performance of several BCH codes with block length 127. (Ref. 9.)

implementation loss of 2.0 dB in both cases. From Figure 5.26, lower left curve, the E_b/N_0 is 9.6 dB for QPSK (uncoded), and from Table 5.7 for the coded system use 6.5 dB. Allow 4 dB of margin. Initially set the satellite G/T at 0 dB/K.

Example 5. A satellite has a $+30$ dBW EIRP downlink at 7.3 GHz. The desired BER is 1×10^{-6}; the modulation is coherent BPSK; FEC is implemented with convolutional encoding and three-bit receiver quantization; the

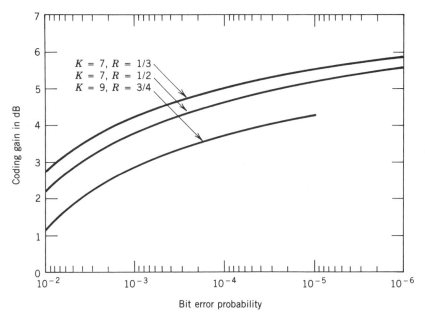

FIGURE 5.32 Coding gain for several convolutional codes with BPSK modulation, AWGN, and three-bit receiver quantization. (Ref. 9.)

bit rate is 45 Mbps and the free space loss is 202.0 dB. What is the terminal G/T assuming a 5-dB margin? See Table 5.11. Assume that the satellite uses a spot beam, and a rate $\frac{1}{2}$ $K = 7$ convolutional code.

Discussion. The required G/T is the sum of 26.18 + 4.9 + 2.0 + 5.0 dB or + 38.08 dB/K. If that value is now substituted for the initial G/T of 0 dB/K, the last entry in the table or "sum" would drop to 0. The value for the required E_b/N_0 was derived first from Figure 5.26, lower left curve, using BER = 1×10^{-6}, thence to Figure 5.32 where we found that the coding gain for $K = 7$, $R = \frac{1}{2}$ is 5.6 dB, we then subtracted. The rainfall loss value was, in a way, arbitrary. As we will show in Chapter 6, 3 dB in this band, at a $10°$ elevation angle, would provide this performance 99.9% of the time for central North America. However, with interleaving using a sufficient interleaving interval (about 4 or 5 sec) and the coding employed would permit us to decrease the excess attenuation due to rainfall to zero. The value would be accommodated in the coding gain. The coding gain used was for AWGN conditions only. It will also provide a fading improvement. Heavy rain manifests itself in fading, which, in the worst case, can be considered a Rayleigh distribution. The incidental losses (i.e., polarization, pointing errors) are good estimates and probably can be improved upon with a real system where firm values can be used.

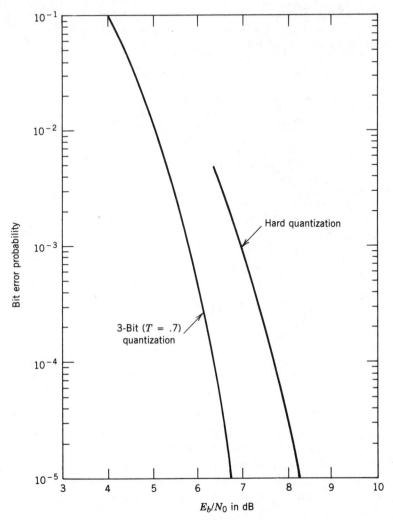

FIGURE 5.33 Bit error probability versus E_b/N_0 performance of a $K = 7$, $R = \frac{1}{2}$ convolutional coding system with DBPSK modulation and an AWGN channel. (Ref. 9.)

5.8 VSAT (VERY SMALL APERTURE TERMINAL)

A very small aperture terminal (VSAT) is a satellite communications terminal with an antenna aperture of 1 m or less. One factor driving the technology in this direction is the networking of personal computers (PCs) by satellite with a DTU (Direct to User) application. Commonly such networking is carried out with an information bit rate of 2400 bps or less. Of course, in the case of the *personal* computer application, low cost is essential.

TABLE 5.8. Summary of the E_b / N_0 Requirements and Coding Gains of $K = 7$, $R = \frac{1}{2}$ Viterbi-Decoded Convolutional Coding Systems with Several Modulation Types at a BER $= 10^{-5}$

Modulation	Number of Bits of Receiver Quantization Per Binary Channel Symbol (see Table 5.9 note)	E_b/N_0 (dB) Required for $P_b = 10^{-5}$	Coding Gain (dB)
Coherent biphase			
BPSK or QPSK	3	4.4	5.2
BPSK or QPSK	2	4.8	4.8
BPSK or QPSK	1	6.5	3.1
Octal-PSK[a]	1	9.3	3.7
DBPSK[a]	3	6.7	3.6
DBPSK[a]	1	8.2	2.1
Differentially[a] coherent QPSK	1	9.0	3.0
Noncoherently demodulated binary FSK	1	11.2	2.1

[a] Interleaving/deinterleaving assumed.

Source. Reference 9.

There are other applications as well. Low cost, geographically dispersed, industrial networks and academic resource sharing are two examples. The terminals may also be used only for news and stock market information dissemination. In this case a central hub station with a larger aperture antenna can transmit to hundreds or even thousands of remote terminals, regionally, nationally, or internationally.

Let's examine the technical question. Can a satellite system be built around the following elementary baseline parameters using today's technology? That is a terminal with a 1 m aperture (39 in) supporting 2400 bps with a BER of 1×10^{-5}.

Consider a sample link at Ku band (14 GHz uplink and 12 GHz downlink). Calculate the free space loss from equation (1.7a) with a 19° elevation angle:

at 14 GHz $L_{dB} = 36.58 + 20 \log 14,200 \text{ MHz} + 20 \log 24,660$ (uplink)

where 24,660 statute miles is the range to the satellite at 19° elevation angle.

$$L_{dB} = 207.5 \text{ dB}$$

at 12 GHz $L_{dB} = 36.58 + 20 \log 12,000 \text{ MHz} + 20 \log 24,660$ (downlink)

$$L_{dB} = 206 \text{ dB}$$

TABLE 5.9. Summary of E_b/N_0 Requirements of Several Coded Communication Systems for a BER $= 10^{-5}$ with BPSK Modulation

Coding Type		Number of Bits Receiver Quantization	Coding[a] Gain (dB)
$K = 7$, $R = \frac{1}{2}$	Viterbi-decoded convolutional	1	3.1
$K = 7$, $R = \frac{1}{2}$	Viterbi-decoded convolutional	3	5.2
$K = 7$, $R = \frac{1}{3}$	Viterbi-decoded convolutional	1	3.6
$K = 7$, $R = \frac{1}{3}$	Viterbi-decoded convolutional	3	5.5
$K = 9$, $R = \frac{3}{4}$	Viterbi-decoded convolutional	1	2.4
$K = 9$, $R = \frac{3}{4}$	Vitervi-decoded convolutional	3	4.3
$K = 24$, $R = \frac{1}{2}$	Sequential-decoded convolutional 20 kbps,[b] 1000-bit blocks	1	4.2
$K = 24$, $R = \frac{1}{2}$	Sequential-decoded convolutional 20 kbps,[b] 1000-bit blocks	3	6.2
$K = 10$, $L = 11$, $R = \frac{1}{2}$	Feedback-decoded convolutional	1	2.1
$K = 8$, $L = 8$, $R = \frac{2}{3}$	Feedback-decoded convolutional	1	1.8
$K = 8$, $L = 9$, $R = \frac{3}{4}$	Feedback-decoded convolutional	1	2.0
$K = 3$, $L = 3$, $r = \frac{3}{4}$	Feedback-decoded convolutional	1	1.1
(24, 12) Golay		3	4.0
(24, 12) Golay		1	2.1
(127, 92) BCH		1	3.3
(127, 64) BCH		1	3.5
(127, 36) BCH		1	2.3
(7, 4) Hamming		1	0.6
(15, 11) Hamming		1	1.3
(31, 26) Hamming		1	1.6

[a] 9.6 dB required for uncoded system.

[b] The same system at a data rate of 100 kbps has 0.5 dB less coding gain.

Notation

$K =$ Constraint length of a convolutional code defined as the number of binary register stages in the encoder for such a code. With the Viterbi-decoding algorithm, increasing the constraint length increases the coding gain but also the implementation complexity of the system. To a much lesser extent the same is also true with sequential and feedback decoding algorithms.

$L =$ Look-ahead length of a feedback-decoded convolutional coding system defined as the number of received symbols, expressed in terms of the corresponding number of encoder input bits, that are used to decode an information bit. Increasing the look-ahead length increases the coding gain but also the decoder implementation complexity.

(n, k) denotes a block code (Golay, BCH, or Hamming here) with n decoder output bits for each block of k encoder input bits.

Receiver   quantization describes the degree of quantization of the demodulator outputs. Without coding and biphase ($0°$ or $180°$) modulation the demodulator output (or intermediate output if the quantizer is considered as part of the demodulator) is quantized to one bit (i.e., the sign if provided). With coding, a decoding decision is based on several demodulator outputs, and the performance can be improved if in addition to the sign the demodulator provides some magnitude information.

Source. Reference 9.

TABLE 5.10. Example 4, Uplink Power Budget

Terminal EIRP	+65 dBW	
Terminal pointing loss	0.5 dB	
Free space loss	199.2 dB	
Satellite pointing loss	0.0 dB	(earth coverage)
Polarization loss	0.5 dB	
Atmos losses	0.5 dB	(Section 4.3.9)
Rainfall (excess attenuation)	0.25 dB	(Section 4.3.9)
		(10° elevation angle)
Isotropic receive level	−135.95 dBW	
G/T	0.0 dB	
Sum	−135.95 dBW	
Boltzmann's constant	−(−228.6 dBW)	
C/N_0	92.65 dB	
−10 log(bit rate)	−70.00 dB	
E_b/N_0	+22.65 dB	
Required E_b/N_0	−9.6 dB	
Implementation loss	−2.0 dB	
Margin	11.05 dB	
Allowable margin	−4 dB	
Excess margin	7.05 dB	

The satellite G/T can be −7.05 dB for the uncoded system.
For the coded system with a 3.1-dB coding gain, the satellite
G/T can be degraded to −10.15 dB [−7.05 dB + (−3.1 dB)]

Assume the satellite $G/T = -6$ dB/K

$$EIRP = +44 \text{ dBW}$$

We will also assume that the satellite is a processing satellite operating in the FDMA mode and each access has a 10 kHz channel assignment, including guard bands, which allows about 3000 accesses in a 36 MHz transponder bandwidth. Statistically we find that no more that 20 percent of the accesses will transmit simultaneously during the busy hour (valid for our sample model only). The satellite EIRP then must be shared among 600 accesses or

$$44 \text{ dBW} - 10 \log 600 = +16.2 \text{ dBW per active access}$$

TABLE 5.11. Example 5, Downlink Power Budget

Satellite EIRP	+30 dBW
Satellite pointing loss	0.5 dB
Footprint error	0.25 dB
Off-footprint center loss	1.0 dB
Terminal pointing loss	0.5 dB
Polarization loss	0.5 dB
Atmospheric losses	0.5 dB
Free space loss	202.0 dB
Rainfall loss	3.0 dB
Isotropic receive level	−178.25 dBW
Terminal G/T	0.0 dB/K
Sum	−178.25 dBW
Boltzmann's constant	−(−228.6) dBW
C/N_0	50.35 dB
$10 \log(BR)$	−76.53 dB
Difference	−26.18 dB
Required	−4.9 dB
Modulation implementation loss	−2.0 dB
Margin	−5.0 dB
Sum	38.08 dB

The modulation is QPSK and with modulation implementation loss, an E_b/N_0 of 11 dB is required for a bit error rate (BER) of 1×10^{-5}.

The VSAT receiving system has a noise temperature (T_{sys}) of 500 K. The gross antenna gain for the 1 m dish at 55 percent efficiency is:

$$12 \text{ GHz} = 38.9 \text{ dB}$$

$$14.2 \text{ GHz} = 40.1 \text{ dB}$$

If we allow the net gain at the input to the VSAT LNA to be 38.0 dB, then the G/T [equation (4.27)]:

$$G/T = G - 10 \log T_{sys}$$

$$= 38.0 - 10 \log 500$$

$$= +11.0 \text{ dB}$$

The next step is to carry out a link budget using these parameters. For the downlink:

Satellite EIRP/carrier	+16.2 dBW
Free space loss (12 GHz)	206.0 dB
Pointing loss	0.5 dB
Polarization loss	0.5 dB
Gaseous absorption loss	1.0 dB
Isotropic receive level	−191.8 dBW
Terminal G/T	+11.0 dB/K
Sum	−180.8 dBW
Boltzmann's constant	−(−228.6) dBW
C/N_0	47.8 dB
−10 log(2400)	−33.8 dB
E_b/N_0	14.0 dB
Required E_b/N_0	−11.0 dB
Margin	3.0 dB

For the uplink:

Terminal EIRP	+40 dBW
Free space loss (14.2 GHz)	207.5 dB
Pointing loss	0.5 dB
Polarization loss	0.5 dB
Gaseous absorption loss	1.0 dB
Isotropic receive level	−169.5 dBW
Satellite G/T	−6.0 dB/K
Sum	−175.5 dB
Boltzmann's constant	−(−228.6 dBW)
C/N_0	53.1 dB
−10 log(2400)	−33.8 dB
E_b/N_0	19.3 dB
Required E_b/N_0	11.0 dB
Margin	8.3 dB

Calculate the transmitter output power that derives an EIRP of +40 dBW. Assume transmission line losses of 2 dB. Use equation (2.6).

$$\text{EIRP}_{\text{dBW}} = P_0 + L_1 + G_a$$

where P_0 is the transmitter power output in dBW; L_1 is the transmission line loss; and G_a is the antenna gain in dB.

$$+40 \text{ dBW} = P_0 - 2 \text{ dB} + 40.1 \text{ dB}$$

and then

$$P_0 = +1.9 \text{ dBW or about } 1.55 \text{ watts}$$

Discussion. The margins on the downlink and uplink would be used to compensate for rainfall attenuation. The downlink margin would probably be on the low side for a number of applications.

A processing satellite with 3000 accesses would require 3000 demodulators and 3000 modulators, a considerable feat with today's technology. With 600 carriers driving a TWT, one would expect a large backoff requirement to keep IM noise to a reasonable value.

Another approach would be the use of a larger "hub" station with a "bent pipe" satellite. Half the transponder would be dedicated to the "hub" uplink. The access bandwidth for the VSATs could be cut to 8 kHz allowing 2000 accesses instead of 3000 as proposed previously. The hub would receive each access, demodulate them, and then combine them in a serial TDM bit stream with a bit rate in the order of 5.0 Mbps, including overhead from framing and stuffing bits. Then half of the satellite transponder EIRP could then be dedicated to the downlink to the VSATs.

The reader with insight will comment that we cannot close the downlink to the VSATs because of the impact of bit rate on the link budget. Instead of -33.8 dB for 10 log(bit rate), we must use $10 \log 5 \times 10^6$ or approximately 67 dB. This is a difference of 67 dB -33.8 db or 33.2 dB. We only compensated for 24.8 dB by increasing the EIRP to $+41$ dBW for the single TDM carrier. We are short 8.4 dB and still maintain the 3 dB margin.

There are several approaches to make up the shortfall of 8.4 dB. One would be to limit the accesses to about $1/4$ of the 2000, or 500, and reduce the bit rate accordingly. Another would be to increase the satellite EIRP, yet keep under the requisite flux denity limits on the earth's surface as required the Radio Regulations. Still another would be to implement an FEC coding scheme to achieve a net coding gain of 5.4 dB.

The TDM downlink would require that each VSAT would be more complex with the added timing and processing requirements to synchronize, demodulate, and demultiplex a serial bit stream in the megabit range. This will add cost to the VSAT.

With present technology the hub approach is probably more attractive for some applications, such as for hotel/motel, rent-a-car, and airline reservation systems. In these cases the hub would be the location for the system CPU (computer or central processing unit). A good reference based on the hub scheme is D. Chakraborty's paper (Ref. 10).

PROBLEMS AND EXERCISES

1. Explain the basic difference between bent pipe digital satellite systems and processing satellite systems.

2. What is the principal advantage of TDMA operation from the point of view of transmitter efficiency? What is the principal design factor in TDMA operation that is not a requirement in FDMA operation?

3. Name factors limiting TDMA data throughput.

4. Explain the rationale for using CR/BTR and its location in a typical TDMA frame.

5. A transponder operating in the TDMA mode has a bandwidth of 80 MHz and 8-ary PSK modulation is used. What maximum link transmission bit rate can be expected? Assume bandwidth-limited operation.

6. A satellite transponder is power limited, operating in the TDMA mode, with an EIRP of +19 dBW, operating with an earth terminal with a G/T of +33 dB/K using BPSK modulation (coherent) with a 6-dB margin and a 2-dB modulation implementation loss. The required E_b/N_0 for a BER of 1×10^{-5} is 9.6 dB (theoretical). What is the maximum bit rate for this power limited case?

7. Explain the difference between open-loop and closed-loop TDMA synchronization systems.

8. Differentiate TASI-based DSI with speech-predictive DSI.

9. Describe the principal degradations of DSI.

10. What is the function of the unique word in a TDMA system?

11. Describe a method of open-loop acquisition in a TDMA system.

12. Give at least two advantages of on-board regenerative repeaters.

13. Name at least five functions that must be carried out by an on-board regenerative repeater.

14. Describe the function and give the rationale for the sync window in SS/TDMA.

15. Explain the advantages of IF on-board switching when compared to conventional SS/TDMA.

16. Give at least two reasons why the band 58–62 GHz was selected for intersatellite links (cross-links).

17. Give at least three advantages in the use of a satellite cross-link architecture in total system design.

18. Give at least two advantages and two disadvantages of implementing a coding (FEC) technique on a digital satellite system.

19. Compare ARQ and FEC as error control techniques in a digital satellite system.

20. The information rate input to a coder is 75 bps and the code rate is $\frac{1}{2}$. What is the output symbol rate?

21. What is one principal difference between block codes and convolutional codes.

22. If a convolutional code has a constraint length of 7 ($K = 7$), what does that tell us about the coder?

23. Give the three-types of decoders that can be used for decoding convolutional codes.

24. Differentiate between hard decisions and soft decisions.

25. What is one form of diversity that can be achieved with a coding system?

26. Name at least three factors that can affect coding gain as it impacts system design.

27. Carry out the link budget for the following uplink. Calculate the terminal EIRP in dBW. The uplink frequency is 6200 MHz; the elevation angle is $10°$; the satellite G/T is -7 dB/K; the data rate is 2400 bps; and the required BER is 1×10^{-6}. The modulation is BPSK. Use reasonable values for the link losses not given. Assume that the satellite has an earth coverage antenna. Use a 3-dB margin.

28. Show how the EIRP in question 27 can be reduced to half by implementation of an FEC scheme. Describe the scheme selected and rework the link budget.

29. Size the HPA and antenna for the uplink of question 28 and again for question 27. Parametrically compare antenna size and transmitter power output for the EIRPs in questions 27 and 28.

30. Size an earth terminal receiving system for 1.544 Mbps for a satellite with a transponder EIRP of $+30$ dBW. Use 4 GHz for the model. Use an elevation angle of $20°$.

REFERENCES

1. O. G. Gabbard and P. Kaul, "Time-Division Multiple Access," IEEE Electronics and Aerospace Systems Convention, October 7–9, 1974, as reproduced in H. L. Van Trees, *Satellite Communications*, IEEE Press, New York, 1979.

2. V. K. Bhargava et al., *Digital Communications by Satellite,*, Wiley, New York, 1981.

3. S. J. Campanella, "Digital Speech Interpolation," COMSAT Tech. Review, Spring 1976, as reproduced in H. L. Van Trees, *Satellite Communications*, IEEE Press, New York, 1979.

4. R. L. Freeman, *Reference Manual for Telecommunications Engineers*, Wiley, New York, 1984.

5. K. Koga, T. Muratani, and A. Ogawa, "On-Board Regenerative Repeaters Applied to Digital Satellite Communications," *Proceedings of the IEEE*, March 1977, as reproduced by H. L. Van Trees, *Satellite Communications*, IEEE Press, New York, 1979.

6. N. R. Edwards, *Satellite Communications Reference Data Handbook*, Computer Sciences Corp., Falls Church, VA, March 1983, under DCA contract DCA100-81-C-0044.

7. R. L. Freeman, *Telecommunication Transmission Handbook*, Wiley, New York, 1982.

8. A. J. Viterbi, "Convolutional Codes and their Performance in Communication Systems," *IEEE Trans. Comm.* **Com-19** (October, 1971).

9. J. P. Odenwalder, *Error Control Coding Handbook*, Linkabit Corporation, San Diego, CA, July 1976 under USAF contract F44620-76-C-0056.

10. D. Chakraboty, "Constraints in Ku-band Continental Satellite Network Design," *IEEE Communications Magazine*, **24**, Aug. 1986, IEEE Press NY.

SYSTEM DESIGN ABOVE 10 GHz

6.1 THE PROBLEM—AN INTRODUCTION

There is an ever increasing demand for radio-frequency spectrum in the industrialized nations of the world. This is due to the information transfer explosion in our society, resulting in a rapid increase in telecommunication connectivity, and the links satisfying that connectivity are required to have ever greater capacity.

The most desirable spectrum to satisfy these needs is the band between 1 and 10 GHz. It is called the "noise window," where galactic and manmade noise are minimum. Atmospheric absorption may generally be neglected in this region.

Congestion in the 1–10 GHz region has forced us to look above 10 GHz for operational frequencies. By careful engineering we have found that frequencies above 10 GHz can give equivalent performance or nearly equivalent performance to those below 10 GHz.

We have arbitrarily selected 10 GHz as a demarcation line. Generally, below 10 GHz, in radiolink design, we can neglect excess attenuation due to rainfall and atmospheric absorption. For frequencies above 10 GHz, excess attenuation due to rainfall and atmospheric absorption can have an overriding importance in radiolink design. In fact, certain frequency bands display so much gaseous absorption that they are unusable for many applications.

The principal thrust of this chapter is to describe techniques for band selection and link design for line-of-sight (LOS) microwave and earth–space–earth links for frequencies above 10 GHz. The chapter also deals with low-elevation-angle space links, and how to deal with the effects resulting from elevation angles under 5° or so.

6.2 THE GENERAL PROPAGATION PROBLEM ABOVE 10 GHz

Propagation of radio waves through the atmosphere above 10 GHz involves not only free space loss but several other important factors. As expressed in

Ref. 1, these are as follows:

(a) The gaseous contribution of the homogeneous atmosphere due to resonant and nonresonant polarization mechanisms.
(b) The contribution of inhomogeneities in the atmosphere.
(c) The particulate contributions due to rain, fog, mist, and haze (dust, smoke, and salt particles in the air).

Under (a) we are dealing with the propagation of a wave through the atmosphere under the influence of several molecular resonances, such as water vapor (H_2O) at 22 and 183 GHz, oxygen with lines around 60 GHz, and a single oxygen line at 119 GHz. These points with their relative attenuation are shown in Figure 6.1.

Other gases display resonant lines as well, such as N_2O, SO_2, O_3, NO_2, and NH_3, but because of their low density in the atmosphere, they have negligible effect on propagation.

The major offender is precipitation attenuation [under (b) and (c)]. It can exceed that of all other sources of attenuation in the atmosphere above 18 GHz. Rainfall and its effect on propagation is covered at length in this chapter.

We will first treat total loss due to absorption and scattering. It will be remembered that when an incident electromagnetic wave passes over an object that has dielectric properties different from the surrounding medium, some energy is absorbed and some is scattered. That which is absorbed heats the absorbing material; that which is scattered is quasi-isotropic and relates to the wavelength of the incident wave. The smaller the scatterer, the more isotropic it is in direction with respect to the wavelength of the incident energy.

We can develop a formula from equation (1.5) to calculate total transmission loss for a given link:

$$\text{Attenuation (dB)} = 92.45 + 20 \log F_{GHz} + 20 \log D_{km} + a + b + c + d + e$$

$$(6.1)$$

where F is in gigahertz and D is in kilometers. Also,

a = excess attenuation (dB) due to water vapor
b = excess attenuation (dB) due to mist and fog
c = excess attenuation (dB) due to oxygen (O_2)
d = sum of the absorption losses (dB) due to other gases
e = excess attenuation (dB) due to rainfall

Notes and comments on equation (6.1):

1. a varies with relative humidity, temperature, atmospheric pressure, and altitude. The transmission engineer assumes that the water vapor content

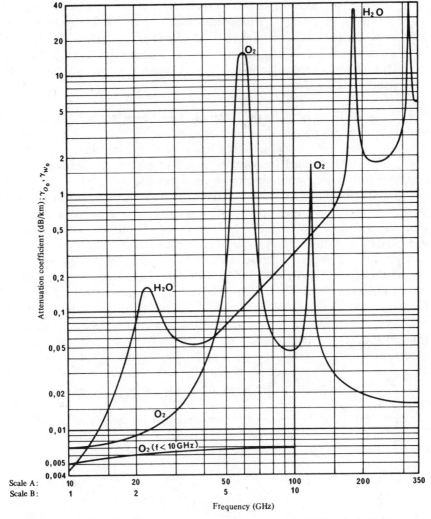

FIGURE 6.1 Atmospheric attenuation by oxygen and water vapor. Use Scale B for oxygen absorption below 10 GHz. Pressure, 1 atm (1013.6 mb); temperature, 20°C; water vapor density, 7.5 g/m³. From CCIR Rep. 719 (Ref. 2). Courtesy of ITU-CCIR, Geneva.

is linear with these parameters and that the atmosphere is homogeneous (actually horizontally homogeneous but vertically stratified). There is a water vapor absorption band about 22 GHz caused by molecular resonance.

2. c and d are assumed to vary linearly with atmospheric density, thus directly with atmospheric pressure, and are a function of altitude (e.g., it is assumed that the atmosphere is homogeneous).

3. *b* and *e* vary with the density of the rainfall cell or cloud and the size of the rainfall drops or water particles such as fog or mist. In this case the atmosphere is most certainly not homogeneous. (Droplets less than 0.01 cm in diameter are considered mist/fog, and more than 0.01 cm, rain). Ordinary fog produces about 0.1 dB/km excess attenuation at 35 GHz, rising to 0.6 dB/km at 75 GHz.

In equation (6.1) terms *b* and *d* can often be neglected; terms *a* and *c* are usually lumped together and called "atmospheric attenuation." If we were to install a 10-km LOS link at 22 GHz, in calculating transmission loss, 1.6 dB would have to be added for what is called atmospheric attenuation but is predominantly water vapor absorption, as shown in Figure 6.1.

It will be noted in Figure 6.1 that there are frequency bands with relatively high levels of attenuation per unit distance, some are rather narrow bands and some fairly wide. For example, the O_2 absorption band covers from about 58 to 62 GHz and with skirts down to 50 and up to 70 GHz. At its peak at about 60 GHz, the sea level attenuation is about 15 dB/km. One could ask "of what use are these bands?" Actually, the 58–62-GHz band is appropriately assigned for satellite cross-links. These links operate out in space far above the limits of the earth's atmosphere, where the terms *a* through *e* may be completely neglected. It is particularly attractive on military cross-links having an inherent protection from earth-based enemy jammers by that significant atmospheric attenuation factor. It is also useful for very-short-haul military links such as a ship-to-ship secure communication system. Again, it is the atmospheric attenuation that offers some additional security for signal intercept (LPI) and against jamming.

On the other hand, Figure 6.1 shows a number of bands that are relatively open. These openings are often called windows. Three such windows are suggested for point-to-point service in Table 6.1.

6.3 RAINFALL LOSS

6.3.1 Basic Rainfall Considerations

Of the factors *a* through *e* in equation (6.1), factor *e*, excess attenuation due to rainfall, is the principal one affecting path loss. For instance, even at 22

TABLE 6.1. Windows for Point-to-Point Service

Band (GHz)	Excess Attenuation due to Atmospheric Absorption (dB/km)
28–42	0.13
75–95	0.4
125–140	1.8

GHz, the water vapor line, excess attenuation due to atmospheric gases accumulates at only 0.165 dB/km, and for a 10-km path only 1.65 dB must be added to free space loss to compensate for water vapor loss. This is negligible when compared to free space loss itself, such as 119.3 dB for the first kilometer at 22 GHz, accumulating thence approximately 6 dB each time the path length is doubled (i.e., add 6 dB for 2 km, 12 dB for 4 km, etc.). Thus a 10-km path would have a free space loss of 139.3 dB plus 1.65 dB added for excess attenuation due to water vapor (22 GHz), or a total of 140.95 dB.

Excess attenuation due to rainfall is another matter. It has been common practice to express rainfall loss as a function of precipitation rate. Such a rate depends on the liquid water content and the fall velocity of the drops. The velocity, in turn, depends on raindrop size. Thus our interest in rainfall boils down to drop size and drop-size distribution for point rainfall rates. All this information is designed to lead the transmission engineer to fix an excess attenuation due to rainfall on a particular path as a function of time and time distribution. This is a method similar to which was used in Chapter 2 for overbuilding a link to accommodate fading.

An earlier approach dealt with rain on a basis of rainfall given in millimeters per hour. Often this was done with rain gauges, using collected rain averaging over a day or even periods of days. For path design above 10 GHz such statistics are not sufficient where we may require path availability better than 99.9% and do not wish to resort to overconservative design procedures (e.g., assign excessive link margins).

As we mentioned, there is a fallacy in using annual rainfall rates as a basis for calculation of excess attenuation due to rainfall. For instance, several weeks of light drizzle will affect the overall long-term path availability much less than several good downpours that are short lived (i.e., 20-min duration). It is simply this downpour activity for which we need statistics. Such downpours are cellular in nature. How big are the cells? What is the rainfall rate in the cell? What are the size of the drops and their distribution?

Hogg (Ref. 3) suggests the use of high-speed rain gauges with outputs readily available for computer analysis. These gauges can provide minute-by-minute analysis of the rate of fall, something lacking with conventional types of gauges. Of course, it would be desirable to have several years statistics for a specific path to provide the necessary information on fading caused by rainfall that will govern system parameters such as LOS repeater spacing, antenna size, and diversity.

Some such information is now available and is indicative of a great variation of short-term rainfall rates from one geographical location to another. As an example, in one period of measurement it was found that Miami, FL has maximum rain rates about 20 times greater than those of heavy showers occurring in Oregon, the region of heaviest rain in the United States. In Miami a point rainfall rate may even exceed 700 mm/hr. The effect of 700 mm/hr on 70 and 48 GHz paths can be extrapolated from Figure 6.2. In the figure the rainfall rate in millimeters per hour extends to 100, which at 100 mm/hr provides an excess attenuation of from 25 to 30 dB/km.

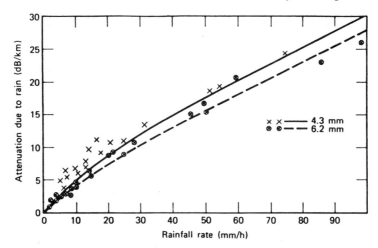

FIGURE 6.2 Measurements made by Bell Telephone Laboratories of excess attenuation due to rainfall at wavelengths of 6.2 and 4.2 mm (48 and 70 GHz) compared with calculated values (Ref. 3, copyright 1968 by AT & T).

When identical systems were compared (Ref. 3) at 30 GHz with repeater spacings of 1 km and equal desired signals (e.g., producing a 30-dB signal-to-noise ratio), 140 min of total time below the desired level was obtained at Miami, FL; 13 min at Coweeta, NC; 4 min at Island Beach, NJ; 0.5 min at Bedford, England; and less than 0.5 min at Corvallis, OR. Such outages, of course, can be reduced by increasing transmitter output power, improving receiver noise factor (NF), increasing antenna size, implementing a diversity scheme, etc.

One valid approach to lengthen repeater sections (space between repeaters) and still maintain performance objectives is to use path diversity. This is the most effective form of diversity for downpour rainfall fading. Path diversity is the simultaneous transmission of the same information on paths separated by at least 10 km, the idea being that rain cells affecting one path will have a low probability of affecting the other at the same time. A switch would select the better path of the two. Careful phase equalization between the two paths would be required, particularly for the transmission of high-bit-rate information.

6.3.2 Calculation of Excess Path Attenuation due to Rainfall
for LOS Paths

When designing radio links (or satellite links), a major problem in link engineering is to determine the excess path attenuation due to rainfall. The adjective "excess" is used to denote path attenuation in *excess* of free space loss [i.e., the terms *a*, *b*, *c*, *d*, and *e* in equation (6.1) are in excess of free space attenuation (loss)].

Before treating the methodology of calculation of excess rain attenuation, we will review some general link engineering information dealing with rain. When discussing rainfall here, all measurements are in millimeters per hour of rain and are point rainfall measurements. From our previous discussion we know that heavy downpour rain is the most seriously damaging to radio transmission above 10 GHz. Such rain is cellular in nature and has limited coverage. We must address the question whether the entire hop is in the storm for the whole period of the storm. Light rainfall (e.g., less than 2 mm/hr), on the other hand, is usually widespread in character, and the path average is the same as the local value. Heavier rain occurs in convective storm cells which are typically 2–6 km across and are often embedded in larger regions measured in tens of kilometers (Ref. 4). Thus for short hops (2–6 km) the path-averaged rainfall rate will be the same as the local rate, but for longer paths it will be reduced by the ratio of the path length on which it is raining to the total path length.

This concept is further expanded upon by CCIR Rep. 593-1 (Ref. 5), where rain cell size is related to rainfall rate. This is shown in Figure 6.3. This concept of rain cell size is very important, whether engineering a LOS link or a satellite link, particularly when the satellite link has a low elevation angle. CCIR Rep. 338-3 (Ref. 6) is quoted in part below:

> Measurements in the United Kingdom over a period of two years...at 11, 20 and 37 GHz on links of 4–22 km in length show that the attenuation due to rain and multipath, which is exceeded for 0.01% of the time and less, increased rapidly with path length up to 10 km, but further increase up to 22 km produced a small additional effect.

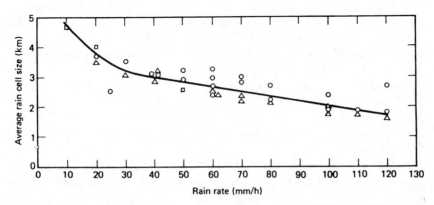

FIGURE 6.3 Average rain cell size as a function of rain rate (CCIR Rep. 593-1; courtesy ITU-CCIR).

The use and application of specific rain cell size was taken from CCIR 1978. We will call this method the "liberal method" because when reviewing CCIR Vol. V of 1982, CCIR Rep. 563-2, we can see that the CCIR has turned more cautious and conservative. Our Figure 6.3 does not appear in the report. Let us partially quote from this report:

> For attenuation predictions the situation is generally more complex (than that of interference scattering by precipitation). Volume cells are known to cluster frequently within small mesoscale areas.... Terrestrial links exceeding 10 km may therefore traverse more than one volume cell within a mesoscale cluster. In addition, since the attenuating influence of the lower intensity rainfall surrounding the cell must be taken into account, any model used to calculate attenuation must take these larger rain regions into account. The linear extent of these regions increases with decreasing rain intensity and may be as large as several tens of kilometers.

One of the most accepted methods of dealing with excess path attenuation A due to rainfall is an empirical procedure based on the approximate relation between A and the rain rate R:

$$A = aR^b \qquad (6.2)$$

where a and b are functions of frequency f and rain temperature T. Allowing a rain temperature of 20°C and for the Laws and Parsons drop-size distribution, Table 6.2 gives the regression coefficients for estimating specific attenuations from equation (6.2).

We note that horizontally polarized waves suffer greater attenuation than vertically polarized waves because large rain drops are generally shaped as oblate spheroids and are aligned with a vertical rotation axis. Thus we use the subscript notation h and v for horizontal and vertical polarizations in Table 6.2 for the values a and b. A is the specific attenuation in dB/km. We obtain a and b from Table 6.2. Next we must obtain a value for the rain rate. This is obtained from local data sources, and we need a value of rain intensity exceeded 0.01% of the time with an integration time of 1 min. If this information cannot be obtained from local sources, an estimate can be obtained by identifying the region of interest from the maps appearing in Figures 6.4, 6.5, and 6.6, then selecting the appropriate rainfall intensity for the specified time percentage from Table 6.3, which gives the 14 regions in the maps. This gives a value for R in equation (6.2). Knowing the frequency and polarization, we calculate A in dB/km from values of a and b from Table 6.2 using equation (6.2). Figure 6.7 and/or the nomogram in Figure 6.8 may also be used to calculate A.

We now turn again to the problem of effective path length or L_{eff} (Ref. 8). This is obtained by multiplying the actual path length L by a reduction factor r. A first estimate to calculate r is given as

$$r = \frac{90}{90 + 4L} \qquad (6.3)$$

TABLE 6.2. Regression Coefficients for Estimating Specific Attenuations in Equation (6.2)[a]

Frequency (GHz)	a_h	a_v	b_h	b_v
1	0.0000387	0.0000352	0.912	0.880
2	0.000154	0.000138	0.963	0.923
4	0.000650	0.000591	1.121	1.075
6	0.00175	0.00155	1.308	1.265
7	0.00301	0.00265	1.332	1.312
8	0.00454	0.00395	1.327	1.310
10	0.0101	0.00887	1.276	1.264
12	0.0188	0.0168	1.217	1.200
15	0.0367	0.0335	1.154	1.128
20	0.0751	0.0691	1.099	1.065
25	0.124	0.113	1.061	1.030
30	0.187	0.167	1.021	1.000
35	0.263	0.233	0.979	0.963
40	0.350	0.310	0.939	0.929
45	0.442	0.393	0.903	0.897
50	0.536	0.479	0.873	0.868
60	0.707	0.642	0.826	0.824
70	0.851	0.784	0.793	0.793
80	0.975	0.906	0.769	0.769
90	1.06	0.999	0.753	0.754
100	1.12	1.06	0.743	0.744
120	1.18	1.13	0.731	0.732
150	1.31	1.27	0.710	0.711
200	1.45	1.42	0.689	0.690
300	1.36	1.35	0.688	0.689
400	1.32	1.31	0.683	0.684

[a] Raindrop size distribution [Laws and Parsons, 1943].
Terminal velocity of raindrops [Gunn and Kinzer, 1949].
Index of refraction of water at 20°C [Ray, 1972].

Values of a and b for spheroidal drops [Fedi, 1979; Maggiori, 1981] based on regression for the range 1 to 150 mm/h.

Source. Reference 7; courtesy of ITU-CCIR, Geneva.

The attenuation exceeded for 0.01% of the time is found from

$$A_{0.01} = aR_{0.01}^b \qquad (6.4)$$

$$A_{eff} = A \times L \times r \qquad (6.5)$$

where A is the value calculated in equation (6.4). Attenuation exceeded for other percentages P can be found by the following power law:

$$A_p = A_{0.01}\left(\frac{P}{0.01}\right)^{-a} \qquad (6.6)$$

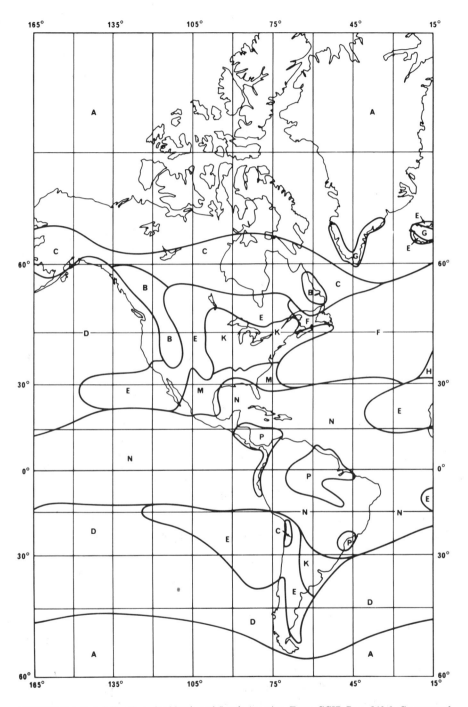

FIGURE 6.4 Rainfall regions for North and South America. From CCIR Rep. 563-2. Courtesy of ITU-CCIR Geneva (Ref. 9).

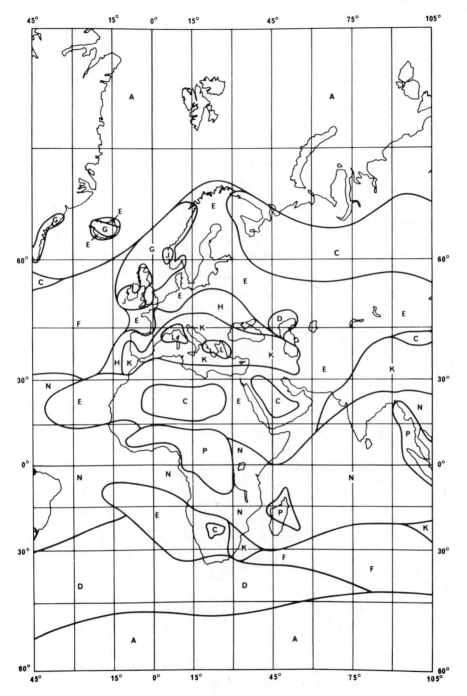

FIGURE 6.5 Rainfall regions for Europe and Africa. From CCIR Rep. 563-2. Courtesy of ITU-CCIR, Geneva (Ref. 9).

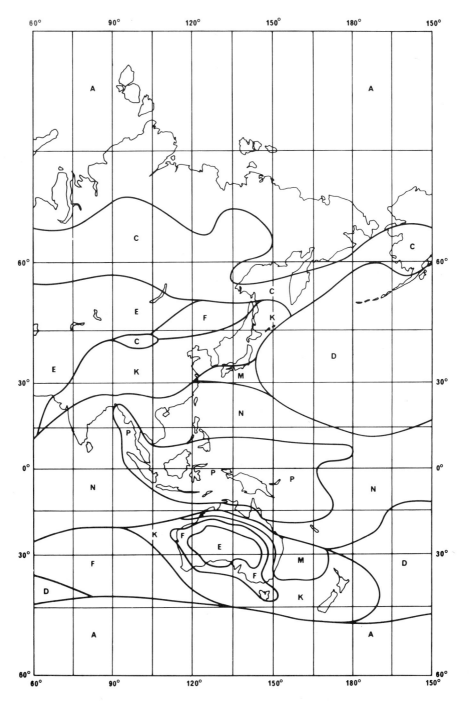

FIGURE 6.6 Rainfall regions for Asia and Oceana. From CCIR Rep. 563-2. Courtesy of ITU-CCIR (Ref. 9).

TABLE 6.3. Rainfall Climatic Regions, Rainfall Intensity Exceeded (mm / h)

Percentage of time (%)	A	B	C	D	E	F	G	H	J	K	L	M	N	P
1.0	—	1	—	3	1	2	—	—	—	2	—	4	5	12
0.3	1	2	3	5	3	4	7	4	13	6	7	11	15	34
0.1	2	3	5	8	6	8	12	10	20	12	15	22	35	65
0.03	5	6	9	13	12	15	20	18	28	23	33	40	65	105
0.01	8	12	15	19	22	28	30	32	35	42	60	63	95	145
0.003	14	21	26	29	41	54	45	55	45	70	105	95	140	200
0.001	22	32	42	42	70	78	65	83	55	100	150	120	180	250

Source. Reference 9.

The coefficient a is given by

$$a = 0.33 \text{ for } 0.001 \leq P \leq 0.01\%$$

$$a = 0.41 \text{ for } 0.01 \leq P \leq 0.1\%$$

Example 1. Consider a path in West Germany 10 km long operating at 30 GHz. Use equation (6.2). Obtain values of a and b and assume vertical polarization. Use Table 6.2 and $a = 0.167$ and $b = 1.000$. Assume a time availability of 99.99% and thus the rainfall intensity exceeded is 0.01%, and from Table 6.3 from climate region E (Figure 6.5), $R = 22$ mm/hr.

Calculate $A_{0.01}$ using equation (6.4):

$$A_{\text{dB}} = 0.167(22)^{1.000}$$

$$A = 3.674 \text{ dB/km}$$

Calculate the effective distance. First calculate the reduction factor r using equation (6.3):

$$r = \frac{90}{90 + (4 \times 10)}$$

$$r = 0.69$$

Calculate the excess attenuation due to rainfall A_{eff} using equation (6.5):

$$A_{\text{eff}} = A \times 10 \times r$$

$$= 3.674 \times 10 \times 0.69$$

$$= 25.35 \text{ dB}$$

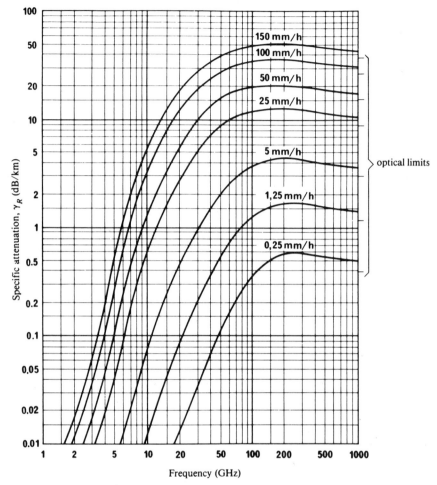

FIGURE 6.7 Specific attenuation A due to rain. Raindrop-size distribution (Laws and Parsons, 1943); Terminal velocity of raindrops (Gunn and Kinzer, 1949); Index of refraction of water at 20°C (Ray, 1972); Spherical drops. From ITU-CCIR, Geneva (Ref. 7).

Example 2. Calculate the excess attenuation due to rainfall for a 15-km path in the Florida panhandle of the United States operating at 18 GHz. Assume horizontal polarization and a time availability of 99.9% (an exceedance of 0.1%).

Turn to Figure 6.4 and the climatic region is N, thence to Table 6.3 and we obtain a rainfall rate of 99 mm/hr for an exceedance of 0.01%. Obtain the values of a and b from Table 6.2 and interpolate between the values of 15 and 20 GHz. $a = 0.05974$ and $b = 1.121$. Use equation (6.4) to calculate A:

$$A = 0.05974(95)^{1.121}$$

$$A_{0.01} = 9.85 \text{ dB/km}$$

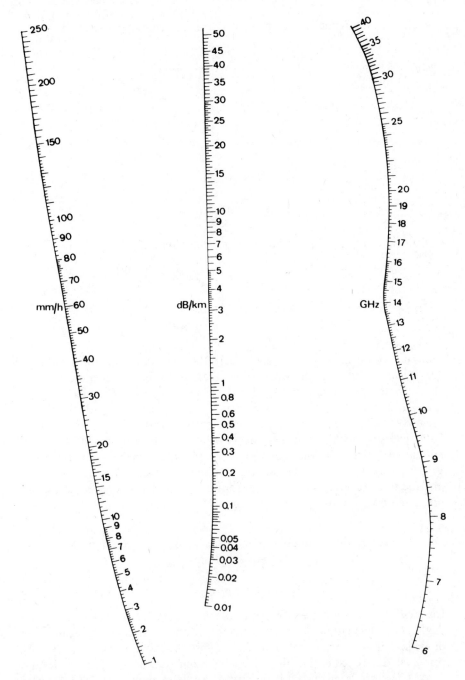

FIGURE 6.8 Specific attenuation A due to rain. Raindrop-size distribution (Laws and Parsons, 1943); Terminal velocity of raindrops (Gunn and Kinzer, 1949); Index of refraction of water at 18°C (Ray, 1972); Spherical drops. From CCIR Rep. 721-1 (Ref. 7). Courtesy of ITU-CCIR, Geneva.

Calculate the reduction factor r, where $L = 15$; use equation (6.3):

$$r = \frac{90}{90 + (4 \times 15)}$$

$$r = 0.6$$

Use equation (6.6) to calculate $A_{0.1}$:

$$A_{0.1} = 9.85(0.1/0.01)^{-0.41}$$

$$A_{0.1} = 3.83 \text{ dB/km}$$

Calculate A_{eff} for the 15-km path; use equation (6.5):

$$A_{eff} = 3.83 \times 0.6 \times 15$$

$$A_{eff(0.1)} = 34.47 \text{ dB}$$

Thus we would have to add 34.47 dB to the free space loss for a path with 99.9% availability. For the worst case situation, to this total value, we would also have to add the fade margin for the same path availability. Conventional space and frequency diversity would mitigate against multipath fading. The use of vertical polarization would reduce the excess attenuation due to rainfall. Path diversity, with a path separation of at least 10 km, would be a major mitigating factor. The value of 10-km separation has been taken from CCIR Vol. V; however, it can be shown that separations of 2 km or more will provide some diversity improvement for rainfall. A more in-depth discussion is presented on path diversity in Section 6.3.4.

6.3.3 Calculation of Excess Attenuation due to Rainfall for Satellite Paths

6.3.3.1 Introduction

Rainfall attenuation on satellite paths is a function of frequency and elevation angle. The calculation of excess attenuation due to rainfall for uplinks and downlinks is somewhat similar to the exercise described in Section 6.3.2. The principal difference is that the path is elevated (e.g., a function of elevation angle). The preferred model is described in this section is basically based on work done by R. K. Crane (1966, 1979) (Ref. 10), embellished by Feldman and characterized by Kaul (ORI, 1980) (Refs. 11 and 14). The reader is also encouraged to review CCIR Rep. 564-2, Section 6 (Ref. 12).

6.3.3.2 Scattering

Raindrops both attenuate and scatter microwave energy along an earth–space path. From the basic Rayleigh scattering criteria (the dimensions of the scatterer are much smaller than the wavelength) and the fact that the median raindrop diameter is approximately 1.5 mm, one would expect that Rayleigh scattering theory should be applied in the frequency range above 10 GHz. However, Rayleigh scattering also requires that the imaginary component of the refractive index be small, which is not the case for water drops. Because of this effect and the wide distribution of raindrop diameters, the Rayleigh scattering theory seems to apply only up to 3 GHz. Above 3 GHz, Mie scattering applies and is the primary technique utilized for calculations of excess attenuation due to rainfall. Mie scattering accounts for the deficiencies of Rayleigh scattering and has proven to be the most accurate technique (Ref. 11).

6.3.3.3 Drop-Size Distribution

Several investigators have studied the distribution of raindrop sizes as a function of rain rate and type of storm activity. The three most commonly used distributions are:

☐ Laws and Parsons (LP) (Section 6.3.2)
☐ Marshall-Palmer (MP)
☐ Joss-thunderstorm (J-T) and drizzle (J-D)

The LP distribution is generally more favored for design purposes because it has been widely tested by comparison to measurements for both widespread (lower rain rates) and convective rain (higher rain rates) and at frequencies above 10 GHz, the LP values give higher values for excess attenuation due to rainfall than the J-T values. In addition, it has been observed that the raindrop temperature is most accurately modeled by the 0°C data rather than 20°C, since for most high-elevation-angle satellite links the raindrops are cooler at high altitudes and warm as they fall to earth.

6.3.3.4 The Global Model (Crane)

The Global Model (Crane) (Ref. 10) uses a specific rain model based entirely on meteorological observations, not attenuation measurements. The rain model, combined with the attenuation estimation, was tested by comparison with attenuation measurements. This procedure was used to circumvent the requirement for attenuation observations over a span of many years. The overall attenuation model is based on the use of independent, meteorologically derived estimates for the cumulative distributions of point rainfall rate, horizontal path averaged rainfall rate, the vertical distribution of rain intensity, and a theoretically derived relationship between specific attenuation and rain rate obtained using median observed drop-size distributions at a number of rain rates.

The first step is to determine the instantaneous point rain rate R_p distribution. The Model provides median distribution estimates for eight rainfall regions, A through H, covering the entire globe. Figures 6.9 and 6.10 give the geographic rain climate regions for the continental and ocean areas of the earth. The continental (contiguous) United States and European portions are further expanded in Figures 6.11 and 6.12, respectively.

The climate regions shown in the figures are very broad. The upper and lower rain rate bounds provided by the nearest adjacent region have a ratio of 3.5 at 0.01% of the year for the CCIR climate region D, for example, which produces a ratio of upper-to-lower bound attenuation values of 4.3 dB at 12 GHz. The attendant uncertainty in the estimated attenuation value can be reduced by using actual rain rate distributions for an area of interest if these long-term statistics are available. Region D for the United States has been further broken down into regions D_1, D_2, and D_3 for convenience.

Once the region of interest has been identified, R_p values may then be obtained from the rain rate distribution curves in Figure 6.13. Figure 6.13a gives the curves for the eight global climate regions designated A through H for 1 min averaged surface rain rate as a function of the percentage of a year that the rain rate is exceeded. For the Region D subdivisions distributions, Figure 6.13b should be used. Numerical values for R_p are provided in Table 6.4.

6.3.3.5 Discussion of the Model

In Section 6.2 we used a path averaged rainfall rate using an effective path averaging factor r. This was a valid approach for a LOS microwave path, but will not apply for the estimation of attenuation on a slant path to a satellite. Here account must be taken of the variation of specific attenuation with height. As we are aware, atmospheric temperature decreases with height, and above some height, the precipitation particles will all be ice particles. Ice and snow do not produce significant attenuation. Only regions in the atmosphere with liquid water precipitation particles are of interest in the estimation of attenuation. The size and number of raindrops per unit volume may vary with height. Weather radar measurements have shown that the reflectivity of a rain volume may vary with height but, on the average, the reflectivity is roughly constant up to the 0°C isotherm and decreases above that height. The rain rate can be considered to be constant to the height of the 0°C isotherm at low rain rates, and this height is used to define the upper boundary of the region of attenuation. However, a high correlation between the 0°C height and the height to which liquid rain drops exist in the atmosphere should not be expected for the higher rain rates because large liquid raindrops are carried aloft above the 0°C height in the strong updraft cores of intense rain cells

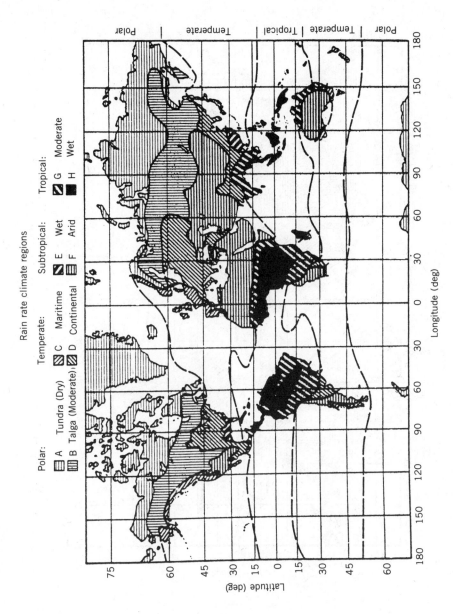

FIGURE 6.9 The global rain rate regions for the continental areas. From Ref. 11.

415

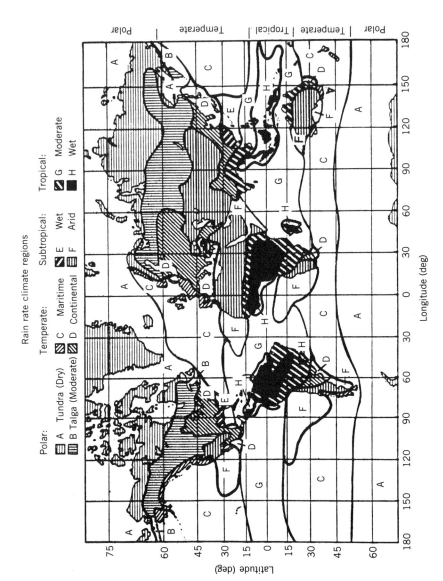

FIGURE 6.10 Global rain rate climate regions for the ocean areas. From Ref. 11.

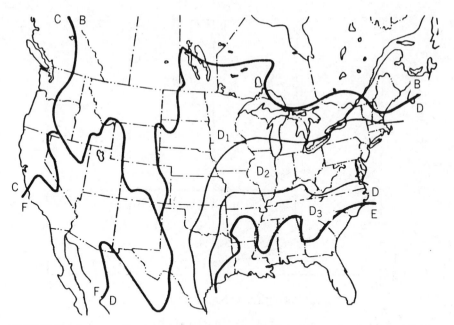

FIGURE 6.11 Rain rate climate regions for the contiguous United States showing the subdivision of region D (Ref. 11).

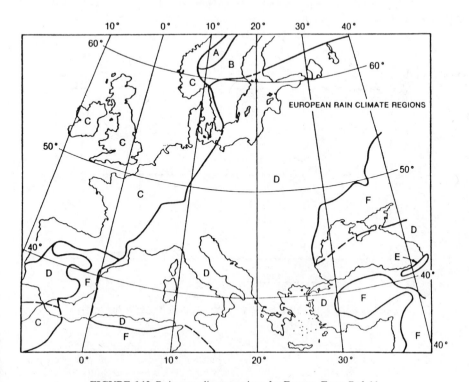

EUROPEAN RAIN CLIMATE REGIONS

FIGURE 6.12 Rain rate climate regions for Europe. From Ref. 11.

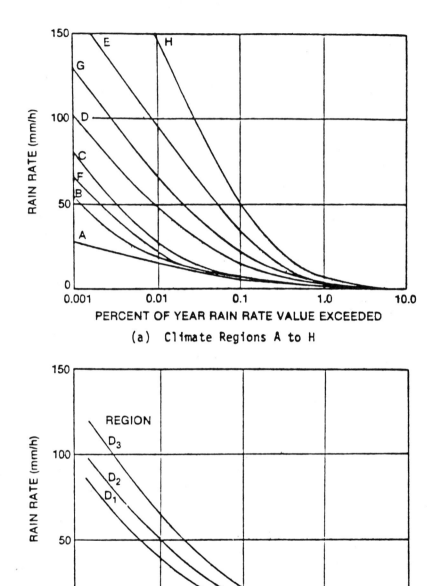

(a) Climate Regions A to H

(b) Climate Regions D divided into three subregions
(D_2 = D above)

FIGURE 6.13 Point rain rate distributions as a function of percentage of year exceeded: (a) climate regions $A-H$; (b) climate region D divided into three subregions. From Ref. 11.

TABLE 6.4. Point Rain Rate Distribution Values (mm / hr) Versus Percent of Year Rain Rate is Exceeded

Percentage of Year	Rain Climate Region										Minutes Per Year	Hours Per Year
	A	B	C	D_1	D_2	D_3	E	F	G	H		
0.001	28	54	80	90	102	127	164	66	129	251	5.3	0.09
0.002	24	40	62	72	86	107	144	51	109	220	10.5	0.18
0.005	19	26	41	50	64	81	117	34	85	178	26	0.44
0.01	15	19	28	37	49	63	98	23	67	147	53	0.88
0.02	12	14	18	27	35	48	77	14	51	115	105	1.75
0.05	8	9.5	11	16	22	31	52	8.0	33	77	263	4.38
0.1	6.5	6.8	72	11	15	22	35	5.5	22	51	526	8.77
0.2	4.0	4.8	4.8	7.5	9.5	14	21	3.8	14	31	1052	17.5
0.5	2.5	2.7	2.8	4.0	5.2	7.0	8.5	2.4	7.0	13	2630	43.8
1.0	1.7	1.8	1.9	2.2	3.0	4.0	4.0	1.7	3.7	6.4	5260	87.66
2.0	1.1	1.2	1.2	1.3	1.8	2.5	2.0	1.1	1.6	2.8	10520	175.3

Source. Reference 11.

(Ref. 11). We must estimate the rain layer height appropriate to the path in question before proceeding with the attenuation calculation, since even the 0°C isotherm height depends on latitude and the general rain conditions.

With the Global Model the average height of the 0°C isotherm for days with rain was taken to correspond to the height expected for 1% of the year. The highest height observed with rain was taken to correspond to the value expected 0.001% of the year, the average summer height of the −5°C isotherm. The latitude dependences of the heights to be expected for surface point rain rates exceed 1% of the year and 0.001% of the year were obtained from the latitude dependences (Oort and Rasmusson 1971) (Ref. 13). The resultant curves are shown in Figure 6.14. The model has a seasonal rms uncertainty for the 0°C isotherm height of 500 m or roughly 13% of the average estimated height. This value of 13% is used to estimate the expected uncertainties associated with Figure 6.14.

The correspondence between the 0°C isotherm height values and the excessive precipitation events showed a tendency toward a linear relationship between R_p and the 0°C isotherm height H_0 for high values of R_p. Since, for high rain rates, the rain rate distribution function displays a nearly linear relationship between R_p and $\log(P)$, where P is the probability of occurrence, the interpolation model used for the estimation of H_0 for P between 0.001% and 1% is assumed to have the form

$$H_0 = a + b \log(P) \qquad (6.7)$$

The relationship was used to provide the intermediate values displayed in

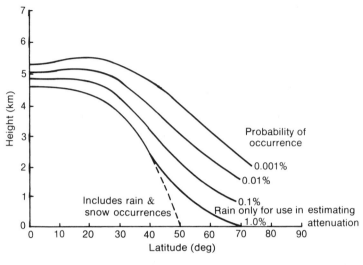

(a) Variable Isotherm

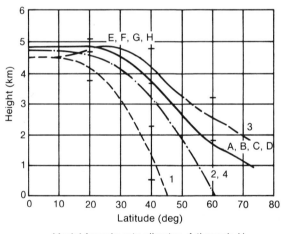

—— Model for rain rate climates *A* through *H*
—·— Annual
---- Seasonal

1—Winter (northern hemisphere)
2—Spring
3—Summer
4—Fall

(b) 0°C Isotherm Height

FIGURE 6.14 Effective heights for computing path lengths through rain events. (*a*) Variable isotherm. (*b*) 0°C isotherm height. From Ref. 11.

Figure 6.14a. In Figure 6.14b the 0°C isotherms are shown for various latitudes and seasons.

6.3.3.6 The Attenuation Model

To estimate the excess attenuation due to rainfall on a satellite link, we first determine the vertical distance between the height (or altitude) of the earth terminal and the 0°C isotherm height or $H_0 - H_g$, where H_g is the earth terminal height for the percentage of the year (or R_p) value of interest. The path horizontal projection distance D can then be calculated by

$$D = \begin{cases} \dfrac{H_0 - H_g}{\tan \theta} & \theta \geq 10° \quad\quad (6.8a) \\[2mm] E\psi \ (\psi \text{ in radians}) & \theta < 10° \quad\quad (6.8b) \end{cases}$$

where H_0 = height of 0°C isotherm
$\quad\quad H_g$ = height of ground terminal
$\quad\quad \theta$ = path elevation angle
$\quad\quad E$ = effective earth radius (8500 km)
$\quad\quad \psi$ = path central angle

and

$$\psi = \sin^{-1}\left\{ \frac{\cos \theta}{H_0 + E}\left[(H_g + E)^2 \sin^2\theta + 2E(H_0 - H_g) \right.\right.$$

$$\left.\left. + H_0^2 - H_g^2 - (H_g + E)\sin \theta \right]^{1/2} \right\} \quad (6.9)$$

The specific attenuation may then be calculated for an ensemble of raindrops if their size and shape densities are known. Experience has shown that adequate results may be obtained using the LP number density model for the attenuation calculations based on Crane (1966) (Ref. 23) and a power-law relationship is fit to calculated values to express the dependence of specific attenuation on rain rate based on Olsen et al. work in 1978 (Ref. 24). The parameters a and b (Section 6.2) in the power-law relationship

$$A = aR_p^b \quad\quad (6.10)$$

where A = specific attenuation (dB/km)
$\quad\quad R_p$ = point rain rate (mm/hr)

[also see equation (6.2)] are both functions of operating frequency. The appropriate values of a and b may be taken from Table 6.5. For consistency, it is recommended that the reader use this table rather than Table 6.2 for satellite uplink and downlink calculations.

TABLE 6.5. Parameters *a* and *b* for Computing Specific Attenuation $A = aR^b$, 0°C, Distribution (Crane, 1966) (Ref. 23)

Frequency f (GHz)	Multiplier $a(f)$	Exponent $b(f)$
1	0.00015	0.95
4	0.00080	1.17
5	0.00138	1.24
6	0.00250	1.28
7.5	0.00482	1.25
10	0.0125	1.18
12.5	0.0228	1.145
15	0.0357	1.12
17.5	0.0524	1.105
20	0.0699	1.10
25	0.113	1.09
30	0.170	1.075
35	0.242	1.04
40	0.325	0.99
50	0.485	0.90
60	0.650	0.84
70	0.780	0.79
80	0.875	0.753
90	0.935	0.730
100	0.965	0.715

6.3.3.7 Calculation of Excess Attenuation due to Rainfall by the Variable Isotherm Height Technique

The variable isotherm height technique is based on the fact that the effective height of the attenuating medium changes depending on the type of rainfall event. It also takes into consideration that various types of rainfall events selectively influence various percentages of time throughout the rainfall cycle. Therefore, a relation exists between the effective isotherm height and the percentage of time that the rain event occurs. This relation has been shown earlier in Figure 6.14*a*. The total attenuation is obtained by integrating the specific attenuation along the path. The resulting equation that is used for estimating the slant path attenuation is

$$A = \frac{aR_p^b}{\cos\theta}\left[\frac{e^{UZb} - 1}{Ub} - \frac{X^b e^{YZb}}{Yb} + \frac{X^b e^{YDb}}{Yb}\right], \qquad \theta \geq 10° \quad (6.11)$$

where U, X, Y, and Z are empirical constants that depend on the point rain rate. These constants are:

$$U = \frac{1}{Z}(e^{YZ} \ln X) \qquad (6.12)$$

$$X = 2.3R_p^{-0.17} \qquad (6.13)$$

$$Y = 0.026 - 0.03 \ln R_p \qquad (6.14)$$

$$Z = 3.8 - 0.6 \ln R_p \qquad (6.15)$$

for lower elevation angles $\theta < 10°$

$$A = \frac{L}{D} aR_p^b \left[\frac{e^{UZb} - 1}{Ub} - \frac{X^b e^{YZb}}{Yb} + \frac{X^b e^{YDb}}{Yb} \right] \qquad (6.16)$$

where

$$L = \left[(E + H_g)^2 + (E + H_0)^2 - (E + H_g)(E + H_0)\cos\psi \right]^{1/2} \qquad (6.17)$$

ψ = path central angle defined above.

The following steps apply the variable isotherm height rain attenuation model to a general earth-to-space path:

STEP 1.   At the satellite earth terminal's geographic latitude and longitude, obtain the appropriate climate region, one of the eight regions (A through H). Use either Figure 6.9, 6.10, 6.11, or 6.12. However, if long-term rain rate statistics are available for the location of the earth terminal, they should be used instead of the model distribution functions.

STEP 2.   Select the probabilities of occurrence (P) covering the range of interest in terms of the percentage of time rain rate is exceeded (e.g., 0.01%, 0.1%, or 1%). This of course will be based on the specified path time availability. If the path availability is specified as 99.9%, P would then equal 0.1% or $100 - 99.9\%$.

STEP 3.   Obtain the terminal point rain rate R_p (mm/hr) using Table 6.4, or long-term measured values, if available, of rain rate versus the percentage of the year rain rate is exceeded at the climate region and probabilities of occurrence (Step 2).

STEP 4.   For a satellite link through the entire atmosphere, obtain the rain layer height from the height of the 0°C isotherm (melting layer) H_0 at the path latitude from Figure 6.14a. The heights will vary correspondingly with the probabilities of occurrence (Step 2). To interpolate, plot $H_0(P)$ vs log P and use a straight line to relate H_0 to P.

STEP 5. Obtain the horizontal path projection D of the oblique path through the rain volume:

$$D = \frac{H_0 + H_g}{\tan \theta}; \qquad \theta \geq 10° \qquad (6.18)$$

$$H_0 = H_0(P) = \text{height (km) of isotherm for probability } P \quad (6.19)$$

$$H_g = \text{height of ground terminal (km)}$$

$$\theta = \text{path elevation angle}$$

STEP 6. Test $D \leq 22.5$ km; if true, proceed to the next step. If $D \geq 22.5$ km, the path is assumed to have the same attenuation value as for a 22.5-km path but the probability of occurrence is adjusted by the ratio of 22.5 km to the path length:

$$\text{new probability of occurrence, } P' = P\left(\frac{22.5 \text{ km}}{D}\right) \quad (6.20)$$

where D = path length projected on surface (> 22.5 km).

STEP 7. Obtain the parameters $a(f)$ and $b(f)$, relating the specific attenuation to the rain rate from Table 6.5.

STEP 8. Compute the total attenuation due to rain using R_p, a, b, θ, D:

$$A = \frac{aR_p^b}{\cos \theta}\left[\frac{e^{UZb} - 1}{Ub} - \frac{X^b e^{YZb}}{Yb} + \frac{X^b e^{YDb}}{Yb}\right]; \qquad \theta \geq 10° \quad (6.21)$$

where A = total path attenuation due to rain (dB)

a, b = parameters relating the specific attenuation to rain rate (from Step 7), $\alpha = aR_p^b$ = specific attenuation

R_p = point rain rate (Step 3)

θ = elevation angle of path

D = horizontal path distance (from Step 5) $Z \leq D \leq 22.5$ km or alternatively, if $D < Z$,

$$A = \frac{aR_p^b}{\cos \theta}\left[\frac{e^{UbD} - 1}{Ub}\right] \quad (6.22)$$

or if $D = 0$, $\theta = 90°$,

$$A = (H - H_g)(aR_p^b) \quad (6.23)$$

Example 3. Calculate the excess attenuation due to rainfall for a downlink operating at 21 GHz where the elevation angle is 10° and the desired path time availability is 99.9%. The earth terminal is in southeastern Minnesota. (45N, 90W)

Procedure. Select the appropriate climate region. Use Figure 6.11 and the climate region is D_1. Turn to Table 6.4 and select the value for R_p for an exceedance of 0.1% (e.g., $100.0 - 99.9$) and this value is $R_p = 11$ mm/hr. Obtain the rain layer height from the height of the 0°C isotherm (melting layer) H_0 at the path latitude (45°N). Use Figure 6.14a and $H_0 = 2.8$ km. Assume $H_g = 0.2$ km.

Calculate D (Step 5); use equation (6.18):

$$D = \frac{28 - 0.2}{\tan 10°}$$

$$= \frac{2.6}{0.176}$$

$$= 14.77 \text{ km}$$

D is less than 22.5 km, thus proceed to step 7. Obtain the parameters $a(f)$ and $b(f)$; use Table 6.5 and interpolate values for 21 GHz: $a = 0.0785$ and $b = 1.098$.

Turn to equations (6.12)–(6.15). Start with (6.15) and proceed backward:

$$Z = 3.8 - 0.6\ln(11)$$

$$Z = 2.36$$

$$Y = 0.026 - 0.03\ln(11)$$

$$Y = -0.046$$

$$X = 2.3(11)^{-0.17}$$

$$X = 1.53$$

$$U = \frac{1}{2.36\left(e^{(-0.04 \times 2.36)}\ln 1.53\right)}$$

$$U = \frac{1}{2.36\left(e^{-0.108} \times 0.425\right)}$$

$$U = \frac{1}{2.36 \times 0.381}$$

$$U = 1.112$$

Apply these calculated values, the values of a and b, and $\theta = 10°$ to calculate

A, the total excess attenuation due to rainfall for the path in question. Use equation (6.21); θ is equal to $10°$:

$$A = \frac{0.0785(11)^{1.098}}{\cos 10°}\left[\frac{e^{1.112 \times 2.36 \times 1.098} - 1}{1.112 \times 1.098} - \frac{1.53^{1.098}e^{-0.046 \times 2.36 \times 1.098}}{-0.046 \times 1.098}\right.$$

$$\left. + \frac{1.53^{1.098}e^{-0.046 \times 14.77 \times 1.098}}{-0.046 \times 1.098}\right]$$

It is often easier to calculate by pieces and simplify factors as follows: $\cos 10° = 0.985$. The factor in front of the brackets is 1.11. The first term inside the parens is $16.74/1.219 = 13.74$. The second term is $(-1.6 \times 0.888)/-0.051 = +27.86$. The third term is $(1.6 \times 0.474)/-0.051 = -14.87$. Simplify:

$$A = 1.11(13.74 + 27.86 - 14.87)$$

$$= 29.67 \text{ dB}$$

This value would be entered into the link budget for the downlink.

6.3.4 Utilization of Path Diversity to Achieve Performance Objectives

Excess attenuation due to rainfall often degrades satellite uplinks and downlinks operating above 10 GHz so seriously that the requirements of optimum economic design and reliable performance cannot be achieved simultaneously. Path diversity can overcome this problem at some reasonable cost compromise. Path diversity advantage is based on the hypothesis that rain cells and, in particular, the intense rain cells that cause the most severe fading are rather limited in spatial extent. Furthermore, these rain cells do not occur immediately adjacent to one another. Thus, the probability of simultaneous fading on two paths to spatially separated earth stations would be less than that associated with either individual path. The hypothesis has been borne out experimentally (Hodge, 1973; Ref. 22).

Let us define two commonly used terms: diversity gain and diversity advantage. Diversity gain is defined (in this context) as the difference between the rain attenuation exceeded on a single path and that exceeded jointly on separated paths for a given percentage of time. Diversity advantage is defined (in this context) as the ratio of the percentage of time exceeded on a single path to that exceeded jointly on separated paths for a given rain attenuation level.

Diversity gain may be interpreted as the reduction in the required system margin at a particular percentage of time afforded by the use of path diversity. Alternatively, diversity advantage may be interpreted as the factor by which

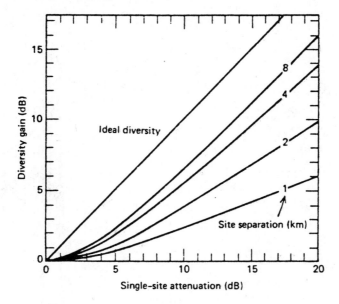

FIGURE 6.15 Diversity gain for various site separations (Ref. 14).

the fade time is improved at a particular attenuation level due to the use of path diversity.

The principal factor to achieve path diversity to compensate for excess attenuation due to rainfall is separation distance. The diversity gain increases rapidly as the separation distance d is increased over a small separation distance, up to about 10 km. Thereafter the gain increases more slowly until a maximum value is reached, usually between about 10 and 30 km. This is shown in Figure 6.15.

The uplink/downlink frequencies seem to have little effect on diversity gain up to about 30 GHz. (Ref. 11). This same reference suggests that for link frequencies above 30 GHz attenuation on both paths simultaneously can be sufficient to create an outage. Therefore, extrapolation beyond 30 GHz is not recommended, at least with the values given in Figure 6.15.

6.4 EXCESS ATTENUATION DUE TO ATMOSPHERIC GASES ON SATELLITE LINKS

The zenith one-way attenuations for a moderately humid atmosphere (e.g., 7.5 g/m^3 surface water vapor density) at various starting heights above sea level are given in Figure 6.16 and in Table 6.6. These curves were computed by Crane and Blood (1979) (Ref. 10) for temperate latitudes assuming the U.S. standard atmosphere, July, 45°N latitude. The range of values shown in

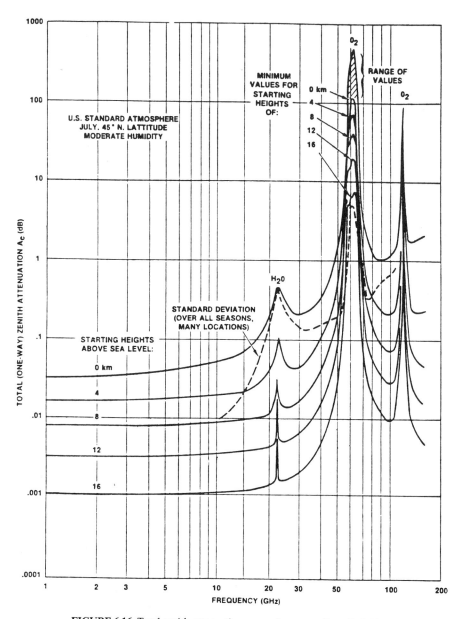

FIGURE 6.16 Total zenith attenuation versus frequency (from Ref. 11).

428

TABLE 6.6. Typical One-Way Clear Air Total Zenith Attenuation Values (7.5 g / m³ H₂O, July, 45°N Latitude, 21°C)

Frequency (GHz)	Altitude				
	0 km	0.5 km	1.0 km	2.0 km	4.0 km
10	0.055	0.05	0.043	0.035	0.02
15	0.08	0.07	0.063	0.045	0.023
20	0.30	0.25	0.19	0.12	0.05
30	0.22	0.18	0.16	0.10	0.045
40	0.40	0.37	0.31	0.25	0.135
80	1.1	0.90	0.77	0.55	0.30
90	1.1	0.92	0.75	0.50	0.22
100	1.55	1.25	0.95	0.62	0.25

Source. Reference 11.

Figure 6.16 refers to the peaks and valleys of the fine absorption lines. The range of values for starting heights above 16 km is even greater.

Figure 6.16 also shows the standard deviation of the clear air zenith attenuation as a function of frequency. The standard deviation was calculated from 220 measured atmosphere profiles spanning all seasons and geographical locations by Crane (1976) (Ref. 15). The zenith attenuation is a function of frequency, earth terminal altitude above sea level, and water vapor content. Compensating for earth terminal altitudes can be done by interpolating between the curves in Figure 6.16.

The water vapor content is the most variable component of the atmosphere. Corrections should be made to the values derived from Figure 6.16 and Table 6.6 in regions that notably vary from the 7.5 g/m³ value given. Such regions would be arid or humid, jungle or desert. This correction to the total zenith attenuation is a function of the water vapor density at the surface p_0 as follows:

$$\Delta A_{c1} = b_p (p_0 - 7.5 \text{ g/m}^3) \tag{6.24}$$

where ΔA_{c1} is the additive correction to the zenith clear air attenuation that accounts for the difference between the actual surface water vapor density and 7.5 g/m³. The coefficient b_p is frequency dependent and is given in Figure 6.17. To convert from the more familiar relative humidity or partial pressure of water vapor, refer to Section 6.4.2.

The surface temperature T_0 also affects the total attenuation because it affects the density of both the wet and dry components of the gaseous attenuation. This relation is (Ref. 10)

$$\Delta A_{c2} = c_T (21° - T_0) \tag{6.25}$$

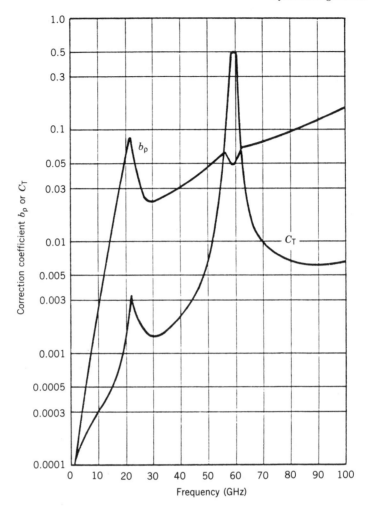

FIGURE 6.17 Water vapor density and temperature correction coefficients. From Ref. 11.

where ΔA_{c2} is an additive correction to the zenith clear air attenuation. Figure 6.17 gives the frequency dependent values for c_T.

The satellite earth terminal elevation angle has a major impact on the gaseous attenuation value for a link. For elevation angles greater than about 5°, the zenith clear air attenuation value A_c is multiplied by the cosecant of the elevation angle θ. The total attenuation for an elevation angle θ is given by

$$A_c = A'_c \csc \theta \qquad (6.26)$$

6.4.1 Example Calculation of Clear Air Attenuation—Hypothetical Location

For a satellite downlink, we are given the following information: frequency, 20 GHz; altitude of earth station, 600 m; relative humidity (RH), 50%; temperature (surface, T_0), 70°F (21.1°C); and elevation angle, 25°. Calculate clear air attenuation.

Obtain total zenith attenuation, A'_c from Table 6.6, interpolate value for altitude: $A'_c = 0.24$ dB.

Find the water vapor density p_0. From Figure 6.18, the saturated partial pressure of water vapor at 70°F is $e_s = 2300$ N/m². Apply formula (6.27) (Section 6.4.2) and

$$p_0 = (0.5)2300/(0.461)(294.1)$$

$$p_0 = 1150/135.6$$

$$p_0 = 8.48 \text{ g/m}^3$$

Calculate the water vapor correction factor ΔA_{cl}. From Figure 6.17 for a frequency of 20 GHz, correction coefficient $b_p = 0.05$, then [equation (6.24)]

$$\Delta A_{cl} = (0.05)(8.48 - 7.5) = 0.05 \text{ dB}.$$

Compute the temperature c_T using equation (6.25). At 20 GHz $c_T = 0.0015$. As can be seen, this value can be neglected in this case.

Calculate the clear air zenith attenuation corrected A'_c:

$$A'_c = 0.24 \text{ dB} + 0.05 \text{ dB} + 0 \text{ dB}$$

$$= 0.29 \text{ dB}$$

Compute the clear air slant attenuation using equation (6.26):

$$A_c = 0.29 \csc 25°$$

$$A_c = 0.29 \times 2.366$$

$$= 0.69 \text{ dB}$$

This value would then be used in the link budget for this hypothetical link.

6.4.2 Conversion of Relative Humidity to Water Vapor Density

The surface water vapor density p_0 (g/m³) at a given surface temperature (T_0) may be calculated from the ideal gas law:

$$p_0 = \frac{(\text{RH})e_s}{0.461 \text{ J g}^{-1}\text{ K}^{-1}(T_0 + 273)} \tag{6.27}$$

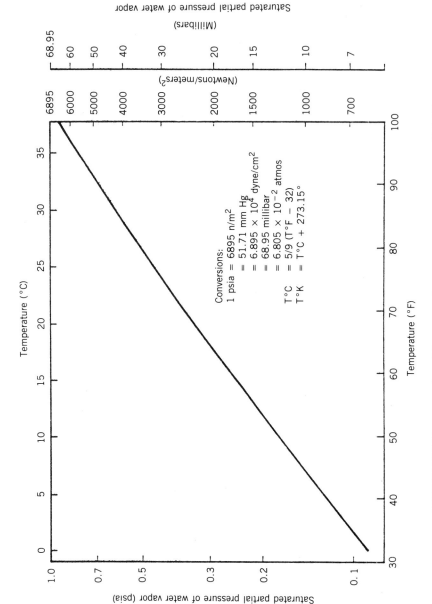

FIGURE 6.18 The saturated partial pressure of water vapor versus temperature (Ref. 11).

where RH is the relative humidity, e_s (N/m^2) is the saturated partial pressure of water vapor which corresponds to the surface temperature T_0 (°C). See Figure 6.18. The relative humidity corresponding to 7.5 g/m^3 at 20°C (68°F) is RH = 0.42 or 42%. (Ref. 11.)

6.5 ATTENUATION DUE TO CLOUDS AND FOG

Water droplets that constitute clouds and fog are generally less than 0.01 cm in diameter (Ref. 11). This allows a Rayleigh approximation to calculate the attenuation due to clouds and fog for frequencies up to 100 GHz. The specific attenuation a_c is, unlike the case of rain, independent of drop-size distribution. It is a function of liquid water content p_1 and can be expressed by

$$a_c = K_c p_1 \quad (\text{dB/km}) \tag{6.28}$$

where p_1 is normally expressed in g/m^3. K_c is the attenuation constant which is a function of frequency and temperature and is given in Figure 6.19. The curves in Figure 6.19 assume pure water droplets. The values for salt water droplets, corresponding to ocean fogs and mists, are higher by approximately 25% at 20°C and 5% at 0°C (Ref. 16).

The liquid water content of clouds varies widely. Stratiform or layered clouds display ranges of 0.05 to 0.25 g/m^3 (Ref. 11). Stratocumulus, which is the most dense of this cloud type, has shown maximum values from 0.3 to 1.3 g/m^3 (Ref. 17). Cumulus clouds, especially the large cumulonimbus and cumulus congestus that accompany thunderstorms, have the highest values of liquid content. Fair weather cumulus clouds generally have a liquid water content of less than 1 g/m^3. Reference 19 reported values exceeding 5 g/m^3 in cumulus congestus and estimates an average value of 2 g/m^3 for cumulus congestus and 2.5 g/m^3 for cumulonimbus clouds.

Care must be exercised in estimating excess attenuation due to clouds when designing uplinks and downlinks. First, clouds are not homogeneous masses of air containing uniformly distributed droplets of water. Actually, the liquid water content can vary widely with location in a single cloud. Even sharp differences have been observed in localized regions on the order of 100 m across. There is a fairly rapid variation with time as well, owing to the complex patterns of air movement taking place within cumulus clouds.

Typical path lengths through cumulus congestus clouds roughly fall between 2 and 8 km. Using equation (6.28) and the value given for water vapor density and the attenuation coefficient K_c from Figure 6.19, an added path loss at 35 GHz from 4 to 16 dB would derive. Fortunately, for the system designer, the calculation grossly overestimates the actual attenuation that has been observed through this type of cloud structure. Table 6.7 provides values that seem more dependable. In the 35 and 95 GHz bands, cloud attenuation, in most cases, is 40% or less of the gaseous attenuation values. One should not

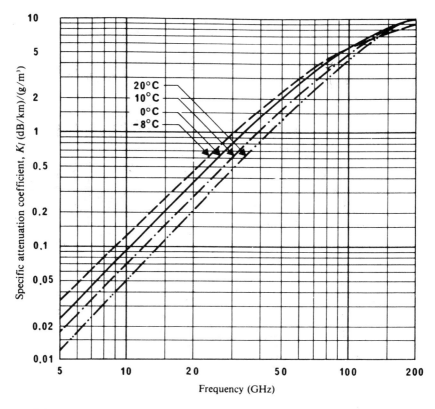

FIGURE 6.19 Attenuation coefficient K_c due to water vapor droplets. From CCIR Rep. 72-11 (Ref. 7).

lose sight of the fact, in these calculations, of the great variability in the size and state of development of the clouds observed. Data from Table 6.7 may be roughly scaled in frequency, using the frequency dependence of the attenuation coefficient from Figure 6.19.

Fog results from the condensation of atmospheric water vapor into water droplets that remain suspended in air. The water vapor content of fog varies from less than 0.4 up to as much as 1 g/m^3.

The attenuation due to fog in dB/km can be estimated using the curves in Figure 6.19. The 10°C curve is recommended for summer, and the 0°C curve should be used for the other seasons. Typical liquid water content values for fog vary from 0.1 to 0.2 g/m^3. Assuming a temperature of 10°C, the specific attenuation would be about 0.08–0.16 dB/km at 35 GHz and 0.45–0.9 dB/km for 95 GHz. In a typical fog layer 50 m thick, a path at a 30° elevation angle would have only 100 m extension through fog, producing less than 0.1 dB excess attenuation at 95 GHz. In most cases, the result is that fog attenuation is negligible for satellite links.

TABLE 6.7. Zenith Cloud Attenuation Measurements

TABLE 6.7. Zenith Cloud Attenuation Measurements

| | Cloud Attenuation (dB) | |
Cloud Type	95 GHz	150 GHz
Stratocumulus	0.5–1	0.1–1
Small, fine weather cumulus	0.5	0.5
Large cumulus	1.5	2
Cumulonimbus	2–7	3–8
Nimbostratus (rain cloud)	2–4	5–7

Source. Reference 7.

6.6 CALCULATION OF SKY NOISE TEMPERATURE AS A FUNCTION OF ATTENUATION

The effective sky noise (see Section 4.3.8.1 and Figures 4.7–4.9) due to the troposphere is primarily dependent on the attenuation at the frequency of observation. Reference 19 shows the derivation of an empirical equation relating specific attenuation (A) to sky noise temperature:

$$T_s = T_m(1 - 10^{(-A/10)})$$ (6.29)

where T_s is the sky noise and T_m is the mean absorption temperature of the attenuating medium (e.g., gaseous, clouds, rainfall) and A is the specific attenuation that has been calculated in the previous subsections. Temperatures are in kelvins. The value

$$T_m = 1.12(\text{surface temperature in K}) - 50 \text{ K}$$ (6.30)

has been empirically determined by Ref. 19.

Some typical values taken in Rosman, NC (Ref. 11) are given in Table 6.8 for rainfall.

Example 4. From Table 6.8 with a total rain attenuation of 11 dB, what is the sky noise at 20 GHz? Assume $T_m = 275$ K.

Use equation (6.29):

$$T_s = 275(1 - 10^{-11/10})$$

$$T_s = 253.16 \text{ K}$$

TABLE 6.8. Cumulative Statistics of Sky Temperature Due to Rain for Rosman, NC at 20 GHz (T_m = 275 K)

Percentage of Year	Point Rain Rate Values (mm/hr)	Average Rain Rate (mm/hr)	Total Rain Attenuation[a] (dB)	Sky Noise Temperature[b] (K)
0.001	102	89	47	275
0.002	86	77	40	275
0.005	64	60	30	275
0.01	49	47	23	274
0.02	35	35	16	269
0.05	22	24	11	252
0.1	15	17	7	224
0.2	9.5	11.3	4.6	180
0.5	5.2	6.7	2.6	123
1.0	3.0	4.2	1.5	82
2.0	1.8	2.7	0.93	53

[a]At 20 GHz the specific attenuation $A = 0.06 R_{av}^{1.12}$ dB/km and for Rosman, NC the effective path length is 5.1 km to ATS-6.

[b]For a ground temperature of 17°C = 63°F, the T_m = 275 K.

Source. Reference 11.

6.7 THE SUN AS A NOISE GENERATOR

The sun is a white noise jammer of an earth terminal when the sun is aligned with the downlink terminal beam. This alignment occurs, for a geostationary satellite, twice a year near the equinoxes, and in the period of the equinox will occur for a short period each day. The sun's radio signal is of sufficient level to nearly saturate the terminal's receiving system, wiping out service for that period. Figure 6.20 gives the power flux density of the sun as a function of frequency. Above about 20 GHz the sun's signal remains practically constant at -188 dBW Hz^{-1} m^{-2} for "quiet sun" conditions.

Reception of the sun's signal or any other solar noise source can be viewed as an equivalent increase in a terminal's antenna noise temperature by an amount T_s. T_s is a function of terminal antenna beamwidth compared to the apparent diameter of the sun (e.g., 0.48°), and how close the sun approaches the antenna boresight. The following formula, taken from Ref. 20, provides an estimate of T_s, when the sun or any other extraterrestrial noise source is aligned in the antenna beam:

$$T_s = \frac{1 - \exp\left[-(D/1.2\theta)^2\right]}{f^2 D^2}\left(\log^{-1}\frac{S + 250}{10}\right) \qquad (6.31)$$

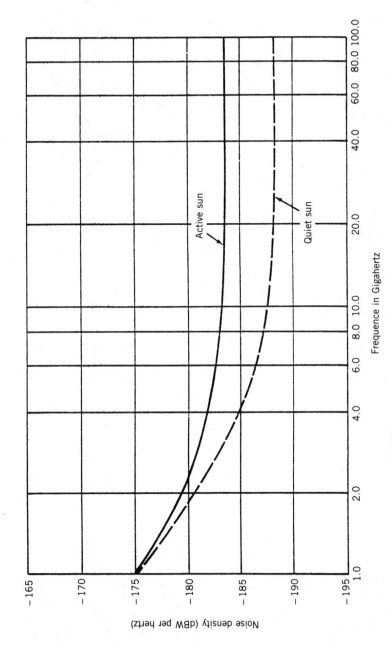

FIGURE 6.20 Values of noise from quiet and active sun. Sun fills entire antenna beam (Ref. 21).

where D = apparent diameter of the sun or 0.48°
  f = frequency in gigahertz
  S = power flux density, dBW Hz^{-1} m^{-2}
  θ = half-power beamwidth of the terminal antenna in degrees

Example 5 An earth station operating with a 20-GHz downlink has a 2-m antenna (beamwidth of 0.5°). What is the maximum increase in antenna noise temperature that would be caused by a quiet sun transit?
  Use formula (6.31):

$$T_s = 8146 \text{ K}$$

6.8 PROPAGATION EFFECTS WITH A LOW ELEVATION ANGLE

As the elevation angle of an earth terminal is lowered, the ray beam penetrates an ever increasing amount of atmosphere. Below about 10°, fading on the downlink signal must be considered. Fading or signal fluctuations apply only to the ground terminal downlink because its antenna is in close proximity to a turbulent medium. The companion uplink satellite path will suffer uplink fluctuation gain degradation only due to scattering of energy out of the path (Ref. 11). Because of the large distance traversed by the uplink signal since leaving the troposphere, the signal arrives at the satellite as a plane wave and with only a small amount of angle-of-arrival effects.

Phase variations must also be expected for the low-elevation-angle condition. Phase variations arise due to the variable delay as the wave passes through a medium with variable refractivity. Phase scintillation can also occur.

6.9 DEPOLARIZATION ON SATELLITE LINKS

Depolarization is an effect wherein a satellite link's wave's polarization is altered by the troposphere. Some texts refer to depolarization as cross-polarization. For the case of a linearly polarized wave passing through a medium, components of the opposite polarization will be developed. In the case of circularly polarized waves, there will be a tendency to develop into an elliptical wave. This is particularly important on frequency reuse systems where the depolarization effectively reduces the polarization isolation and can tend to increase crosstalk on the signal.

Depolarization on a satellite link can be caused by rain, ice, snow, multipath effects, and refractive effects.

PROBLEMS AND EXERCISES

1. Give at least two causes of excess attenuation on satellite paths that must be taken into account for satellite systems operating above 10 GHz.

2. Give the two frequency bands between 10 and 100 GHz where excess attenuation due to atmospheric gases is high, one of which is excessive.

3. Give some uses one might make of these high attenuation bands.

4. Argue why cumulative annual rainfall rates may not be used for calculation of excess attenuation due to rainfall and where we must use point rainfall rates.

5. Name at least four ways a system design engineer can build a link margin (obviously for rain and gaseous attenuation, maintenance margin).

6. In early attempts to build in sufficient margin on satellite and LOS links operation above 10 GHz, it was found that the required margins were excessively large because we integrated excess attenuation per kilometer along the entire path (the entire path in the atmosphere for satellite links). Describe how statistics on rain cell size assisted to better estimate excess attenuation due to rainfall.

7. Calculate the specific attenuation per kilometer for a path operating at 30 GHz on a LOS basis with a time availability for the path of 99.9%. Neglect path length considerations, of course. The path is located in northeastern United States. Carry out the calculation for both horizontal and vertical polarizations.

8. Calculate the excess attenuation due to rainfall for a LOS path operating at 50 GHz operating in central Australia. The path length is 20 km and the desired time availability is 99.99%. Assume vertical polarization.

9. Calculate the excess attenuation due to rainfall for a LOS path 25 km long with an operating frequency of 18 GHz and the desired path availability (propagation reliability) is 99.99%. The path is located is the state of Massachusetts.

10. Name five ways to build a rainfall margin for the path is question 9.

11. Calculate the excess attenuation due to rainfall for a satellite path with a 20° elevation angle for a 21-GHz downlink. The earth station is located in southern Minnesota and the desired time availability for the link is 99%.

12. An earth station is to be located near Bonn, FRG and will operate at 14 GHz. The desired uplink time availability is 99.95% and the subsatellite point is 10°W. What is the excess attenuation due to rainfall?

13. An earth station is to be installed in Diego Garcia with an uplink at 44 GHz. The elevation angle is 15° and the desired path (time) availability is 99%. What value of excess attenuation due to rainfall should be used in the link budget?

14. There is an uplink at 30 GHz and the required excess attenuation due to rainfall is 15 dB. Path diversity is planned. Show how the value of excess attenuation due to rainfall for a single site can be reduced for site separations of 1, 2, 4, and 8 km.

15. For an earth station, calculate the excess attenuation due to atmospheric gases for a site near sea level. The site is planned for 30/20 GHz operation. The elevation angle is 15°. The relative humidity is 60% and the surface temperature is 70°F.

16. Calculate the sky noise contribution for the attenuation of gases calculated in question 15. Calculate the sky noise temperature due to the excess attenuation due to rainfall from question 13.

REFERENCES

1. H. J. Liebe, "Atmospheric Propagation Properties in the 10 to 75 GHz Region: A Survey and Recommendations," ESSA Technical Report ERL 130-ITS 91, Boulder, CO 1969.
2. CCIR Rep. 719, "Recommendations and Reports of the CCIR," 1978, XIV Plenary Assembly, Kyoto, 1978, Vol. 5.
3. D. C. Hogg, "Millimeter-Wave Propagation through the Atmosphere," *Science* (1968).
4. R. K. Crane, "Prediction of the Effects of Rain on Satellite Communication Systems," *Proceedings of the IEEE*, **65**, 456–474 (1977).
5. CCIR Rep. 593-1, "Recommendations and Reports of the CCIR," 1978, XIV Plenary Assembly, Kyoto, 1978, Vol. V.
6. CCIR Rep. 338-3, "Recommendations and Reports of the CCIR," 1982, XV Plenary Assembly, Geneva, 1982, Vol. V.
7. CCIR Rep. 721-1, "Recommendations and Reports of the CCIR," 1982, XV Plenary Assembly, Geneva, 1982, Vol. V.
8. CCIR Rep. 338-4, "Recommendations and Reports of the CCIR," 1982, XV Plenary Assembly, Geneva, 1982, Vol. V.
9. CCIR Rep. 563-2, "Recommendations and Reports of the CCIR," 1982, XV Plenary Assembly, Geneva, 1982, Vol. V.
10. R. K. Crane and D. W. Blood, "Handbook for the Estimation of Microwave Propagation Effects—Link Calculations of Earth-Space Paths," Environmental Research and Technology Report, No. 1, DOC No. P-7376-TRL, U.S. Department of Defense, 1979.
11. R. Kaul, R. Wallace, and G. Kinal, *A Propagation Effects Handbook for Satellite Systems Design: A Summary of Propagation Impairments on 10–100 GHz Satellite Links, with Techniques for System Design*, ORI Inc., Silver Spring, MD, 1980 (NTIS N80-25520).
12. CCIR Rep. 564-2, "Recommendations and Reports of the CCIR," 1982, XV Plenary Assembly, Geneva, 1982, Vol. V.
13. A. H. Oort and E. M. Rasmusson, "Atmospheric Circulation Statistics," NOAA Professional Paper No. 5, U.S. Department of Commerce, 1971.
14. R. L. Freeman, *Telecommunication Transmission Handbook*, 2nd ed, Wiley, New York, 1981.
15. R. K. Crane, "An Algorithm to Retrieve Water Vapor Information from Satellite Measurements," NEPRF Tech. Report 7-76, Final Report, Project No. 1423, Environmental Research and Technology, Inc., Concord, MA, 1976.

16. K. L. Koester and L. H. Kosowsky, "Millimeter Wave Propagation in Ocean Fogs and Mists," Proceedings of IEEE Antenna Propagation Symposium, 1978.

17. B. J. Mason, *The Physics of Clouds*, Clarendon Press, Oxford, UK, 1971.

18. H. K. Weickmann and H. J. Kaumpe, "Physical Properties of Cumulus Clouds," Journal of Meteorology, Vol. 10, 1953.

19. K. H. Wulfsberg, "Apparent Sky Temperatures at Millimeter-Wave Frequencies," Physical Science Research Paper No. 38, Air Force Cambridge Research Lab., No. 64-590, 1964.

20. J. W. M. Baars, "The Measurement of Large Antennas with Cosmic Radio Sources," *IEEE Trans. Ant. Prop.*, **AP-21** (4) (1973).

21. S. Perlman et al., "Concerning Optimum Frequencies for Space Vehicle Communications," *IRE Trans. Mil. Electronics*, **Mil-4** (2–3) (1960).

22. D. B. Hodge, "The Characteristics of Millimeter Wavelength Satellite-to-Ground Space Diversity Links," IEE Conference No. 98, London, April 1978.

23. R. K. Crane, "Microwave Scattering Parameters for New England Rain," MIT Lincoln Lab. Technical Report 426, AD 647798, 1966.

24. R. L. Olsen, D. V. Rogers, and D. B. Hodge, "The aR^b Relation in Calculation of Rain Attenuation," *IEEE Trans. Ant. Prop.*, **AP-26**, 1978.

RADIO TERMINAL DESIGN—A SYSTEM APPROACH

7.1 INTRODUCTION

This chapter deals with the design of radio terminals and the operation of associated subsystems. A modular approach is taken. The details of the design of these modules can be researched from the appropriate literature. The concern here is the functions of the necessary subassemblies or modules of a terminal and their interfaces. Modulation/demodulation has been discussed in Chapters 2, 3, 4, and 5. The text relies heavily on block diagrams. The objective is to describe the basic components of a terminal and how they relate to one another and to the system overall. Generic terminals are described for radiolinks [line-of-sight (LOS) microwave], diffraction/troposcatter, and earth station terminals, both digital and analog.

7.1.1 The Basic Terminal

For nearly every application in the point-to-point service a radio terminal consists of a transmitting and receiving subsystem as shown in Figure 7.1. On the transmit side we can expect to find a modulator, an upconverter, and some sort of power amplifier (sometimes called an HPA or high-power amplifier). On the receive side, a low-noise amplifier (LNA), a downconverter, and a demodulator. In most cases a common antenna system is used for receive and transmit. A service channel, which may be separate (as shown in Figure 7.1) or composite with the information baseband, is used for site-to-site operational coordination and/or for system-wide coordination. The information baseband, which may be either analog or digital in makeup, carries the electrical signal information that originated at a source and is destined for a sink. The intermediate frequency for many if not most implementations is 70 MHz; 140 MHz is used in certain applications (CCIR Rec. 403-3). Some new equipments use 600, 700, and 1200 and 1300 MHz. F_0 in Figure 7.1 is the operating frequency.

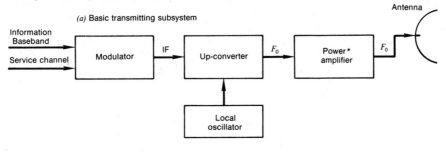

(a) Basic transmitting subsystem

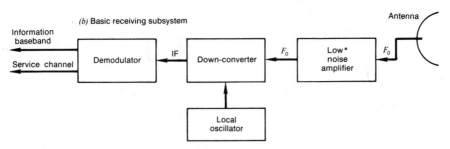

(b) Basic receiving subsystem

FIGURE 7.1 Basic elements of a radio terminal (*a*) Basic transmitting subsystem. (*b*) Basic receiving subsystem. * Optional on microwave LOS radiolinks.

7.2 ANALOG LINE-OF-SIGHT RADIOLINK TERMINALS AND REPEATERS

7.2.1 Basic Analog Radiolink Terminal

Analog microwave LOS systems in the context presented here provide broadband LOS communication on a point-to-point basis. A method of sizing these terminals was presented in Chapter 2. A block diagram of a typical terminal is shown in Figure 7.2, which is an obvious outgrowth of that shown in Figure 7.1.

The great majority of analog microwave systems use frequency modulation (FM), although there are some being fielded today that use single-sideband techniques, which are considerably more bandwidth conservative.

FM systems have been favored because their design is simple, the technology is mature, and there is the advantage of trading off bandwidth for reduced thermal noise owing to the very nature of wide deviation FM.

Turning to the block diagrams of Figure 7.2, a terminal, on the transmit side, consists of baseband-conditioning equipment, an FM modulator, IF equipment, an upconverter, a power amplifier, and an antenna. On the receive side, there is an antenna, downconverter, IF equipment, demodulator, and baseband-conditioning equipment. An optional LNA may be inserted ahead of the downconverter to improve system gain and noise performance. Figure

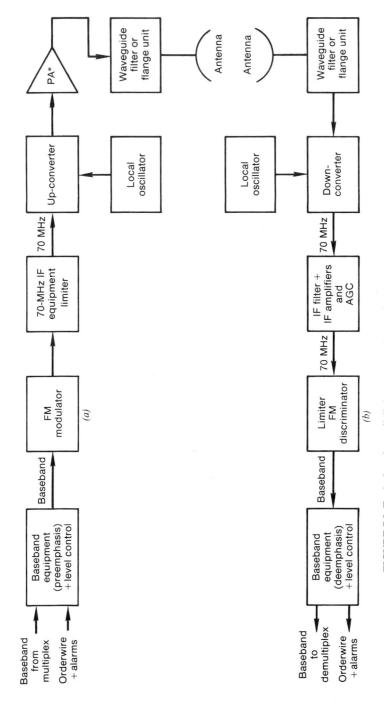

FIGURE 7.2 Typical analog radiolink (*a*) transmitter and (*b*) receiver subsystems. * PA = power amplifier, either traveling wave tube (TWT) or solid-state amplifier (SSA).

7.2 was drawn to show functionality. In an actual terminal, much equipment is shared by function, although the transmit and receive signal paths are separate. For instance, in most installations there is a common antenna and the modulator/demodulator is shown as a common assembly as is the baseband equipment.

FM radio terminals commonly transmit two types of broadband signals: a composite FDM waveform or television. In certain implementations both waveforms may be transmitted simultaneously, again on a frequency division basis. A typical FDM signal is shown in Figure 7.3. For a further description of FDM, the reader may wish to read Chapter 3 of Ref. 1, and for video transmission, Chapter 12 of the same reference.

On the transmit side, looking in from the baseband input, the baseband equipment conditions the input signal, combines the service channel (orderwire + alarms) with the baseband signal, and applies the necessary pre-emphasis. Pre-emphasis/de-emphasis is described in Section 2.6.5. The service channel, its makeup, and function are described subsequently.

The output of the baseband equipment is fed to the FM modulator, which frequency modulates the signal. The key parameter in FM modulation is deviation. The calculation of peak deviation is described in Section 2.6.4. The FM modulator frequency modulates a 70-MHz IF carrier or other IF carrier that is upconverted to IF. The IF (intermediate frequency) is usually 70 MHz. In certain equipment designed for very-broad-band operation, a higher-frequency IF is used, such as 140 MHz.

The output of the FM modulator feeds the IF equipment consisting of an amplifier and filters. The IF is then fed to an upconverter, which consists of a mixer and appropriate filters. The local oscillator provides the required injection frequency to the mixer for the upconversion process. Upconversion translates the IF to the operational radio frequency F_0. The filters in the upconverter output remove the local oscillator frequency and its harmonics and out-of-band spurious emissions. The output of the mixer feeds a power amplifier, which brings the signal level up to the required output.

The power output of LOS radiolink transmitters is rather modest and is commonly 1 W or 0 dBW. Some transmitters have no more than 100 mW output and some as much as 10 W.

The output of the PA (power amplifier) is further filtered to attenuate spurs (spurious emissions) and fed to the antenna. Antennas will be discussed at length in the following subsection.

On the receive side of the terminal, the RF signal (F_0) is fed to the downconverter. As was pointed out earlier, in some installations a LNA may be inserted here to improve performance. (See Figure 2.10.) The downconverter carries out the reverse operation of the upconverter; it translates F_0 to the IF and feeds the resulting signal to the IF equipment consisting of a filter and an amplifier. The output of the IF equipment feeds an FM demodulator, which, in many cases, consists of a limiter and a discriminator. The output of the demodulator is fed to the baseband equipment. This consists of a deem-

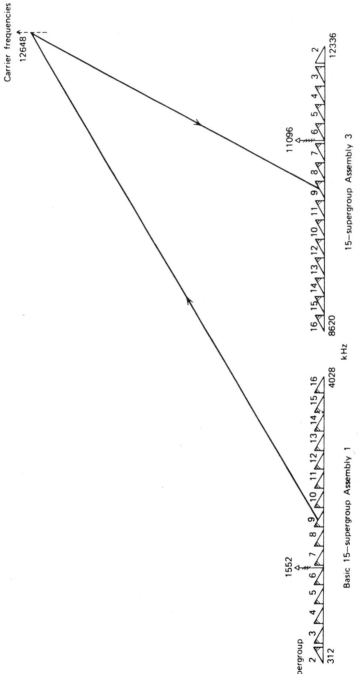

FIGURE 7.3 A typical broadband frequency division multiplex (FDM) configuration for the simultaneous transmission of 1800 voice channels. The carrier frequency, in this case 12,648 MHz, is used in the final frequency translation stage in the FDM equipment. The two vertical arrows are the FDM (not radio) pilot tones, which are used for level regulation and FDM alarms.

phasis network and the necessary circuits to separate the service channel from the information baseband signal.

7.2.2 The Antenna Subsystem

For conventional microwave LOS radiolinks, whether digital or analog, the antenna subsystem offers more room for tradeoff to meet minimum performance requirements than any other subsystem. Basically, the antenna subsystem looking outward from the transmitter must have:

- ☐ Transmission line (coaxial cable or waveguide)
- ☐ An antenna: a reflecting surface or device
- ☐ An antenna feed: a feed horn or other feeding device

In addition the antenna subsystem may have:

- ☐ Circulators or isolators
- ☐ Directional coupler(s)
- ☐ Phaser(s)
- ☐ Passive reflectors
- ☐ Radome
- ☐ A mounting device

7.2.2.1 Antennas

Below about 700 MHz, antennas used for point-to-point radiolinks are often yagis and are fed with coaxial transmission lines. Above 700 MHz, some form of parabolic reflector-feed arrangement is used; 700 MHz is no hard and fast dividing line. Above 2000 MHz, the transmission line is usually a waveguide. As was previously pointed out, the same antenna is used for both transmission and reception. The essential requirements imposed on an antenna relate to the following characteristics:

(a) Antenna gain in the direction of the main beam. For LOS radiolinks, antenna gains of over 45 dB should be avoided because the half-power beamwidth (i.e., less than 1°) results in greatly increased requirements for tower and mounting stability and rigidity.
(b) Half-power beamwidth, which affects requirements for antenna and tower design.
(c) Sidelobe attenuation to reduce or prevent interference to/from other systems using the same frequency or adjacent frequencies.

The power radiated from or received by an antenna depends on its aperture area. The power gain G of an antenna over the area A relative to an isotropic antenna can be expressed by

$$G = \frac{4\pi\eta A}{\lambda^2} \tag{7.1}$$

where A = area of the aperture in the same units as λ

η = efficiency of the antenna aperture, usually 55% for LOS radiolinks

λ = wavelength of the operating frequency (F_0)

The antenna gain in decibels is

$$G_{\mathrm{dB}} = 20\log F_{\mathrm{MHz}} + 20\log D_{\mathrm{ft}} + 10\log \eta - 49.92 \tag{7.2}$$

For an antenna with 55% efficiency the gain in decibels is

$$G_{\mathrm{dB}} = 20\log F_{\mathrm{MHz}} + 20\log D_{\mathrm{ft}} - 52.5 \text{ dB} \tag{7.3}$$

where F = frequency (F_0) in megahertz

D = aperture diameter (e.g., for a parabolic dish, the diameter of the dish) in feet

Directivity is another term commonly used to describe antenna performance. Directivity is the antenna lobe pattern that actually determines the antenna gain. An antenna may radiate in any direction, but it usually suffices to know the directivity in the horizontal and vertical planes.

Beamwidth is another important parameter. Radiation patterns for antennas are often plotted in a form (simplified) as shown in Figure 7.4. The center of the graph represents the location of the antenna, and the field strength is plotted along radial lines outward from the center (on polar graph paper). The line at 0° is the direction of maximum radiation or what we have previously

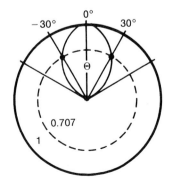

FIGURE 7.4 A typical (simplified) antenna pattern.

called the ray beam or main beam. For this simplified example at 30° either side of center, the voltage has dropped to 0.707 of its maximum value. The decibel ratio of this voltage to the maximum is $20 \log(E_{max}/E) = 20 \log(1/0.707) = 3$ dB. These 3-dB points are considered to be a measure of the antenna directivity. In this simplified case, the antenna beamwidth $\theta_u = 2 \times 30° = 60°$. These diagrams are usually plotted directly in decibels rather than in terms of field strength.

VSWR (voltage standing wave ratio) is another important parameter used to describe antenna performance. It deals with the impedance match of the antenna feed point to the feed line or transmission line. The antenna input impedance establishes a load on the transmission line as well as on the radiolink transmitter and receiver. To have the RF energy produced by the transmitter radiated with minimum loss or the energy picked up by the antenna passed to the receiver with minimum loss, the input or base impedance of the antenna must be matched to the characteristic impedance of the transmission line or feeder.

Mismatch gives rise to reflected waves on the transmission line or standing waves. These standing waves may be characterized by voltage maxima (V_{max}) and minima (V_{min}) following each other at intervals of one-quarter wavelength on the line: $VSWR = V_{max}/V_{min}$. A similar parameter is the reflection coefficient (ρ) which is the ratio of the amplitude of the reflected wave to that of the incident wave. Both VSWR and ρ are representative of the quality of impedance match. They are related by

$$\rho = \frac{VSWR - 1}{VSWR + 1} \tag{7.4}$$

Return loss (RL) is another mismatch parameter. It is the decibel difference between the power incident upon a mismatched discontinuity and the power reflected from the discontinuity. Return loss can be related to the reflection coefficient by

$$RL_{dB} = 20 \log(1/\rho) \tag{7.5}$$

Obviously, we would want a return loss as high as possible, in excess of 30 dB, and the reflected power as low as possible. VSWR should be less than 1.5 : 1.

Front-to-back ratio is still another measure of antenna performance. It is the ratio of the power radiated from the main ray beam to that radiated from the back lobe of the antenna. This is illustrated in Figure 7.4 where there is a small lobe extending from the back of the antenna. The ratio is expressed in decibels. For example, if an antenna radiates 20 times the power forward than back, its front-to-back ratio is 13 dB. Parabolic reflector antennas attain front-to-back ratios of 50–60 dB. The more efficient horn-reflector antennas can achieve as much as 70 dB. Figure 7.5 is a nomogram to determine gain of parabolic reflector antennas as a function of reflector diameter in feet and frequency in gigahertz.

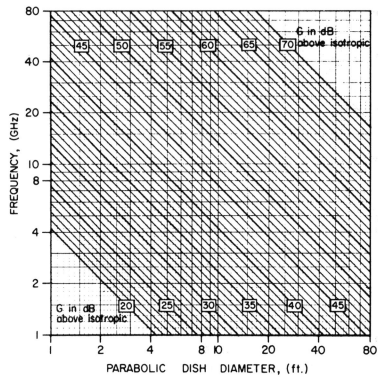

FIGURE 7.5 Parabolic reflector antenna gain nomogram. From MIL-HDBK-416 (Ref. 3).

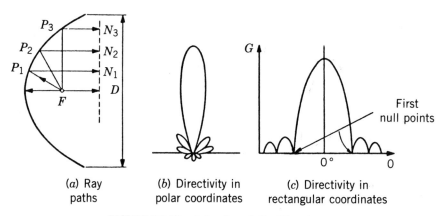

| (a) Ray paths | (b) Directivity in polar coordinates | (c) Directivity in rectangular coordinates |

FIGURE 7.6 Directivity of parabolic dish antennas.

The operation of a parabolic reflector antenna is shown in Figure 7.6. The feed point is located at the focus F of the parabola. The drawing represents a cross section through a paraboloid of revolution. For large circular apertures (i.e., those whose diameter is large compared to the wavelength) with uniform illumination, the beamwidth can be calculated from the following expression:

$$\theta = 142/\sqrt{G} \text{ (in degrees)} \qquad (7.6)$$

where

$$\sqrt{G} = \log^{-1}(G/20) \qquad (7.7)$$

with G expressed in decibels.

Example 1. An antenna has a gain of 38 dB, what is its half-power beamwidth?
Use equation (7.7) to calculate $\sqrt{G} = 79.4$. Now use this value in equation (7.6):

$$\theta = 142/79.4 = 1.79°$$

In practice, parabolic dishes are never illuminated uniformly, but the illumination tapers off toward the outer edge, reducing the overall gain somewhat. The taper acts to reduce the sidelobes, improving the front-to-back ratio, and reducing the potential for interference.

Antenna feeds are commonly waveguide horns (Figure 7.7), but dipole elements are sometimes used as radiators from about 300 MHz to approximately 3 GHz. Three types of parabolic antennas are illustrated in Figure 7.7. In Figure 7.7a, two types of feed horns are shown, the "button hook" type and the front-feed type. Such antennas permit only one polarization when a

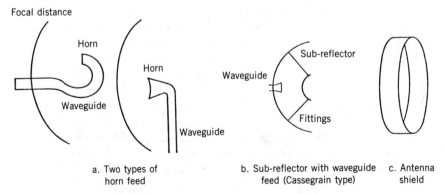

a. Two types of horn feed

b. Sub-reflector with waveguide feed (Cassegrain type)

c. Antenna shield

FIGURE 7.7 Parabolic antennas and related devices.

rectangular waveguide feed is used, and both horizontal and vertical polarizations with a square waveguide feed. Bandwidths are usually sufficient to cover several hundred megahertz with a fairly linear response.

The more efficient "Cassegrain" feed is shown in Figure 7.7*b*. This antenna is described in the earth station antenna subsection. Figure 7.7*c* shows a shield mounted around the edge of the dish to suppress sidelobes and back lobes. The inside surface of the shield is often lined with absorbing material to prevent reflections.

7.2.2.2 RF Transmission Lines

Two types of transmission line are used in radiolink terminals and repeaters: coaxial cable and waveguide. Coaxial cable, in general, is easier to install. Its loss increases exponentially with frequency, and, as a result, its upper limit of application is in the range of 2–3 GHz. Figures 7.8 and 7.9 give data on coaxial cable loss versus frequency.

There are a number of important parameters to be considered for the application of coaxial cable as a transmission line. Probably the most important for the system engineer is attenuation or loss as shown in Figures 7.8 and 7.9. Loss varies with ambient temperature. The reference value in the figures is 24°C (75°F). Figure 7.10 shows how loss varies with ambient temperature.

VSWR is another important parameter as previously described. VSWR can effectively increase transmission line loss. Such additional loss is called mismatch loss. This is shown in Figure 7.11. However, the effect is quite small for normal operating conditions.

The power rating of the line is another important parameter. Typically, peak power ratings limit the amplitude modulation or pulsed usage, while average power ratings limit the CW usage. The peak power rating is a function of the insulation material and structure between the inner and outer conductors. Voltage breakdown is independent of frequency but varies with line pressure (see subsequent discussion) and type of pressurizing gas. Voltage breakdown can result in permanent damage to the cable.

Waveguide is superior to coaxial cable in attenuation characteristics, particularly at the higher frequencies, and will handle higher power levels. For the lower frequencies (e.g., below about 3 GHz), the choice between coaxial cable and waveguide is economic, not only for the cost of the transmission line, but its installation. There are three types of waveguide in common use: rectangular, elliptical, and circular. Rectangular waveguide is that which is most commonly associated with microwave installations. However, generally, elliptical (flex) or circular waveguides are favored because of their low-loss properties. For ease of installation, elliptical waveguide, often called "flex," is the most commonly used for installations operating below 20 GHz.

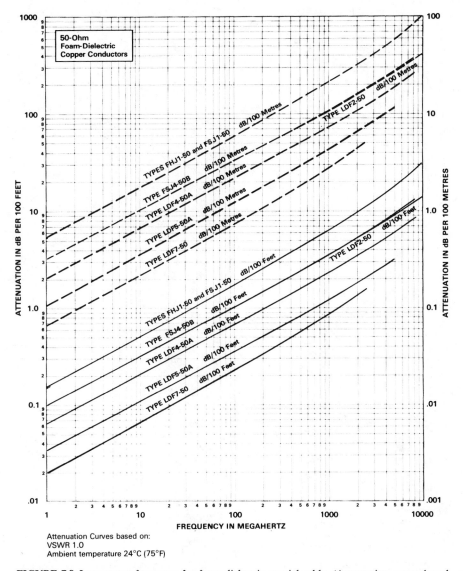

FIGURE 7.8 Loss versus frequency for foam dielectric coaxial cable. (Attenuation curves based on: VSWR 1.0, ambient temperature 24° C (75° F). Courtesy of the Andrew Corporation (Ref. 2).

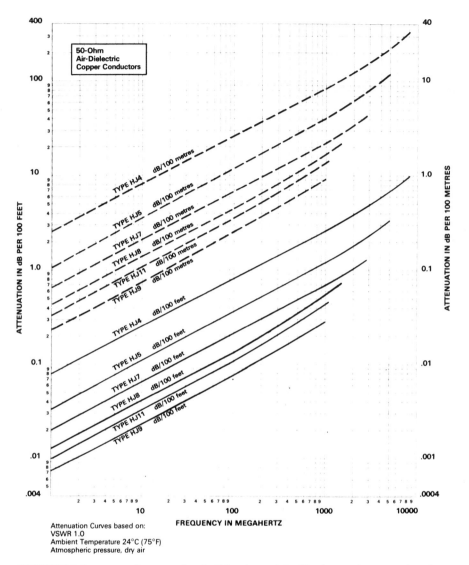

FIGURE 7.9 Loss versus frequency for air dielectric coaxial cable. Attenuation curves based on: VSWR 1.0, ambient temperature 24° C (75° F), atmospheric pressure, dry air. Courtesy of the Andrew Corporation (Ref. 2).

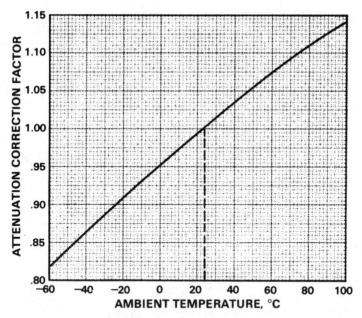

FIGURE 7.10 Variation of attenuation with ambient temperature. Courtesy of the Andrew Corporation (Ref. 2).

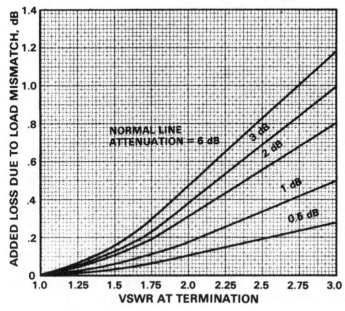

FIGURE 7.11 Effect of load VSWR on transmission loss. Courtesy of the Andrew Corporation (Ref. 2).

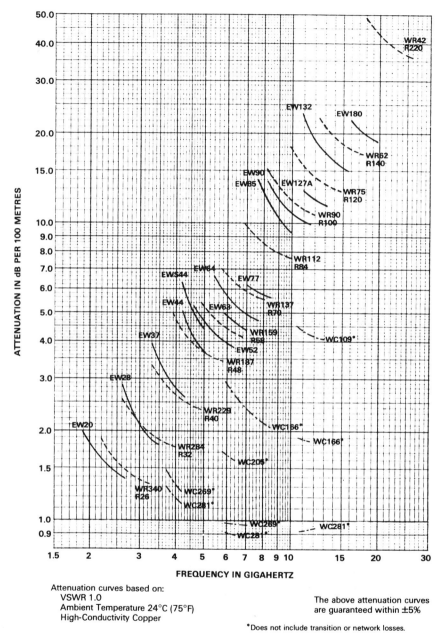

FIGURE 7.12 Loss versus frequency for several common waveguide types (metric units). Attenuation curves based on: VSWR 1.0, ambient temperature 24° C (78° F); high-conductivity copper. Courtesy of the Andrew Corporation (Ref. 2).

Circular waveguide displays minimum loss and is particularly suited for long vertical waveguide runs to tower-mounted antennas.

Most of the performance parameters applicable to coaxial cable are also applicable to waveguide. Figure 7.12 gives loss versus frequency for a number of the more commonly used waveguide types. Waveguide types are abbreviated: WR for rectangular, EW for elliptical, and WC for circular.

All air-dielectric waveguides, coaxial cables, and rigid lines are maintained under dry gas pressure to prevent electrical performance degradation. If a constant positive pressure is not maintained, "breathing" can occur with temperature variations. This permits moisture to enter the line causing increased loss, increased VSWR, and a path for voltage breakdown. One pressurizer/dehydrator can usually serve a number of waveguide installations in a common location.

7.2.2.3 Waveguide Devices—Separating and Combining Elements— Filters and Directional Couplers

By means of separating/combining networks, groups of several transmitters and receivers are connected to the same antenna. These include circulators, isolators, branching network, and combining networks. Figure 7.13 shows the various applications of these devices.

A waveguide circulator is used to couple two or three microwave radio equipments to a single antenna. A circulator consists essentially of three basic waveguide sections combined into a single assembly and is commonly a four-port device. The center section is a ferrite nonreciprocal phase shifter. An external permanent magnet causes the ferrite material to exhibit phase-shifting characteristics. Normally, an antenna transmission line is connected to one arm and either three radio equipments or two equipments and a shorting plate are connected to the other three arms. Attenuation in a clockwise direction from arm to arm is low, on the order of 0.5 dB, whereas in the counterclockwise direction it is high, on the order of 20 dB. Figure 7.13a shows a typical application of circulators in a microwave radiolink antenna subsystem.

Bridge networks consist of filter networks and four-arm bridge elements such as 3-dB directional couplers or "magic tees." Two bridge elements are connected by two identical filters to produce a separating-filter element. Of the four ports of a separating-filter element, one is connected to the equipment terminal and another to a termination. The two remaining ports are connected to neighboring separating-filter elements or one of two ports is connected to the antenna transmission line or a termination. (See Figures 7.13b and 7.13c.)

Branching networks (Figure 7.13d) connect multiple equipments to a single antenna by means of bandpass filters. Polarization filters (Figure 7.13e) combine/separate polarizations from/to a common antenna.

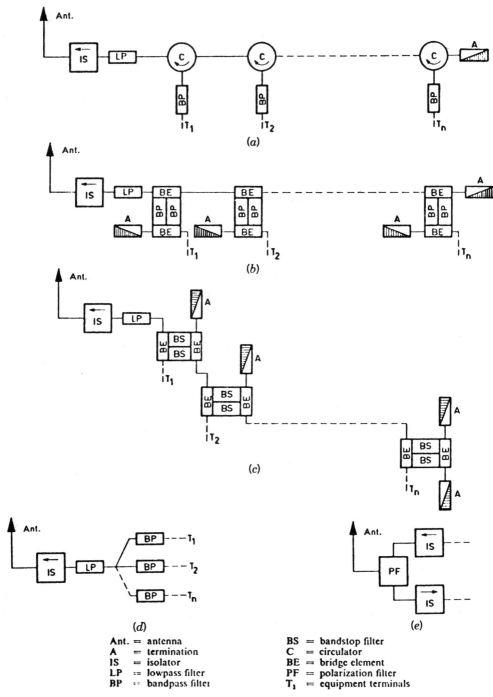

FIGURE 7.13 Application of waveguide devices as separating and combining elements: (*a*) circulator networks; (*b*) bridge-type network; (*c*) bridge-type network; (*d*) branching networks; (*e*) polarization filters. From MIL-MDBK-416 (Ref. 3).

458

A load isolator is a ferrite waveguide component that provides isolation between a single source and its load. A typical source is a HPA and the load is the antenna. These are commonly used in troposcatter installations. They reduce the ill effects of higher VSWRs and serve to protect the transmitter from high values of reflected power, and in some instances have to be cooled. Owing to the ferrite material with its associated permanent magnetic field, ferrite load isolators have a unidirectional property. Energy traveling toward the antenna is relatively unattenuated, whereas energy traveling back from the antenna undergoes fairly severe attenuation. The forward and reverse attenuations are on the order of 1 and 40 dB, respectively.

A directional coupler is a power splitter. It is a relatively simple waveguide device that divides the power on a transmission line, usually the power to/from the antenna. A 3-dB power split device divides the power in half; such a device could be used, for instance, to permit radiation of the power from a transmitter on two different antennas. A 20- or 30-dB power split has an output that serves to sample the power on a transmission line. It is most commonly used for VSWR measurements by measuring the forward and reverse power.

7.2.3 Analog Radiolink Repeaters

Radiolink repeaters amplify the signal along the radio route providing gain on the order of 110 dB. For *analog* systems there are three types of repeaters that can carry out this function:

- □ Baseband repeaters
- □ IF repeaters
- □ RF repeaters

For an FDM/FM system, a block diagram of a typical baseband or demodulating repeater is shown in Figure 7.14. A baseband repeater is required, if, at the repeater relay point, there will be drops and inserts. Such repeaters are often located at or near a telephone switching center. All or part of the baseband may be dropped or inserted. A baseband repeater will insert the same amount of noise into the system as a terminal facility. To reduce some of the noise inserted by the accompanying FDM equipment, through-group and through-supergroup techniques are used for routing through traffic. Most of the gain obtained with this type of repeater is obtained at the IF and through the demodulation/remodulation process.

An IF repeater may be used when there are no drops and inserts at the repeater facility. An IF repeater inserts less noise into the system because there are less modulation/demodulation steps required to carry out the repeating process. An IF repeater eliminates two modulation steps. It simply translates the incoming RF signal to IF with the appropriate local oscillator and mixer,

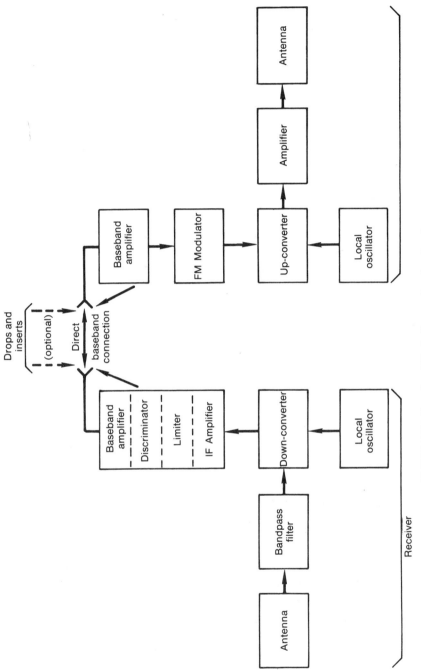

FIGURE 7.14 Functional block diagram for a baseband repeater.

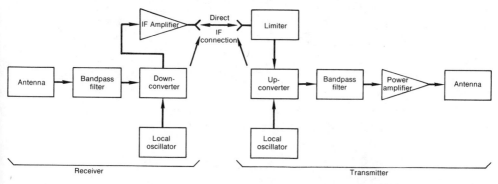

FIGURE 7.15 Functional block diagram of an IF repeater.

amplifies the derived IF, and then upconverts it to a different RF frequency. The upconverted frequency may then be amplified by a TWT or SSA*. Figure 7.15 is a simplified functional block diagram of a typical IF repeater.

With an RF repeater amplification is carried out directly at RF frequencies. The incoming RF signal is amplified, translated in frequency, and then amplified again. RF repeaters are seldom used. They are troublesome in their design with such things as sufficient selectivity, limiting and automatic gain control, and methods to correct group delay.

7.2.4 Diversity Combiners

Diversity combiners are used to combine signals from two or more diversity paths. They are also used in some hot-standby applications. Hot-standby operation improves link availability by switching in standby equipment when an on-line unit fails. The combiner removes at least one of the switching requirements. Hot standby is discussed in Section 7.2.5.

7.2.4.1 Classes of Diversity Combiners

There are two generic classes of diversity combiners: predetection and postdetection. Of course, this classification is made in accordance with where the combining function takes place, before or after detection.

A simplified functional block diagram of a predetection combiner is shown in Figure 7.16. In this case the combining is carried out at IF and phase control circuitry is required to maintain signal coherency of the two (or more) signal paths. If selection combining is used (Section 7.2.4.2), however, this control is unnecessary, since only one signal at a time is on line.

*SSA = Solid state amplifier.

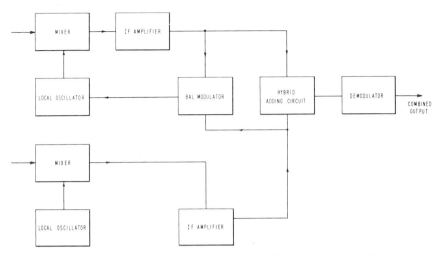

FIGURE 7.16 Functional block diagram of a typical predetection combiner.

Most systems today use postdetection or baseband combining. Figure 7.17 is a functional block diagram of a typical postdetection combiner.

7.2.4.2 Methods of Combining*

There are three general methods of combining: selection combining, equal gain combining, and maximal ratio combining. Figure 7.18 shows a simplified block diagram of each of these combining methods in a typical receiving system.

Combiner performance characteristics are comparatively illustrated in Figure 7.19. The following discussion compares the three types of combiners, and it will be assumed that in each case:

☐ Signals add linearly; noise adds in an rms manner.
☐ All receivers have equal gain.
☐ All receivers have equal noise outputs; the noise is random in character.
☐ The desired output signal-to-noise power ratio S_o/N_o is a constant.

For the case of the selection combiner, only one receiver at a time is used. The output signal-to-noise ratio is equal to the input signal-to-noise ratio of the output of the selected receiver at the time. Curve (a) refers in Figure 7.19.

---

* Based on abridged material from Ref. 4.

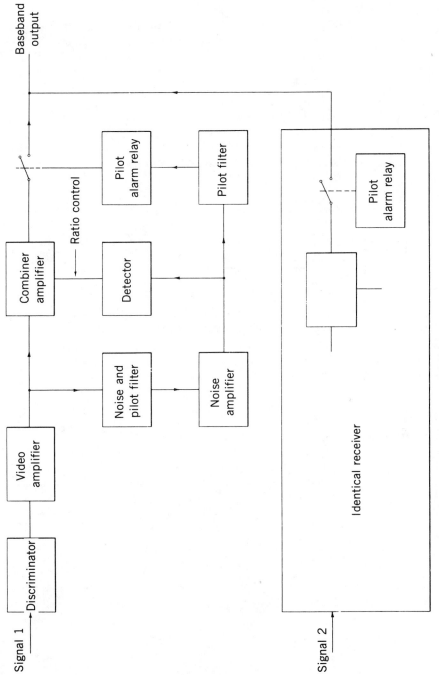

FIGURE 7.17 Functional block diagram of a typical postdetection (baseband) combiner.

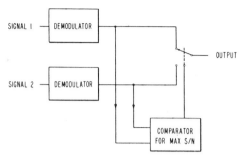

(a) SELECTION COMBINER (POST-DETECTION)

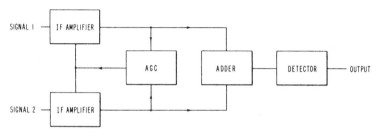

(b) EQUAL GAIN COMBINER (PRE-DETECTION)

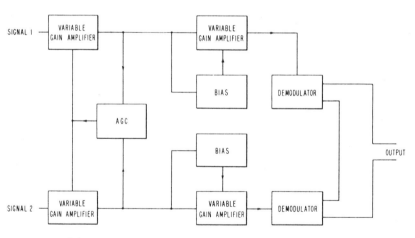

(c) MAXIMAL RATIO COMBINER (PRE-DETECTION)

FIGURE 7.18 Simplified functional block diagrams of the three basic methods of combining.

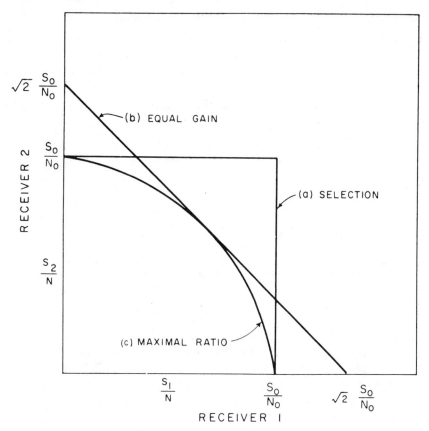

FIGURE 7.19 Combiner performance characteristics.

With the equal gain combiner, the different inputs are only added and when two receivers are in use

$$\frac{S_o}{N_o} = \frac{S_1 + S_2}{\sqrt{N_1^2 + N_2^2}} = \frac{S_1 + S_2}{\sqrt{2}N} \tag{7.8}$$

and is illustrated in curve (b) in Figure 7.19. S and N in Figure 7.19 are the signal and noise levels, respectively, for receivers 1 and 2.

The maximum ratio or ratio-squared combiner uses a relative gain change between the output signals of the receivers in use. For example, if the stronger signal has unity output and the weaker signal has an output proportional to G, then

$$\frac{S_o}{N_o} = \frac{S_1 + GS_2}{\sqrt{N_1^2 + G^2N_2^2}} = \frac{S_1 + GS_2}{N\sqrt{1 + G^2}} \tag{7.9}$$

Maximizing the above expression by differentiating and equating to zero yields

$$G = \frac{S_2}{S_1} \qquad (7.10)$$

in other words this signal gain is adjusted to be proportional to the ratio of the input signals. Then

$$\left(\frac{S_o}{N_o}\right)_{max} = \sqrt{\frac{S_1^2 + S_2^2}{N}} \qquad (7.11)$$

$$\left(\frac{S_o}{N_o}\right)_{max}^2 = \left(\frac{S_1}{N}\right)^2 + \left(\frac{S_2}{N}\right)^2 \qquad (7.12)$$

which is the equation of a circle and is illustrated in Figure 7.19 by curve (c).

7.2.4.3 Comparison of Combiners

The preceding discussion indicates that the maximal ratio combiner utilizes the best features of the two other combiners. When one signal is zero, the maximal ratio combiner acts as a selector combiner. When both signals are of equal level, it acts as an equal-gain combiner. It will also be noted that the maximum ratio combiner yields the best output for any combination of S/N since the curve of operation shows that the optimum output occurs for lower values of input S/N as evidenced by its proximity to the origin.

Figure 7.20 graphically illustrates another comparison based on the difference in output to be expected from each of the three combiners as the number of independent diversity paths increases. If the signals on each of the paths are assumed to be Rayleigh distributed, then a statistical analysis of the output from each of these combiners results in the comparison illustrated in the figure. For quadruple diversity, which was discussed in Chapter 3, the *average* signal-to-noise ratio of the output using the selection combiner is about 3 dB better than the nondiversity case, the equal-gain combiner is about a 5.25 dB improvement, and the maximal ratio combiner yields about a 6-dB improvement.

If the comparison of the combiners is done on the basis of a time distribution of the output signal-to-noise ratio rather than on the basis of average value (Figure 7.20), a new significance of combiner operation is revealed. Consider that a time distribution is the percentage of a time interval where the signal amplitude exceeds particular signal levels. In Figure 7.21, the Rayleigh distribution has a time distribution where the signal amplitude is approximately 5.2 dB above the median value for 10% of the time intervals of the measurement, whereas, for 90% of that interval, the signal amplitude

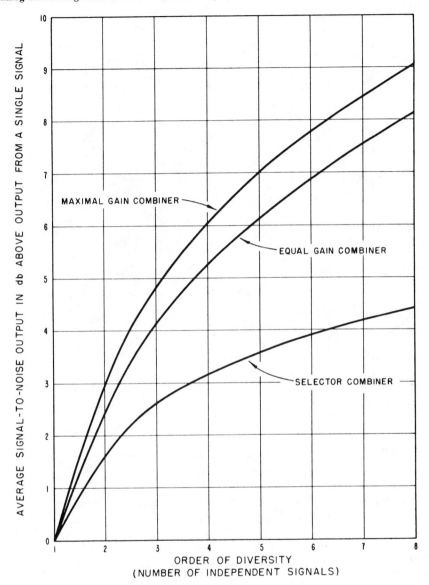

FIGURE 7.20 Signal-to-boise improvement in a diversity system.

exceeded a level that is 8.2 dB below the median. Where Rayleigh distributed signals are combined according to the principles used in these three types of combiners, the output can be predicted as a function of the type of combiner and the order of diversity. Figure 7.21 shows the time distribution comparison for dual diversity with each combiner type and Figure 7.22 is for quadruple diversity. In the case of dual diversity, the signal-to-noise ratio is exceeded

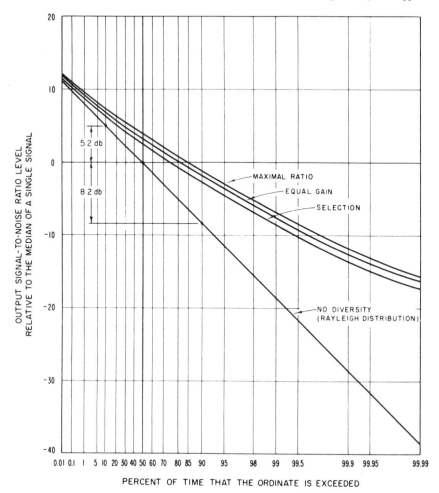

FIGURE 7.21 Comparison of the time distribution of the three types of combiners using dual diversity.

99.9% of the time is improved about 15 dB and for quadruple diversity, at least 23 dB.

7.2.5 Hot-Standby Operation

Radiolinks commonly provide transport of multichannel telephone service and/or point-to-point broadcast television on high-priority backbone routes. A high order of route reliability is essential. Route reliability depends on path reliability or link time availability (propagation) and equipment/system reliability. We have discussed path reliability and equipment reliability in Chapter 2. Redundancy is one way to achieve equipment reliability to minimize

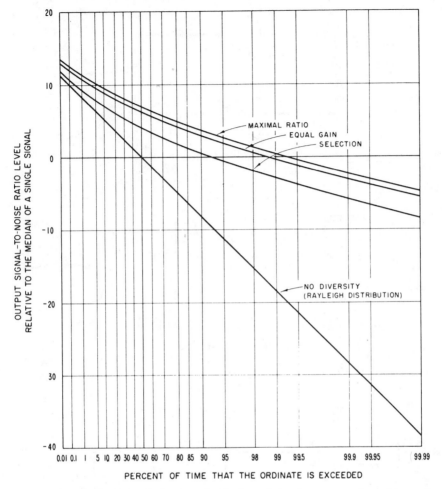

FIGURE 7.22 Comparison of the time distribution of the three types of combiner using quadruple diversity.

downtime and maximize link availability regarding equipment degradation or failure.

One straightforward way to achieve redundancy effectively is to provide a parallel terminal/repeater system. Frequency diversity effectively does just this. With this approach all equipment is active and operated in parallel with two distinct systems carrying the same traffic. This is expensive, but necessary, if a high order of link reliability is desired. Here we mean route reliability.

Often the additional frequency assignments to permit operation in frequency diversity are not available. When this is the case, the equivalent equipment reliability may be achieved by the use of a hot-standby configuration. Figure 7.23 illustrates the hot-standby concept.

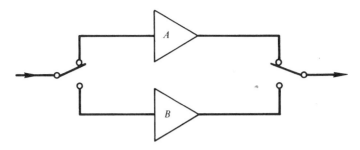

FIGURE 7.23 The concept of hot-standby protection.

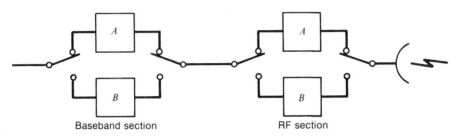

Baseband section   RF section

FIGURE 7.24 Hot-standby radio.

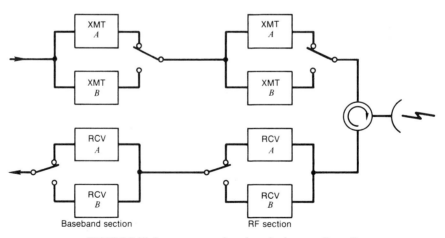

Baseband section   RF section

FIGURE 7.25 Separate transmit and receive hot-standby radio.

The equipment marked *A* and *B* in Figure 7.23 could be single modules, shelves or groups of modules, or whole equipment racks. On a complex equipment, such as a radiolink terminal, whether digital or analog, more than one set of protection can exist.

On a radiolink terminal such as shown in Figure 7.24, the sections marked baseband and RF are both hot-standby protected. The operation of the protection system of these two sections, however, is independent. The protection system is broken down further in that the protection for the transmit and receive paths operate independently, as shown in Figure 7.25. It should be noted that the switches ahead of each set of modules have been replaced by a signal splitter. This technique allows the signal to be fed into each set of modules simultaneously.

As the expression indicates, hot standby is the provision of parallel redundant equipment such that this equipment can be switched in to replace the operating on-line equipment nearly instantaneously when there is a failure in the operating equipment. The switchover can take place in the order of microseconds or less. The changeover of a transmitter and/or receiver line can be brought about by a change, over/under a preset amount, in one of the following values: for a transmitter,

☐ Frequency
☐ RF power
☐ Demodulated baseband (radio) pilot level

and for a receiver,

☐ AGC voltage
☐ Squelch
☐ Received pilot level
☐ Degraded bit error rate for a digital system
☐ Frame misalignment for a digital system

(*Note*: digital systems do not use pilot tones.)

Hot-standby-protection systems provide sensing and logic circuitry for the control of waveguide switches (or coaxial switches where appropriate), in some cases IF switches as well as baseband switches on transmitters, and IF and baseband output signals on receivers. The use of a combiner on the receiver side is common with both receivers on line at once.

There are two approaches to the use of protection equipment. These are called one-for-one and one-for-*n*. One-for-one operation provides one full line of standby equipment for each operational system. See Figure 7.26. One-for-*n* provides only one full line of equipment for *n* operational lines of equipment, where *n* is greater than one. Figure 7.27 illustrates a typical one-for-four configuration.

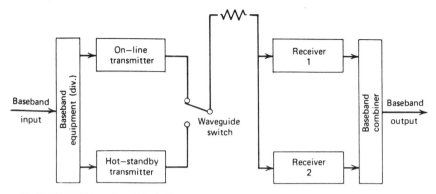

FIGURE 7.26 Functional block diagram of a typical one-for-one hot-standby configuration.

One-for-one is more expensive but provides a higher order of reliability. Its switching system is comparatively simple. One-for-n is more economic, with only one line of spare equipment for several operational lines. It is less reliable (i.e., suppose there were equipment failures in two lines of operational equipment of the n lines), and switching is considerably more complex.

Figure 7.28 shows a typical digital hot-standby configuration with space diversity.

7.2.6 Pilot Tones

On analog radiolinks a radio continuity pilot tone (or tones) is inserted on a link-by-link basis. These pilot tones are usually independent of multiplex pilot tones. The pilot tone is used for:

☐ Gain regulation
☐ Monitoring (fault alarms)
☐ Frequency comparison
☐ Measurement of level stability
☐ Control of diversity combiners

The last application involves the simple sensing of continuity by a diversity combiner. The presence of the continuity pilot tone tells the combiner that a particular diversity path is operative. The problem arises from the fact that most diversity combiners are postdetection and use noise as the means to determine the path contribution to the combined output. The path with the least noise, as in the case of the maximal ratio combiner, provides the greatest path contribution.

If, for some reason, a diversity path were to fail, it would be comparatively noiseless and would provide 100% contribution. Thus, there would be no

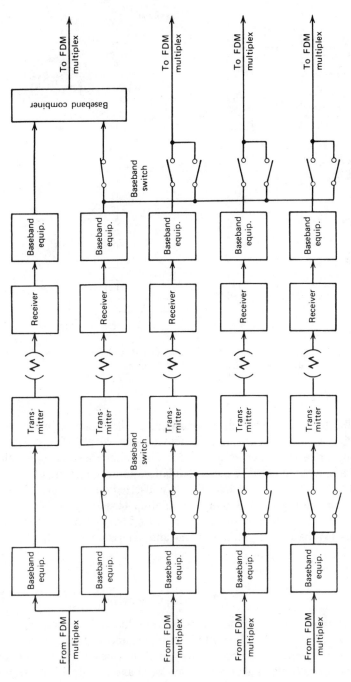

FIGURE 7.27 Functional block diagram of a one-for-n hot-standby configuration, where $n = 4$.

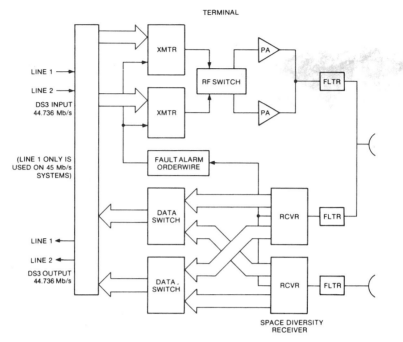

FIGURE 7.28 A typical hot-standby (one-for-one) digital implementation using space diversity. Ref. 5; courtesy of Rockwell International, Collins Transmission Division.

signal output. To avoid such a condition, a pilot tone is inserted prior to modulation at the radiolink transmitter and deleted after the combiner. The pilot tone presence indicates to the combiner that the path is a valid path. Pilots are reinserted anew at each modulation point in a string of radiolinks in tandem. It should be noted that pilots are inserted out-of-band, generally 10% higher than the highest baseband modulating frequency. Table 7.1 provides recommended pilot tone frequencies and rms deviation produced by the pilot for 11 FDM baseband configurations and television. This table was taken from CCIR Rec. 401-2.

7.2.7 Service Channels

Service channels are separate facilities from the information baseband (on analog radiolinks) but transmitted on the same carrier. Service channels operate in frequency slots (on analog radiolinks) below the carrier for FDM telephony operation and above the video baseband for those links transmitting video (and associated but separated aural channels). Service channels are used for maintenance, link and network coordination, and, in these cases, may be

TABLE 7.1. Radio Continuity Pilots Recommended by the CCIR

System Capacity (Channels)	Limits of Band Occupied by Telephone Channels (kHz)	Frequency Limits of Baseband (kHz)[a]	Continuity Pilot Frequency (kHz)	Deviation (r.m.s.) Produced by the Pilot (kHz)[b,e]
24	12–108	12–108	116 or 119	20
60	12–252	12–252	304 or 331	25, 50, 100[c]
	60–300	60–300		
120	12–552	12–552	607[d]	25, 50, 100[c]
	60–552	60–552		
300	60–1300	60–1364	1499, 3200,[f] or 8500[f]	100 or 140
600	60–2540	60–2792	3200 or 8500	140
	64–2660			
960 ⎫ 900 ⎭	60–4028 ⎫ 316–4188 ⎭	60–4287	4715 or 8500	140
1260 ⎫	60–5636 ⎫		⎧ 6199	100 or 140
	60–5564 ⎬	60–5680	⎨	
1800 ⎭	316–5564 ⎭		⎩ 8500	140
1200	312–8204	300–8248	9023	100
	316–8204			
2700	312–12,388	308–12,435	13,627	100
	316–12,388			
Television			⎧ 8500	140
			⎩ 9023[g]	100

[a] Including pilot or other frequencies which might be transmitted to line.

[b] Other values may be used by agreement between the Administrations concerned.

[c] Alternative values dependent on whether the deviation of the signal is 50, 100, or 200 kHz (Recommendation 404-2).

[d] Alternatively 304 kHz may be used by agreement between the Administrations concerned.

[e] This deviation does not depend on whether or not a pre-emphasis network is used in the baseband.

[f] For compatibility in the case of alternate use with 600-channel telephony systems and television systems.

[g] The frequency 9023 kHz is used for compatibility purposes between 1800 channel telephony systems and television systems, or when the establishment of multiple sound channels so indicates.

Source. Reference 6.

called "orderwire(s)." They may also be assigned to carry network status data and fault information from unmanned radio relay sites. For FDM telephony links, service channels commonly occupy the band from 300 Hz to 12 kHz of the transmitted baseband, allowing three nominal 4-kHz channel operation.

Digital radiolinks utilize specific time slots in the digital bit stream for service channels. At each terminal and repeater local digital service channels are dropped and inserted. An express orderwire can also be accommodated. This, of course, requires a reconstitution of the information bit stream at the terminal locations to permit the insertion of service channel information. Another method is to use a separate carrier, either analog or digitally modulated.

7.2.8 Alarm and Supervisory Subsystems

Many radiolink sites are unattended, especially repeater sites. To ensure improved system availability, it is desirable to know the status of unattended sites at a central or manned location. This is accomplished by means of a fault-reporting system. Commonly, such fault alarms are called status reports. The radiolink sites originating status reports are defined as reporting stations. A site that receives and displays such reports is defined as a supervisory location. This is the standard terminology of the industry. Normally, supervisory locations are those terminals that terminate a radiolink section. Status reports may also be required to be extended over a wire circuit to a remote location, often a maintenance center.

The following functions at a radiolink site, which is a reporting location (unmanned), are candidate functions for status reports:

Equipment Alarms

Loss of receive signal
Loss of pilot (at receiver)
High noise level (at receiver)
Power supply failure
Loss of modulating signal
TWT overcurrent
Low transmitter output
Off-frequency operation
Hot-standby actuation

Site Alarms

Illegal entry
Commercial power failure
Low fuel supply
Standby power unit failure
Standby power unit on-line
Tower light status

Additional Fault Information for Digital Systems

Loss of BCI (bit count integrity)
Loss of sync
Excessive BER

Often alarms are categorized into "major" (urgent) and "minor"(nonurgent) in accordance with their importance. For instance, a major alarm would be one where the fault would cause the system or link to go down (cease operation) or seriously deteriorate performance. A major alarm may be audible as well as visible on the status panel. A minor alarm may then show only as an indication on the status panel. On military equipment alarms are referred to as BITE (built-in-test equipment).

The design intent of alarm or BITE systems is to make all faults binary: a tower light is either on or off; the RSL (receive signal level) has dropped below a specified level, -100 dBW, for example; the transmitter power output is 3 dB below its specified output; or the noise on a derived analog channel is above a certain level in picowatts. By keeping all functions binary, using relay closure (or open) or equivalent solid-state circuitry, the job of coding alarms for transmission is made much easier. Thus all alarms are of a "go/no-go" nature.

7.2.8.1 Transmission of Fault Information

On analog systems common practice today is to transmit fault information in a voice channel associated with the service channel groupings of voice channels (Section 7.2.7). Binary information is transmitted by VF telegraph equipment using FSK or tone-on, tone-off (see Section 8.13 of Reference 1). Depending on the system used, 16, 18, or 24 tone channels may occupy the voice channel assigned. A tone channel is assigned to each reporting location (i.e., each reporting location will have a tone transmitter operating on the specific tone frequency assigned to it). The supervisory location will have a tone receiver for each reporting (unmanned site under it supervision) location.

At each reporting location the fault or BITE points previously listed are scanned every so many seconds, and the information from each monitor or scan point is time division multiplexed in a simple serial bit stream code. The data output from each tone receiver at the supervisory location represents a series of reporting information on each remote unmanned site. The coded sequence in each case is demultiplexed and displayed on the status panel.

A simpler method is the tone-on, tone-off method. Here the presence of a tone indicates a fault in a particular time slot; in another method it is indicated by the absence of a tone. A device called a fault-interrupter panel is used to code the faults so that different faults may be reported on the same tone frequency.

On digital radiolinks fault information in a digital format is stored and then inserted into one of the service channel time slots, a time slot is reserved for each reporting location.

7.2.8.2 Remote Control

Through a similar system to that previously described, which operates in the opposite direction, a supervisory station can control certain functions at reporting locations via a voice frequency telegraph tone line (for analog systems) with a tone frequency assigned to each separate reporting location on the span. If only one condition is to be controlled, such as turning on tower lights, then a mark condition could represent lights on and a space for lights off. If more than one condition is to be controlled, then coded sequences are used to energize or deenergize the proper function at the remote reporting location.

There is an interesting combination of fault reporting and remote control that particularly favors implementation on long spans. Here only summary status is normally passed to the supervisory location, that is, a reporting location is either in a "go" or "no-go" status. When a "no-go" is received, that reporting station in question is polled by the supervisory location and detailed fault information is then released. Polling may also be carried out on a periodic basis to determine detailed minor alarm fault data.

On one digital radiolink equipment (Ref. 5) the following operational support system maintenance functions and performance monitoring is listed below:

Maintenance Functions

Alarm Surveillance

> Alarm reporting
> Status reporting
> Alarm conditioning
> Alarm distribution
> Attribute report

Control Functions

> Allow–inhibit local alarms
> Operate alarm cutoff
> Allow–inhibit protection switching
> Operate release protection switch
> Remove–restore service
> Restart processor
> Preemptive switching—override an existing protective switch
> Activate restore lockout (lockout prevents switching)
> Local remote control (i.e., inhibits local operation)
> Operate–release loopback
> Command (control) verification (i.e., set status point)
> Completion acknowledgment (i.e., completed the command received earlier)

Performance Monitoring

Report performance monitoring data such as BER, sync error, error second, etc.

Inhibit–allow performance monitoring data (i.e., collect but don't send unless asked)

Start–stop performance monitoring data

Initialize (reset) performance monitoring data storage registers

7.2.9 Antenna Towers

7.2.9.1 General

Two types of towers are used to mount antennas for radiolink systems: guyed and self-supporting. However, other manmade and natural structures should also be considered or at least taken advantage of. Among these are siting on the highest hill or ridge feasible, or leasing space on tall buildings or on TV towers.

One of the most desirable construction materials for towers is hot-dipped galvanized steel. Guyed towers are often preferred because of overall economy and versatility. Although guyed towers have the advantage that they can be placed closer to the equipment shelter or building than self-supporting types, the fact that they need a larger site may be a disadvantage where land values are high. The larger site is needed because additional space is required for installing guy anchors. Table 7.2 shows approximate land area needed for several tower heights.

Tower and foundation design are dependent four main factors: (1) soil bearing capability at the specific location; (2) the size and number of parabolic antennas and their location on the tower (i.e., these antenna reflectors act as wind sail devices); (3) the meteorological conditions to be expected; and (4) the maximum tower twist and sway that can be tolerated under worst conditions of wind loading, and ice loading, where applicable.

Tower loading is the resultant of all forces acting on the tower. Design of the tower must be such that with all antennas and other required items mounted on the tower, it will, when subjected to maximum specified wind and ice loading conditions, resist deflection or twisting beyond a specified amount.

Since the net result of all the forces acting on the tower are also, in effect, transmitted to the foundation, it in turn must be capable of distributing the force over a large enough area and depth so as not to exceed the soil bearing pressure at any point and also to resist movement in any direction. The depth of the foundation will be governed by the tower load and soil bearing characteristics, but, in colder climate, it is necessary to extend the depth of the foundation below the frost line or to firm ground.

TABLE 7.2. Minimum Land Area Required for Guyed Towers

Tower Height (ft)	Area Required[a] (ft)				
	80% Guyed	75% Guyed	70%Guyed	65% Guyed	60% Guyed
60	87 × 100	83 × 96	78 × 90	74 × 86	69 × 80
80	111 × 128	105 × 122	99 × 114	93 × 108	87 × 102
100	135 × 156	128 × 148	120 × 140	113 × 130	105 × 122
120	159 × 184	150 × 174	141 × 164	132 × 154	123 × 142
140	183 × 212	178 × 200	162 × 188	152 × 176	141 × 164
160	207 × 240	195 × 226	183 × 212	171 × 198	159 × 184
180	231 × 268	218 × 252	204 × 236	191 × 220	177 × 204
200	255 × 296	240 × 278	225 × 260	210 × 244	195 × 226
210	267 × 304	252 × 291	236 × 272	220 × 264	204 × 236
220	279 × 322	263 × 304	246 × 284	230 × 266	213 × 246
240	303 × 350	285 × 330	267 × 308	249 × 288	231 × 268
250	315 × 364	296 × 342	278 × 320	254 × 282	240 × 277
260	327 × 378	308 × 356	288 × 334	269 × 310	249 × 288
280	351 × 406	330 × 382	309 × 358	288 × 332	267 × 308
300	375 × 434	353 × 408	330 × 382	308 × 356	285 × 330
320	399 × 462	375 × 434	351 × 406	327 × 376	303 × 350
340	423 × 488	398 × 460	372 × 430	347 × 400	321 × 372
350	435 × 502	409 × 472	383 × 442	356 × 411	330 × 381
360	447 × 516	420 × 486	393 × 454	366 × 424	339 × 392
380	471 × 544	443 × 512	414 × 478	386 × 446	357 × 412
400	495 × 572	465 × 536	425 × 502	405 × 468	375 × 434
420	519 × 599	488 × 563	456 × 527	425 × 490	393 × 454
440	543 × 627	510 × 589	477 × 551	444 × 513	411 × 475

Tower Height (ft)	Area Required (acre)				
	80% Guyed	75% Guyed	70% Guyed	65% Guyed	60% Guyed
60	0.23	0.21	0.19	0.17	0.15
80	0.38	0.34	0.30	0.26	0.23
100	0.56	0.50	0.44	0.39	0.34
120	0.77	0.69	0.61	0.53	0.46
140	1.03	0.91	0.80	0.70	0.61
160	1.31	1.16	1.03	0.90	0.77
180	1.63	1.45	1.27	1.11	0.96
200	1.99	1.76	1.55	1.35	1.16
210	2.18	1.93	1.70	1.48	1.27
220	2.38	2.11	1.85	1.61	1.39
240	2.81	2.49	2.18	1.90	1.63
250	3.04	2.69	2.36	2.05	1.76
260	3.27	2.89	2.54	2.21	1.90
280	3.77	3.33	2.92	2.54	2.18
300	4.30	3.80	3.33	2.89	2.49
320	4.87	4.30	3.77	3.27	2.81
340	5.48	4.84	4.24	3.65	3.15
350	5.79	5.11	4.48	4.88	3.33
360	6.12	5.40	4.73	4.10	3.52
380	6.79	5.99	5.25	4.55	3.90
400	7.50	6.62	5.79	5.02	4.30
420	8.24	7.27	6.36	5.52	4.73
440	9.03	7.96	6.96	6.03	5.17

[a] Preferred area is a square using the larger of minimum area. This will permit orienting tower in any desired position.

Source. Reference 1.

TABLE 7.3. Soil Bearing Characteristics

Material	Maximum Allowable Bearing Value (lb ft^{-2})
Bedrock (sound) without laminations	200,000
Slate (sound)	70,000
Shale (sound)	20,000
Residual deposits of broken bedrock	20,000
Hardpan	20,000
Gravel (compact)	10,000
Gravel (loose)	8,000
Sand, coarse (compact)	8,000
Sand, coarse (loose)	6,000
Sand, fine (compact)	6,000
Sand, fine (loose)	2,000
Hard clay	12,000
Medium clay	8,000
Soft clay	2,000

Source. Reference 3

The soil bearing capability, usually expressed as a pressure in pounds per square foot, is a determining factor in the design of a tower foundation. Table 7.3 gives the maximum soil bearing values for various types of soil conditions. It should be noted that the designation of the various soil conditions is arbitrary in nature, so the table should be used only as a rough guide for preliminary estimates. Soil borings taken at the area in question are normally required for final design.

The following information should be provided by the communication system engineer in order for the tower designer to properly design a suitable tower and foundation:

1. Size and type of antennas required and the type of radomes, if any.

2. Azimuth and elevation angles for each antenna.

3. Amount of adjustment required for antenna alignment after installation; $\pm 5°$ in both azimuth and elevation is usually sufficient.

4. Height above ground level (AGL) for each antenna.

5. Location of tower with respect to the building/shelter and their corresponding orientations.

6. Beamwidths of the antennas and permissible twist and sway of the tower that can be tolerated under maximum expected wind and ice loading.

7. Type and number of waveguide runs required and power cabling for feedhorn and radome heaters.

8. Requirements for tower lighting, obstruction lights, lightning protection and grounding systems, painting.

9. Requirements for platforms to allow access to antennas for maintenance and/or alignment.

10. Required means of access up the tower including required personnel safety features.

11. Provisions for future antenna requirements.

12. Climatic conditions of the area involved.

13. Local restrictions and/or other constraints that may be involved.

14. Soil conditions.

Tower twist and sway is one of the most important requirements to be specified. As any other structure, a radiolink tower tends to twist and sway due to wind loads and other natural forces. Considering the narrow beamwidths of radiolink antennas, with only a little imagination one can see that only a very small deflection of a tower or antennas will cause the radio ray beam to fall out of the reflection face of the distant end corresponding antenna or move the beam out on the far end transmit side of the link.

Twist and sway, therefore, must be limited. Table 7.4 sets certain limits. The table has been taken from EIA RS-222B (Ref. 7). From the table we can see that angular deflection and tower movement are functions of wind velocity. It should also be noted that the larger the antenna, the smaller the beamwidth, besides the fact that the sail area is larger. Thus the larger the antenna and the higher the frequency of operation, the more we must limit the deflection.

To reduce twist and sway, tower rigidity must be improved. One generality we can make is that towers that are designed to meet required wind load or ice load specifications are sufficiently rigid to meet twist and sway tolerances. One way to increase rigidity is to increase the number of guys, particularly at the top of the tower. Under certain circumstances doubling the number of guys is warranted.

A number of external safety requirements are imposed on towers. A primary consideration is to prevent excessive hazards to air commerce. Therefore, antenna towers must be marked in such a way as to make them conspicuous when viewed from aircraft. The type of marking to be used depends in part on the height of the structure, its location with respect to nearby objects, and its proximity to aircraft traffic routes near landing areas. In the United States the requirements and specifications for marking and lighting potential hazards to air navigation have been established through joint cooperation of the Federal Aviation Agency (FAA), Federal Communications Commission (FCC), Department of Defense (DoD), and appropriate branches of the broadcasting and aviation industries. When conducting the preliminary site survey, it is advisable to determine the prevailing ordinances concerning such structures and to discuss them with local government and building

TABLE 7.4. Nominal Twist and Sway values for Microwave Tower–Antenna–Systems[a]

	Tower-Mounted Antenna			Tower-Mounted Passive Reflector	
A	B	C	D	E	F
Total Beamwidth of Antenna or Passive Reflector between Half-Power Points (°)	Limits of Movement of Antenna Beam with Respect to Tower ($\pm$°)	Limits of Tower Twist or Sway at Antenna Mounting Point ($\pm$°)	Limits of Movements of Passive Reflector with Respect to Tower ($\pm$°)	Limits of Tower Twist at Passive Reflector Mounting Point ($\pm$°)	Limits of Tower Sway at Passive Reflector Mounting Point ($\pm$°)
14	0.75	4.5	0.2	4.5	4.5
13	0.75	4.5	0.2	4.5	4.3
12	0.75	4.5	0.2	4.5	3.9
11	0.75	4.5	0.2	4.5	4.6
10	0.75	4.5	0.2	4.5	3.3
9	0.75	4.5	0.2	4.5	3.9
8	0.75	4.2	0.2	4.5	2.6
7	0.6	4.1	0.2	4.5	2.3
6	0.5	4.0	0.2	4.3	2.1
5	0.4	3.4	0.2	3.7	1.8
4	0.3	3.1	0.2	3.3	1.6
3.5	0.3	2.9	0.2	2.9	1.4
2.0	0.3	2.3	0.1	2.5	1.2
2.5	0.2	1.9	0.1	2.1	1.0
2.0	0.2	1.5	0.1	1.7	0.9
1.5	0.2	1.1	0.1	1.2	0.6
1.0	0.1	0.9	0.1	0.9	0.5
0.75[b]	0.1	0.7	0.1	0.7	0.4
0.5[b]	0.1	0.4	0.1	0.4	0.2

[a] The values are tabulated as a guide for systems design and are based on the values that have been found satisfactory in the operational experience of the industry. These data are listed for reference only.

[b] These deflections are extrapolated and are not based on experience of the industry.

Notes

1. Half-power beamwidth of the antenna to be provided by the purchaser of the tower.
2. a. The limits of beam movement resulting from an antenna mounting on the tower are the sum of the appropriate figures in columns B and C.
   b. The limits of beam movement resulting from twist when passive reflectors are employed are the sum of the appropriate figures in columns D and E.
   c. The limits of beam movement resulting from sway when passive reflectors are employed are twice the sum of the appropriate figures in columns D and F.
   d. The tabulated values in columns, D, E, and F are based on a vertical orientation of the antenna beam.
3. The maximum tower movement shown above (4.5°) will generally be in excess of that actually experienced under conditions of 20 lb ft^{-2} wind loading.
4. The problem of linear horizontal movement of a reflector–parabola combination had been considered. It is felt that in a large majority of cases, this will present no problem. According to tower manufacturers, no tower will be displaced horizontally at any point on its structure more than 0.5 ft/100 ft of height under its designed wind load.
5. The values shown correspond to 10 dB gain degradation under the worst combination of wind forces at 20 lb ft^{-2}. This table is meant for use with standard antenna-reflector configurations.
6. Twist and sway limits apply to 20 lb ft^{-2} wind load only, regardless of survival or operating specifications. If there is a requirement for these limits to be met under wind loads greater than 20 lb ft^{-2}, such requirements must be specified by the user.

Source. Reference 7. Courtesy Electronics Industry Association, Washington, DC.

authorities. Inside the continental United States consult the latest issues of Government Rules and Regulations, FCC Form 715, FCC Fule Part 17, and FAA Standards for marking and lighting obstructions to air navigation, with all revisions.

Both the FCC and FAA lighting specifications are set forth in terms of the heights of antenna structures. The requirements for towers and obstructions are determined from FAA Specifications set forth in the latest edition of AC 70/7460-1( ). The specifications further stipulate that placement of lights on either square or rectangular towers shall be such that at least one top or side light be visible from any angle of approach. When a flashing beam is required, it shall be equipped with a flashing mechanism capable of producing not more than 40 nor less than 12 flashes per minutes with a period of darkness equal to one-half the luminous period.

In the United States the FCC requires that tower lighting be exhibited during the period from sunset to sunrise unless otherwise specified. At unattended radiolink installations, a dependable automatic obstruction-lighting control device will be used to control the obstruction lights in lieu of manual control. This requirement is met in microwave installations by employing a tower-lighting kit. These kits apply power to the lights when the north skylight intensity is less than approximately 35 foot candles and disconnect the power when the north skylight intensity is greater than approximately 58 foot candles.

7.2.10 Waveguide Pressurization

All external waveguide runs should be pressurized with dry air or an inert gas such as nitrogen. While standard tanks of nitrogen may be used for this purpose, the use of automatic dehydrator-pressurization equipment is generally preferred. The pressure should not exceed the manufacturer's recommended values for the particular waveguide involved. This is generally in the order of 2–10 lb in.^{2} Pressure windows are available to isolate pressurized from unpressurized portions of the waveguide.

7.3 DIGITAL MICROWAVE (RADIOLINK) TERMINALS

Figure 7.29 is a functional block diagram of a typical digital radio terminal. Starting from the bottom of the figure on the transmit line (left side) there are a number of baseband processing functions. The line code converter takes the standard PCM line codes, which have been implemented for good baseband transmission properties, and converts the code usually to NRZ format. Then the resulting code is scrambled by means of a PRBS (pseudo-random-binary sequence). This action tends to remove internal correlation among symbols such as long strings of 1s or 0s. The resulting modulation with the PRBS provides an output with a constant power spectrum.

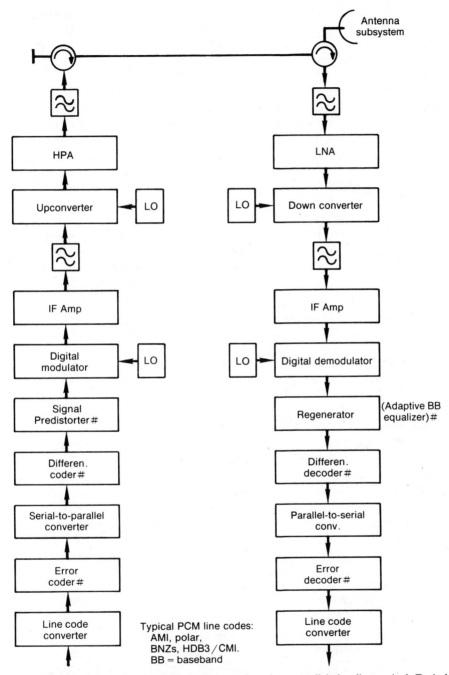

FIGURE 7.29 Typical functional block diagram of a microwave digital radio terminal. Typical PCM line codes: AMI, polar, BNZS, HDB3/CMI. BB = baseband, # = optional.

The signal may then be channel coded for forward error correction, although this is not often done on terrestrial radiolinks. Differential coding/decoding is one method to remove phase ambiguity at the receive end. A coherent system would not require this function. However, a phase coherence would be required.

A serial-to-parallel converter divides the serial bit stream into two components for the I and Q inputs of a phase modulator (See Figure 2.38.) Multilevel coders (and decoders) are required for 8-ary PSK and for so-called QAM modulation schemes. (Refer to Section 2.11.3.) in the cases of 16-QAM, 32-QAM, and 64-QAM more than two amplitude levels are used per orthogonal modulation. Therefore, the digital data have to be converted to multilevel logic. For instance, 64-QAM requires two eight-level modulation signals.

The predistorter is designed to compenstate for the distortion imparted on the signal by the power amplifier.

The digital modulator, of course, carries out the modulation function. Modulation to achieve spectral efficiency was discussed in Section 2.11.3. The output of the modulator is then amplified, filtered, and passed to the upconverter. The upconverter translates this signal to the operating frequency of the terminal. The output of the upconverter is then fed to the HPA, which amplifies the signal to the desired output level. The HPA may be a TWT or a SSA. Commonly, power outputs of HPAs for microwave LOS operation are 0.1, 1, or, in some cases, 10 W. The output of the HPA generally incorporates a bandpass filter to reduce spurious and harmonic out-of-band signals. The signal is then fed to the antenna subsystem for radiation to the distant end.

The received signal, starting from the antenna downward in Figure 7.29, is fed from the antenna subsystem through a bandpass filter through the LNA to a downconverter. In many installations the LNA can be omitted. This decision is driven by economics. If the link can tolerate the additional thermal noise, some savings can be made on the installation. Common downconverters display noise figures of the 7–11 dB, whereas GaAs FET LNAs display noise figures of from 1 to 3 dB, and above 20 GHz, 3.5–5 dB. One alternative is to use a low-noise mixer where the noise figure is on the order of 2.5–5 dB. We will remember from Chapter 2 that the system thermal noise is gated by this first active device.

The downconverter translates the incoming signal to the IF, which is commonly 70 or 140 MHz. These are the more common IF frequencies used in the Industry, and it does mean that other IF frequencies cannot be used, such as 300, 600, and 700 MHz.

The output of the downconverter is then fed through a filter to an IF amplifier prior to inputting the digital demodulator. Excellent linearity throughout the receive chain is extremely important, particularly on high-bit-rate equipments using higher-order modulation schemes such as 64-QAM.

Demodulation and regeneration are probably the most important elements in the digital microwave receiver. The principal purpose, of course, is to achieve a demodulator serial bit stream output that is undistorted and comparatively free of intersymbol interference.

Coherent demodulation is commonly used. Reference phase has to be maintained. The vector status of the modulated signal is compared with that of the carrier. However, since the carrier is not available in the received modulation signal, it has to be reproduced from it. One method that can be used for an AM-SSB signal is to employ a squaring process where a discrete spectral component is available at twice the frequency of the carrier from which a carrier can then be derived. Another method is to use a Costas-loop for carrier regeneration.

After demodulation, the data clock at the incoming symbol rate is regenerated from the baseband digital bit stream. The clock is usually derived in a similar manner as the carrier. A phase lock loop (PLL) is synchronized to the spectral component occurring at the clock rate. Here timing jitter is a major impairment which should be prevented or minimized. Specifications on jitter are called out in CCITT Rec. G.703 (Ref. 8).

The demodulated signal is then regenerated by means of the regenerated clock and a sample-and-hold circuit is applied to regenerate the actual signal. For modulation schemes using M-PSK, the I and Q channel sequences are regenerated separately. The regenerative repeater may also incorporate a baseband equalizer to reduce signal distortion. Methods of reducing distortion on a digital radiolink were discussed in Chapter 2.

For the case of I and Q bit streams, the combining of these two bit streams into a single serial bit stream is accomplished in the parallel-to-serial converter. If FEC is implemented, the signal is then error decoded and the resulting signal is then conditioned for line transmission in the line code converter.

7.4 TROPOSPHERIC SCATTER / DIFFRACTION INSTALLATIONS

Tropospheric scatter/diffraction installations must be configured in such a way as to (1) meet path performance requirements and (2) be econominally viable. All tropospheric scatter installations use some form of diversity, often quadruple diversity. Antennas usually have considerably larger apertures than their LOS radiolink counterparts. Antenna apertures range from 9 to 120 ft. Towers are seldom used. Transmitter power ranges from 1 to 50 kW, and unlike LOS installations, low-noise receiving systems are the rule.

A typical quadruple-diversity tropospheric scatter terminal layout is shown in Figure 7.30. It is made up of identifiable subsystems as follows:

☐ Antennas, duplexer, and transmission lines
☐ Modulator/exciters and power amplifiers

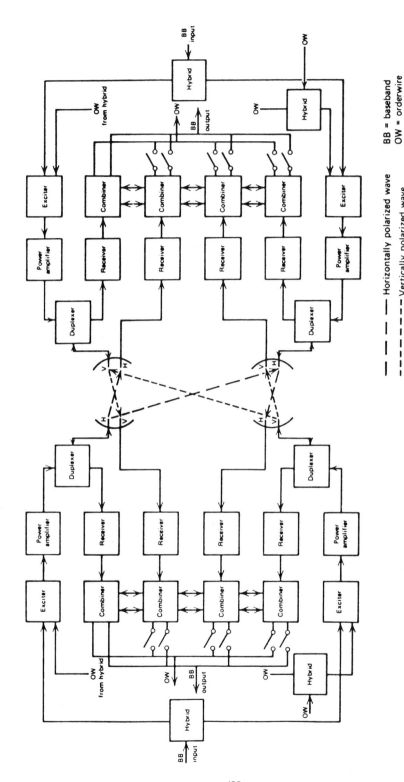

FIGURE 7.30 Simplified functional block diagram of a quadruple-diversity tropospheric scatter configuration for analog operation.

☐ Preselectors, receivers and threshold extension devices (optional)

☐ Diversity combiners

Through the proper selection and sizing of these devices, viable tropospheric/diffraction systems can be implemented for links up to 250 statute miles (400 km).

Tropospheric scatter links commonly operate in the following frequency bands:

350–450 MHz

755–985 MHz

1700–2400 MHz

4400–5000 MHz

The reader should also consult CCIR Rec. 388 and CCIR Reps. 285-2 and 286 (Ref. 9).

7.4.1 Antennas, Transmission Lines, Duplexer, and Related Transmission Line Devices

The antennas most commonly used in tropospheric scatter/diffraction installations are broadband, high-gain parabolic reflector devices. The antennas described here are similar in many respects to those discussed in Section 7.2.2, but have higher gain, in some cases better efficiency, therefore, are larger and considerably more expensive. In some installations they are the major cost driver.

As we have discussed previously in this text, the gain of this type of antenna is a function of the reflector diameter or aperture; 1 or 2 dB additional gain can be obtained by improved efficiency, particularly through the selection of the feed type. Improved antenna feeds illuminate the reflector more uniformly and reduce spillover.

It is desirable, but not always practical, to have the two antennas (space diversity) of a terminal spaced not less than 100 wavelengths apart to ensure proper space diversity operation. Antenna spillover (i.e., radiated energy in the sidelobes and back lobes) must be reduced not only to improve radiation efficiency but also to minimize interference with your own installation, with full duplex operation, and with other services. The first sidelobe should be down (attenuated) as referenced to the main beam at least 23 dB, and the remaining sidelobes should be down at least 40 dB from the main lobe. Antenna alignment is extremely important because of the narrow antenna beamwidths. These beamwidths are usually less than 2° and often less than 1° at the half-power (3-dB) points.

A good VSWR is also important, not only from the standpoint of improving system efficiency, but also because the resulting reflected power with a poor VSWR may damage components further back in the transmission chain. Often, load isolators are required to minimize damaging effects of reflected waves (e.g., reflected power). In high-power tropospheric scatter systems these devices may even require a cooling system to remove that the heat generated by the reflected energy.

A load isolator is a ferrite device with approximately 0.5-dB insertion loss. The forward wave (e.g., the energy radiated toward the antenna) is attenuated 0.5 dB; the reflected wave (e.g., the energy reflected back from the antenna) is attenuated more than 20 dB.

Another important consideration in planning a tropospheric scatter/diffraction antenna system is polarization (see Figure 7.30). For a common antenna the transmit wave should be orthogonal to the receive wave. This means that if the transmitted signal is horizontally polarized, the receive signal should be vertically polarized. The polarization is established by the feed device, usually a feed horn. The primary reason for using opposite polarizations is to improve isolation, although the correlation of fading on diversity paths may also be reduced. A figure commonly encountered for isolation between polarizations on a common antenna is 35 dB.

In selecting and laying out transmission lines for tropospheric scatter installations, it should be kept in mind that losses must be kept to a minimum. That additional fraction of a decibel is much more costly in tropospheric scatter/diffraction installations than in radiolink installations. The tendency, therefore, is to use waveguide on most tropospheric scatter installations because it displays lower loss than coaxial cable. Waveguide is universally used above 1.7 GHz.

Transmission line runs should be less than 200 ft (60 m). The attenuation of the line should be kept under 1 dB, whenever possible, from the transmitter to the antenna and from the antenna to the receiver, respectively. To minimize reflective losses, the VSWR of the line should be 1.05 : 1 or better when the line is terminated in its characteristic impedance. Figure 7.12 shows some waveguide types commercially available.

The duplexer is a transmission line device which permits the use of a single antenna for simultaneous transmission and reception. For tropospheric scatter/diffraction application, a duplexer is a three-port device (See Figure 7.31) so tuned that the receiver leg appears to have an admittance approaching (ideally) zero at the receiving frequency. To establish this, sufficient separation in frequency is required between transmit and receive frequencies. Figure 7.31 is a simplified block diagram of a duplexer. The insertion loss of the duplexer in each direction should be less than 0.5 dB. Isolation between the transmiter port and the receiver port should be better than 30 dB. High-power duplexers are usually factory tuned. It should be noted that some textbooks call the duplexer a diplexer.

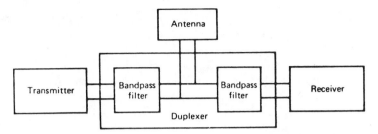

FIGURE 7.31 Simplified functional block diagram of a duplexer.

7.4.2 Modulator–Exciter and Power Amplifier

The type of modulation used on tropospheric scatter/diffraction links is commonly FM, but digital tropo is now taking on a particularly important role in military systems. We should also keep in mind as this discussion develops that tropospheric scatter/diffraction systems are high gain low-noise extensions of radiolink (LOS microwave) systems.

The tropospheric scatter/diffraction transmitter is made up of a modulator–exciter and a power amplifier (see Figure 7.30). Power outputs have been fairly well standardized at 1, 2, 10, 20, and 50 kW. For most commercial applications the 50-kW installation is often not feasible from an economic view point. Installations that are 2 kW or below are usually air-cooled. Those above 2 kW are liquid-cooled, usually with a glycol–water solution using a heat exchanger. When klystrons are used in high-power amplifiers, they are about 33% efficient. Thus a 10-kW klystron will require at least 20 kW of heat exchange capacity.

The transmitter frequency stability (long term) should be at least $\pm 0.001\%$ when FM is to be employed. Other modulation waveforms, such as SSB, require considerably better frequency stability/accuracy. Spurious emission should be down better than 80 dB below the carrier output level. A link using FM will require pre-emphasis/deemphasis as described in Section 2.6.5.

For FDM/FM configurations, the baseband modulating signal consists of FDM groups and supergroups. For many such installations, the baseband is made up of the band 60–552 kHz (CCITT supergroups 1 and 2). However, CCITT subgroup A, 12–60 kHz is often used as well. For longer-route tropospheric scatter systems, the link design engineer may tend to limit the number of voice channels, selecting a baseband configuration that lowers the highest modulating frequency to be transmitted as much as possible. This increases the overall system gain by reducing the bandwidth of the transmitted signal.

Conventionally, on FM systems, the modulator injects an RF pilot tone that is used for alarms at both ends of the link as well as to control far end combiners; 60 kHz is a common pilot tone for U.S. military systems. CCIR recommends 116 or 119 kHz for 24-channel systems, 304 or 331 kHz for 60-channel systems, and 607 (or 304) kHz for 120-channel systems (see CCIR Rec. 401-2) (Ref. 10).

The modulator also has a service channel input (FM operation). This is covered in CCIR Rec. 400-2 (Ref. 10). It recommends the use of the band 300–3400 Hz. U.S. military systems often multiplex more than one service channel in the band 300–12,000 Hz. One of these service channels is commonly assigned to transmit fault and alarm information.

The output of the power amplifier should have a low-pass filter to attenuate second harmonic output by at least 40 dB and third harmonic output by at least 50 dB.

7.4.3 FM Receiver Group

The receiver group in an FM tropospheric scatter installation usually consists of two or four identical receivers in dual- or quadruple-diversity configurations, respectively. Receiver outputs are combined in maximal ratio square combiners or in other types of combiners such as selection combiners. (See Section 7.2.4.) A simplified functional block diagram of a typical quadruple-diversity receiving system is shown in Figure 7.32.

Unlike their microwave LOS counterpart, tropo receiving systems universally use low-noise front ends or LNAs. Table 7.5 gives some typical receiver

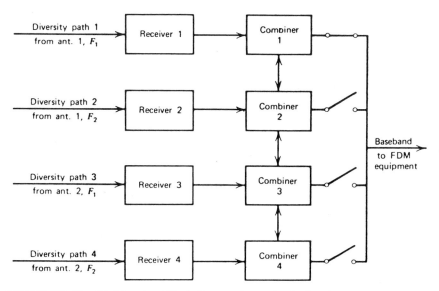

FIGURE 7.32 Simplified functional block diagram of a quadruple-diversity troposcatter receiving system. F_1 = frequency 1 and F_2 = frequency 2.

TABLE 7.5. Typical Receiver Front End NFs

Frequency Band (MHz)	Type	NF (dB)
350–450	Bipolar	1.0
775–985	Bipolar	1.0
1700–2400	GaAs FET	1.1
4400–5000	GaAs FET	1.5

TABLE 7.6. IF Bandwidth Required as a Function of Channel Capacity and Per-Channel Deviation

	Number of Channels			
	60	120	180	240
Baseband limits, kHz				
Lower (f_b)	12 or 60	12 or 60	60	60
Upper (f_m)	252 or 300	504 or 552	804	1052
Baseband bandwidth B_b, kHz	240	492	744	992
$10 \log B_b/b_c$	18.9	22.0	23.8	25.1
RMS noise loading ratio, NLR, dBm0	7.8	10.8	12.6	13.8
NLR × pf	11.58	16.38	20.07	23.17
For $\delta = 100$ kHz				
Peak carrier deviation ΔF, kHz	1158	1638	2007	2317
IF bandwidth B_{IF}, kHz	2820 or 2916	4284 or 4380	5622	6738
For $\delta = 140$ kHz				
Peak carrier deviation ΔF, kHz	1622	2293	2809	3244
IF bandwidth B_{IF}, kHz	3748 or 3844	5594 or 5690	7226	8592
For $\delta = 200$ kHz				
Peak carrier deviation ΔF, kHz	2317	3276	4014	4634
IF bandwidth B_{IF}, kHz	5138 or 5234	7560 or 7656	9636	11373

Source. From MIL-HDBK-417 (Ref. 12).

1. See Section 2.6.4.2.

2. ΔF is the peak carrier deviation in hertz; δf is the RMS per channel deviation in hertz; pf is the baseband signal peak factor, 13.5 dB; NLR is the RMS noise loading ratio in decibels; b_c is the usable voice channel bandwidth, 3100 Hz.

front end noise figures. Table 7.6 gives several maximum IF bandwidths for several voice channel configurations.

Another method to improve FM tropospheric scatter performance, particularly on long paths is to use threshold extension. The FM improvement threshold of a receiver can be "extended" by using a more complex and costly demodulator, called a "threshold extension demodulator." With a modulation index of 3 or better, the amount of improvement that can be expected using threshold extension over conventional receivers is on the order of 7 dB. For example, if the FM improvement threshold for a receiver without extension were -126.2 dBW, then with extension it would be -133.2 dBW.

Threshold extension works on an FM feedback principle, which reduces the equivalent instantaneous deviation, thereby reducing the required IF bandwidth (B_{if}), which in turn effectively lowers the receiver noise threshold. A typical receiver with a threshold extension module may employ a tracking filter that instantaneously tracks the deviation with a steerable bandpass filter having a 3-dB bandwidth of approximately four times the top baseband frequency. The control voltage for the filter is derived by making a phase comparison between the feedback signal and the IF input signal.

7.4.4 Diversity Operation

Some form of diversity is mandatory in a tropospheric scatter receiving system, primarily to mitigate Rayleigh (short-term) fading. Most systems employ quadruple diversity. There are several ways to obtain some form of quadruple diversity. One of the most desirable ways is shown in Figure 7.30, where both frequency and space diversity are utilized. For the frequency-diversity section, the design engineer must consider frequency separation, sufficient to reduce signal correlation between paths enough for the desired performance enhancement. As a minimum, at least 2% separation in frequency is required. That means that at 4 GHz we would calculate the frequency separation as 0.02×4000 MHz or 80 MHz. The ideal separation is 10%, but such a separation is difficult to obtain from regulating authorities.

Space diversity is easier to implement and is used universally on troposcatter systems. The physical separation of the antennas is the critical parameter. Separation is normally in the horizontal plane with a separation distance from 100 to 150 wavelengths.

Frequency diversity, although very desirable, often may not be permitted due to a shortage of frequency assignments in the desired band or due to RFI considerations. Another form of diversity to obtain the four orders of diversity making up quadruple diversity is based on polarization or what some call "polarization diversity." However, this is actually another form of space diversity and has been found not to provide a complete additional order of diversity. Nevertheless, it will often make do when the additional frequencies are not available to implement frequency diversity.

Polarization diversity is usually used in conjunction with conventional space diversity. The four space paths are achieved by transmitting signals in the horizontal plane from one antenna and in the vertical plane from the second antenna. On the receiving end two antennas are used, each antenna having dual-polarized feed horns for receiving signals in both planes of polarization. The net effect is to produce four signal paths that are relatively independent.

A discussion of diversity combiners is given in Section 7.2.4.

7.4.5 Isolation

An important factor in tropospheric scatter/diffraction installation design is the isolation between the emitted transmit signal and the receiver input. Normally we refer to the receiver sharing a common antenna feed with the transmitter.

For this discussion we can say that a nominal receiver input level is -80 dBm. If a tropo transmitter has an output power of 10 kW or $+70$ dBm, and transmission line losses are negligible, then overall isolation should be greater than 150 dB (e.g., 80 + 70).

To achieve the isolation necessary such that the transmitted signal interferes in no way with receiver operation during full duplex operation, the following items contribute to the required isolation when there is sufficient frequency separation between transmitter and receiver:

- ☐ Polarization
- ☐ Duplexer
- ☐ Receiver preselector
- ☐ Transmit filters
- ☐ Normal isolation from receiver conversion to IF

7.5 SATELLITE COMMUNICATIONS, TERMINAL SEGMENT

7.5.1 Functional Operation of a "Standard" Earth Station

7.5.1.1 The Communication Subsystem

Figure 7.33 is a simplified functional block diagram of an earth station showing the communication subsystem only. We shall use this figure to trace a signal through the equipment chain from antenna to baseband. Figure 7.34 is a more detailed functional block diagram of a typical earth station. By "standard" we can assume typically INTELSAT A, B, or C service or regional/national domestic satellite service.

The operation of an earth station communication subsystem in the FDMA/FM mode really varies little from that of an LOS radiolink system as

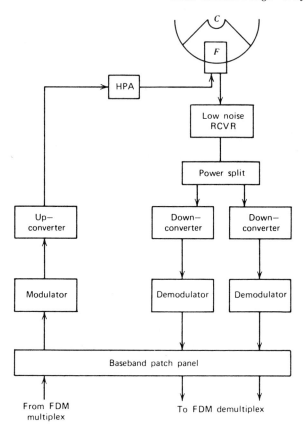

FIGURE 7.33 Simplified functional block diagram of an earth station communication subsystem. F = feed, HPA = high-power amplifier, C = Cassegrain subreflector.

shown in Figure 7.1. The variances are essentially these:

☐ Use of low-noise front ends on receiving systems, cryogenically cooled for large earth stations (e.g., INTELSAT A) and GaAs FETs for smaller terminals

☐ An HPA with a capability of from 200 watts to 8000 watts output

☐ Larger high-efficiency antennas, feeds

☐ Careful design to achieve as low a noise as possible

☐ Use of a signal-processing technique that allows nearly constant transmiter loading (e.g., spreading waveform), FM systems

☐ Use of threshold extension demodulators in some cases (FM systems only)

☐ Use of forward error correction on many digital systems, and above 10 GHz with interleaving to mitigate rainfall fading

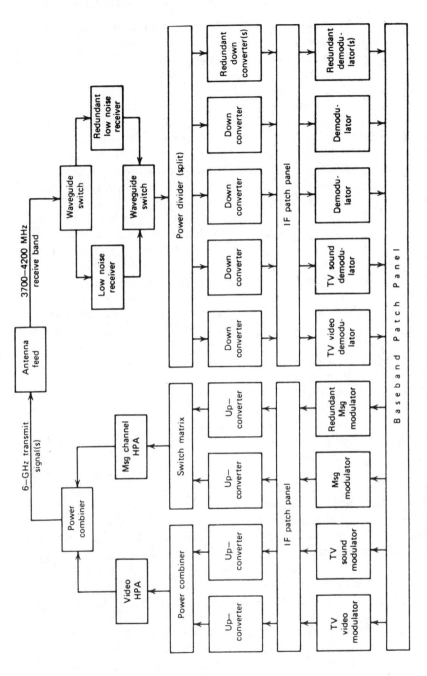

FIGURE 7.34 More detailed functional block diagram of a communication subsystem, typical of an INTELSAT Standard A earth station.

Now let us trace a signal through the communication subsystem typical of Figure 7.33. On the transmit side the FDM baseband is fed from the multiplex equipment through the baseband patch facility to the modulator. A spreading waveform is added to the very low end of the baseband to achieve constant loading. The baseband signal is then shaped with a pre-emphasis network (see Section 2.6.5). The baseband so shaped frequency modulates a carrier, and the resultant is then upconverted to a 70-MHz IF. Patching facilities usually are available at the IF to loop back through the receiver subsystem or through a test receiver for local testing or troubleshooting. The 70-MHz IF is then fed to an upconverter, which translates the IF to the output frequency (6 or 14 GHz). The signal is then amplified by the HPA, filtered by a low-pass filter, directed to the feed, and radiated by the antenna.

For reception, the signal derives from the feed and is fed to a low-noise receiver. In the case of an INTELSAT Standard A earth station, the low-noise receiver looks at the entire 500-MHz band (i.e., in the case of 4-GHz operation, the band from 3700 to 4200 MHz), amplifying this broadband signal 20–40 dB. When there are long waveguide runs from the antenna to the equipment building, the signal is amplified still further by a low-level TWT or SSA called a driver. The low-noise receiver is placed as close as possible to the feed to reduce ohmic noise contributions to the system. On INTELSAT Standard A earth stations, the LNA is commonly cryogenically cooled, usually with liquid helium, achieving a physical temperature of about 12–17 K. The equivalent noise temperature of the receiver, in this case, is on the order of 17–20 K. Smaller earth station facilities use GaAs FET LNAs with a noise temperature typically from 70 to 90 K at 4 GHz. The decision of which type of LNA to use is driven by the G/T requirements.

The comparatively high-level broadband receive signal is then fed to a power split. There is one output from the power split for every downconverter–demodulator chain. In addition, there is often a test receiver available as well as one or several redundant receivers in case of failure of an operational receiver chain. It should be kept in mind that every time the broadband incoming signal is split into two equal-level paths, there is a 3-dB loss due to the split, plus an insertion loss of the splitter. A splitter with eight outputs will incur a loss of something in the order of 10 dB.

A downconverter is required for each receive carrier, and there will be at least one receive carrier from each distant end. Each downconverter is tuned to its appropriate carrier and converts the signal to the 70-MHz IF. In some instances dual conversion is used.

The 70-MHz IF is then fed to the demodulator on each receive chain. The resulting demodulated signal, the baseband, is reshaped in the deemphasis network (see Section 2.6.5) and spreading waveform signal is removed. The resulting baseband output is then fed to the baseband patch facilities and thence to the demultiplex equipment. In many large facilities, threshold extension demodulators are used in lieu of conventional demodulators to achieve a *station margin* (link margin). Threshold extension techniques are described in Section 7.4.3.

It should be noted that the term often used to describe service is "message" service, which connotes common telephony service. When we see the terms "message upconverter" or "message demodulator," they refer to message service which is actually telephony service. This is in opposition to the equipment used to carry video picture, TV sound, or program channel traffic.

7.5.1.2 The Antenna Subsystem

The antenna subsystem is one of the most important component parts of an earth station, since it provides the means of radiating signals to the satellite and/or collecting signals from the satellite. The antenna not only must provide the gain necessary to allow proper transmission and reception, but also must have the radiation characteristics that discriminate against unwanted signals and minimize interference into other satellite or terrestrial systems.

Earth stations most commonly use parabolic dish reflector antennas or derivatives thereof. Dish diameters range from 1 to 30 m.

The sizing of the antenna and its design are driven more by the earth station required G/T than the EIRP. The gain is basically determined by the aperture (e.g., diameter) of the dish; but improved efficiency can also add to the gain in the order of 0.5–2 dB. For this reason the Cassegrain feed technique is almost always used on larger earth terminal installations. Smaller military terminals also resort to the use of Cassegrain. In some cases efficiencies as high as 70% and more have been reported. The Cassegrain feed working into a parabolic dish configuration (Figure 7.35) permits the feed to look into cool space as far as the spillover from the subreflector is concerned.

Antennas of generally less than 30-ft diameter often use a front-mounted feed horn assembly in the interest of economy. The most common type is the prime focus feed antenna (Figure 7.36). This is a more lossy arrangement, but, since the overall requirements are more modest, it is an acceptable one.

In order to keep interference levels on both the up and downlinks to acceptable levels, antenna sidelobe envelope limits (in dB) of $-32 - 25 \log \theta$ relative to the main beam maximum level, have been internationally adopted; θ is the angular distance in degrees from the main beam lobe maximum.

Figure 7.35 shows the functional operation of a Cassegrain fed antenna. Such an antenna consists of a parabolic main (prime) reflector and a hyperbolic subreflector. Here, of course, we refer to truncated parabolic and hyperbolic surfaces. The subreflector is positioned between the focal point and the vertex of the prime reflector. The feed system is situated at the focus of the subreflector, which also determines the focal length of the system. Spherical waves emanating from the feed are reflected by the subreflector. The wave then appears to be emanating from the virtual focus. These waves are then, in turn, reflected by the primary reflector into a plane wave traveling in the direction of the axis of symmetry. The size of the aperture (diameter) of the prime reflector determines the gain. The gain for a parabolic dish reflector antenna can be calculated from equation (4.51) (Section 4.6.4).

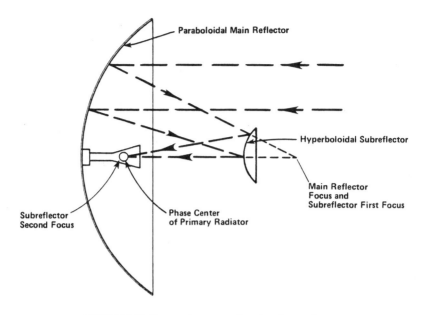

FIGURE 7.35 Cassegrain antenna functional operation.

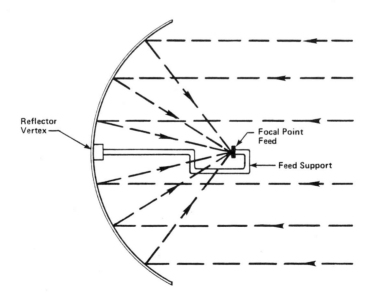

FIGURE 7.36 Prime focus antenna, functional operation.

It can be shown that a uniform field distribution over the reflector gives the highest gain of all constant phase distributions. The aperture efficiency can be shown to be a function of the following:

☐ Phase error loss
☐ Illumination loss due to a nonuniform amplitude distribution over the aperture
☐ Spillover loss
☐ Cross-polarization loss
☐ Blockage loss due to the feed, struts, and subreflector
☐ Random errors over the surface of the reflector (e.g., surface tolerance)

Nearly all very high gain reflector antennas are of the Cassegrain type. Within the main beam the antenna behaves essentially like a long focal-length front-fed parabolic reflector. Slight shaping of the two reflector surfaces can lead to substantial gain enhancement. Such a design also leads to more uniform illumination of the main reflector and less spillover. Typically efficiencies of Cassegrain type antennas are from 65 to 70%, which is at least 10% above most front-fed designs. It also permits the LNA to be placed close to the feed, if desired. The ability of the antenna to achieve these characteristics rests largely with the feedhorn design. The feedhorn radiation pattern has to provide uniform aperture illumination and proper tapering at the edges of the aperture.

The simplest type of feed for the antenna is a waveguide, which can be either open ended or terminated with a horn. Both rectangular and circular waveguide have been used with the circular considered superior, since it produces a more uniform illumination pattern over the aperture and provides better cross-polarization loss characteristics. The illumination pattern should taper to a value of -10 dB at an angle corresponding to the edge of the reflector (subreflector). This results in an asymmetric feed radiation pattern causing a loss in efficiency, increased cross-polarization losses, and moderate reflector spillover. Cross-polarization loss can be reduced by careful selection of waveguide radius, while improvement in efficiency can be achieved by using a corrugated horn. The positive aspect to 10 dB taper is to reduce sidelobe levels.

Smaller earth stations with less stringent G/T requirements resort to using the less expensive prime-focus-fed parabolic reflector antenna. The functional operation of this antenna is shown in Figure 7.36. For intermediate aperture sizes, this type of antenna has excellent sidelobe performance in all angular regions except the spillover region around the edge of the reflector. Even in this area a sidelobe suppression can be achieved that will satisfy FCC/CCIR pattern requirements. The aperture efficiency for apertures greater than about 100 wavelengths is around 60%. Therefore, it represents a good compromise choice between sidelobes and gain. For aperture sizes less than approximately

40 wavelengths, the blockage of the feed and feed support structure raises sidelobes with respect to the peak of the main beam such that it becomes exceedingly difficult to meet the FCC/CCIR sidelobe specification. However, the CCIR specification can be met since it contains a modifier that is dependent on the aperture size.

7.5.1.2.1 POLARIZATION

By use of suitable geometry in the design of an antenna feedhorn assembly, it is possible to transmit a plane electric wavefront in which the E and H fields have a well defined orientation. For linear polarization of the wave front, the convention of vertical or horizontal electric (E) field is adopted. The generation of linearly polarized signals is based on the ability of a length of square section waveguide to propagate a field in the $TE_{1,0}$ and $TE_{0,1}$ modes, which are orthogonally oriented. By exciting a short length of square waveguide in one mode by the transmitted signal and extracting signals in the orthogonal mode for the receiver, a means is provided for cross-polarizing the transmitted and received signals. The square waveguide is flared to form the antenna feedhorn. Similarly, by exciting orthogonal modes in a section of circular waveguide, a left- and right-hand rotation is imparted to the wavefront, providing two orthogonal circular polarizations.

The waveguide assembly used to obtain this dual polarization is known as a diplexer or an "orthomode junction." In addition to generating the polarization effect, the diplexer has the advantage of providing isolation between the transmitter and the receiver ports that could be in the range of 50 dB if the antenna presented a true broadband match. In practice, some 25–30 dB or more of isolation can be expected.

For linear polarization, discrimination between received horizontal and vertical fields can be as high as 50 dB, but diurnal effects coupled with precipitation can reduce this value to 30 dB. To maintain optimum discrimination, large antennas are equipped with a feed-rotating device driven by a polarization-sensing servo loop. Circularly polarized fields do not provide much more than 30 dB discrimination, but they tend to be more stable and do not require polarity tracking. This makes circular polarization more suitable for systems that include mobile terminals. One of the important parameters of antenna performance is how well cross-polarization is preserved across the operating spectrum.

Polarization discrimination can be used to obtain frequency reuse. Alternate transponder channel spectra are allowed to symmetrically overlap and are provided with alternate polarization. The channelization schemes of INTELSAT V and VI are typical. The number of transponders in the satellite is effectively doubled by this process. Polarization discrimination can also provide a certain degree of interference isolation between satellite networks if the nearest satellite (in terms of angular orbit separation) uses orthogonal uplink and downlink polarizations. (From Ref. 11.)

7.5.1.2.2 ANTENNA POINTING AND TRACKING

Satellites orbiting the earth are in motion. They can be in geostationary orbits and inclined orbits. Those in geostationary orbits appear to be stationary with respect to a point on earth. Those that are in inclined orbits are in motion with respect to a point on earth. All satellite terminals working with this latter class of satellite require a tracking capability.

Even though we have said that geostationary satellites appear stationary relative to a point on earth, they do tend to drift in small suborbits (figure eights). However, even with improved satellite station keeping, the narrow beamwidths encountered with large earth station antennas such as the INTELSAT Standard A 30-m (100-ft) require precise pointing and subsequent tracking by the earth station antenna to maximize the signal on the satellite and from the satellite. The basic modes of operation to provide these capabilities are:

☐ Manual pointing

☐ Programmed tracking (open-loop tracking)

☐ Automatic tracking (closed-loop tracking)

Pointing deals with "aiming" the antenna initially on the satellite. Tracking keeps it that way. Programmed tracking (open-loop tracking) may assume both duties. With programmed tracking, the antenna is continuously pointed by interpolation between values of a precomputed time-indexed ephemeris. With adequate information as to the actual satellite position and true satellite terminal position, pointing resolutions are in the order of 0.03–0.05°.

Manual pointing may be effective for initial satellite acquisition or "capture" for later active tracking (closed-loop tracking). It is also effective for wider beamwidth antennas, where the beamwidth is sufficiently wide to accommodate the entire geostationary satellite suborbit. Midsized installations may require a periodic trim up, and some smaller installations need never be trimmed up, assuming, of course, that the satellite in question is keeping good station keeping.

We will now discuss three types of active or closed loop tracking: monopulse, step-track and conscan.

Monopulse Tracking. Monopulse is the earliest form of satellite tracking, and today it is probably still the most accurate. In monopulse tracking, multiple antenna feed elements are used to obtain multiple received signals. The relative signal levels received by the various feed elements are compared to provide azimuth and elevation-angle-pointing error signals. The error signals are used to control the servo system, which operates the antenna drive motors.

Monopulse has taken its name from radar technology, and it derives from the fact that all directional information is obtained from a single radar pulse.

Beam switching or mechanical scanning is not necessary for its operation. In a three-channel monopulse system, RF signals are received by four antenna elements, usually horn feeds, located symmetrically around the boresight axis as an integral part of the antenna feed system. The multiple RF receive signals derived from these horns are combined in a beam-forming network (hybrid comparators) to produce sum and difference signals simultaneously in orthogonal planes. Figure 7.37 is a functional block diagram of the front end of a typical monopulse tracking system. The sum of the four-element radiation pattern is characterized by a single beam whose maximum lies on the antenna axis. The difference patterns are characterized by a null on the antenna axis

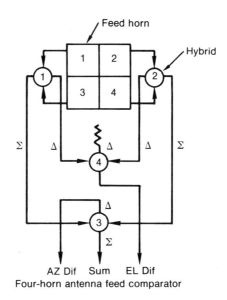

Four-horn antenna feed comparator

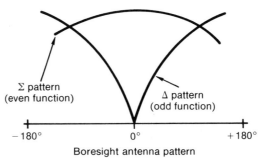

Boresight antenna pattern

Sum and difference processing and
resulting antenna patterns

FIGURE 7.37 Functional operation of three-channel monopulse tracking. From Ref. 11.

with the lobes on opposite sides in antiphase. Because the difference radiation pattern is in phase with respect to the sum pattern on one side of the antenna axis and out of phase on the other side, bearing angle sense information can be derived. By applying the sum signal to one input of a detector and the difference signal to the other input, an error voltage is produced which is proportional to the angle off-center and whose polarity is determined by the direction off-center.

In large earth stations such as INTELSAT Standard A, the sum and the two difference RF signals are kept separate. They are downconverted and applied to a three-channel tracking receiver to generate azimuth and elevation error signals for the antenna servo drive system. In other satellite terminals employing the monopulse technique, the azimuth and elevation difference signals are commutated onto a single channel, often by ferrite switches controlled by a digital scan generator. The commutated output of the ferrite switches is added to the communications (sum) channel by a directional coupler. Since the difference and sum signals are phase coherent, this has the effect of amplitude modulating the sum signal. This latter implementation of monopulse reduces the needed equipment from a three-channel tracking receiver to a single-channel receiver.

The satellite signal used for tracking is usually the satellite beacon channel. Thus, in this single-channel monopulse tracking design, the modulated sum channel is amplified by an LNA, downconverted, and demodulated in the beacon receiver. The signals are next decommutated and phase-detected to obtain error voltages that are used to drive the antenna so that the azimuth and elevation difference signals are minimized. Figure 7.38 is a functional block diagram of a typical satellite terminal monopulse tracking receiver subsystem.

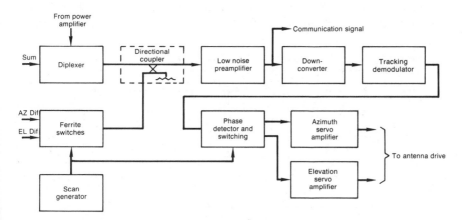

FIGURE 7.38 Functional block diagram of a satellite terminal monopulse tracking receiver subsystem.

The complexity and cost of monopulse feeds arise from the fact that they must be packaged into a small volume and, at the same time, provide low mutual coupling among the units without obstructing the illumination characteristics of the communications feed horn.

With monopulse tracking the beam scanning can be performed at almost any arbitrarily high rate, thus providing the potential for high tracking rates. On the other hand, with step track the tracking rate is limited by the dynamics of offsetting the antenna.

Step-Track. Despite the limitation of step tracking, it is cost effective where there are low dynamic tracking requirements such as with geostationary antennas of medium aperture size. It does not require the complex feed arrangement of the monopulse system and requires only simple, low-cost electronics. The only input signal required from the satellite terminal is the AGC voltage or other DC signal proportional to the received RF signal level such as a signal level indication from a communications demodulator or from a beacon receiver. The output of the step-track processor algorithm can be as simple as a periodic step function for each antenna axis or as complex as a pseudorandom sinusoid. This type of output applied to the antenna servo drive subsystem will result in smooth, continuous antenna motion. Figure 7.39 is a simplified functional block diagram of a step-track system.

Step tracking is a considerably lower-cost tracking system when compared to its monopulse counterpart. It is also less accurate. It lends itself to midsize satellite terminal installations operating with geostationary satellites and to tracking some inclined satellites with relatively slow relative orbital motion.

In the step-track technique the antenna is periodically moved a small amount along each axis, and the level of the received signal is compared to its previous level. A microprocessor, or part of the terminal control processor, provides processing to convert these level comparisons into input signals for the servo system, which will drive the antenna in directions which maximize the received signal level. In contrast to the monopulse technique, which

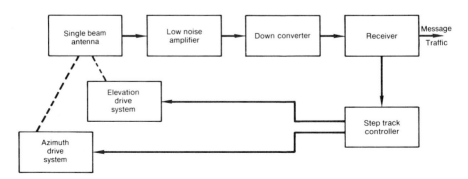

FIGURE 7.39 Functional block diagram of a step-track system.

seeks the null of the antenna difference pattern, step tracking seeks the signal peak. Locating a beam maximum can never be as accurate as finding a sharp null as with the monopulse tracking technique.

Conical Scan Tracking (Conscan). Conical scan tracking is a refinement of the old antenna lobing technique used in World War II radars. Some of these original tracking radars used an array of radiating elements that could be switched in phase to provide two beam positions for the lobing operation. The radar operator observed on his or her display the same target side by side, which were the returns of the two beam positions. When the target was on-axis, the two pulses were of even amplitude, and when moved off-axis, the two pulses became unequal. To track a remote target, all the operator did was to maintain a balance between the two pulses by steering the antenna correctly.

This lobing technique was refined to a continuous rotation of the beam around the target, which is the basis of conscan tracking. Angle-error-detection circuitry is provided to generate error voltage outputs proportional to the tracking error and with a phase or polarity to indicate the direction of the error. The error signal actuates a servosystem to drive the antenna in the proper direction to null the error to zero.

One method to accomplish this continuous beam scanning is by mechanically moving the antenna feed, since the antenna beam will move off axis as the feed is moved off the focal point. The feed is typically moved in a circular path around the focal point, causing a corresponding movement of the antenna beam in a circular path around the satellite to be tracked.

The feed scan motion may be either by a rotation or a nutation. A rotating feed turns as it moves with a circular motion, causing the polarization to rotate. A nutating feed does not rotate with the plane of polarization during the scan; it has an oscillatory movement of the axis of a rotating body or wobble. This is sometimes accomplished in the subreflector. The rotation or wobble modulates the received signal. The percentage of modulation is proportional to the angle tracking error, and the phase of the envelope function relative to the beam-scanning position contains direction information. In other words, this modulation is compared in phase with quadrature reference signals generated by the nutating mechanism to obtain error direction, and the amplitude of the modulation is proportional to the magnitude of the error. Figure 7.40 shows the conscan tracking technique and Figure 7.41 is a simplified functional block diagram of a conscan subsystem.

Conscan tracking may be used to track geosynchronous or polar orbit satellites with high or low target dynamics. Its principal advantages are low cost, that it requires only one RF channel, and it has a single-beam feed. Its principal disadvantages are that it requires four pulses or short duration continuous signal to obtain tracking information, is subject to mechanical reliability problems due to continuous feed rotation, and is subject to tracking loss due to propeller and other low-rate modulation sources.

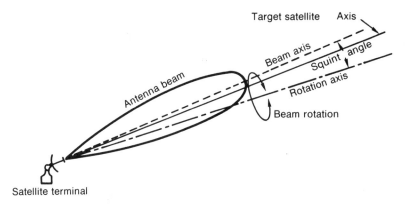

FIGURE 7.40 Conscan tracking operation.

Program Tracking (Open-Loop Tracking). Program tracking is a processor-based tracking system where the processor calculates the required azimuth and elevation angles as a function of time. Such a processor is called an ephemeris processor. Ephemeris (plural: ephemerides) refers to a tabulation of satellite locations referenced to a time scale. The ephemeris processor uses an algorithm that calculates the relative direction of the satellite with respect to the terminal on a continuous and real-time basis. It contains in memory the forecast satellite location with respect to time, which requires periodic updating every 30 or 60 days. For fixed terminal sites, site location latitude and longitude is programmed into the processor at installation; for mobile terminals the processor requires continuous positional updates, often provided by an inertial navigation system.

Since program tracking is an open-loop process, it is subject to several sources of error that are automatically corrected in autotrack (closed-loop) systems. Some of these error sources are atmospheric refractions; structural

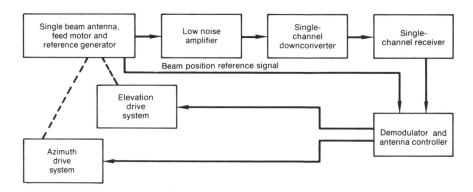

FIGURE 7.41 Simplified functional block diagram of a conical scanning subsystem.

deformation due to wind, ice, and gravity loads on the antenna; misalignment of mechanical axes; errors in axis angle measuring devices; and errors in input data relating to terminal position, satellite ephemeris, and absolute time.

In practice, a method of closed-loop tracking is usually included as well, since the accuracy required in the program tracking system is greatly relaxed if the last few tenths of a decibel in tracking accuracy are not required. In such cases program tracking is primarily used as an aid to initial satellite acquisition. However, for a geostationary satellite, acquisition causes little difficulty. In most cases a "look-up table," such as shown in Figure 4.5, suffices. Program tracking, therefore, usually has application for terminals that have to rapidly acquire nongeostationary satellites or rapidly slew from one satellite to another, as in the case of many military applications. Military ephemeris processors may have in memory ephemeris data for up to 20 satellites.

PROBLEMS AND EXERCISES

1. Draw a simplified functional block diagram of a basic radiolink terminal (as a minimum, nine components).

2. Give at least three reasons why frequency modulation is used so widely for radiolinks.

3. List similarities and differences between analog radiolink terminals and digital radiolink terminals.

4. Calculate the gain of a radiolink antenna with a 4-ft aperture. Assume an efficiency of 55%.

5. Why is sidelobe attenuation important on an antenna? Look at the antenna as both a transmitting device and a receiving device.

6. If the return loss of antenna is to be 30 dB, what is the equivalent reflection coefficient?

7. If an antenna has a gain of 40 dB, what is the beamwidth at the 3-dB points?

8. A radiolink transmitter operates at 7 GHz. Why would I rather use waveguide for the transmission line than coaxial cable? Select a waveguide type and calculate the loss for a 120-ft run?

9. Where would a circulator be used in a radiolink terminal?

10. What is the function of a load isolator?

11. Why would an IF repeater not be appropriate for digital radiolinks?

12. Why would maximal ratio combiners be favored in analog diversity application?

13. Compare one-for-one and one-for-n hot-standby operation.

14. Name at least three parameters/values that would cause hot-standby switchover for a transmitter chain and for a receiver chain.

15. Of what use are pilot tones on an analog radiolink?

16. Why would I want to specify limits to twist and sway for a radiolink tower?

17. Earth stations commonly use an LNA in receiving system front ends. Why is an LNA seldom used on LOS radiolinks?

18. Compare coherent and noncoherent detection for digital radiolinks.

19. Compare typical LOS radiolinks with troposcatter links as far as equipment is concerned.

20. How is sufficient isolation achieved on a troposcatter terminal when the same antenna transmits and receives simultaneously?

21. What is the function of a duplexer in a tropospheric scatter antenna system?

22. Draw a simplified functional block diagram of a satellite TV receive only (TVRO) installation.

23. Compare a prime focus antenna with a Cassegrain antenna.

24. What is the function of an orthomode junction?

25. Name and compare the two basic methods of tracking a satellite.

26. Why is step tracking attractive, where can it be used, and what are some of its limitations?

27. What tracking options does one have to track a highly inclined satellite in a low orbit (e.g., less than one-quarter synchronous)?

REFERENCES

1. R. L. Freeman, *Telecommunication Transmission Handbook*, 2nd ed, Wiley, New York, 1982.

2. "Andrew Antenna Systems," Catalog No. 32, Andrew Corporation, Orland Park, IL 60462, 1983.

3. "Design Handbook for Line-of-Sight Microwave Communication Systems," MIL-HDBK-416, U.S. Department of Defense, Washington, DC, November 1977.

4. A. P. Barkhausen et al., "Equipment Characteristics and Their Relationship to Performance for Tropospheric Scatter Communication Circuits," Technical Note 103, U.S. National Bureau of Standards, Boulder, CO, January 1962.

5. "Transmission Systems Engineering Symposium," Rockwell International, Collins Transmission Systems Division, Dallas, TX, September 1985.

6. CCIR Rec. 401-2, "Recommendations and Reports of the CCIR," 1982, Vol. IX, XVth Plenary Assembly, Geneva, 1982.

7. EIA RS-222B, Electronic Industries Association, Washington, DC, 1965.

8. CCITT Rec. G.703, "Recommendations and Reports from The International Telegraph and Telephone Consultive Committee (CCITT)," VIIth Plenary Assembly, Geneva, November 1980, Vol. III.

9. CCIR Rec. 388 and CCIR Reports 285-2 and 286, "Recommendations and Reports of the CCIR," 1982, XVth Plenary Assembly, Geneva, 1982, Vol. IX.

10. CCIR Rec. 400-2 and 401-2, "Recommendations and Reports of the CCIR," 1982, XVth Plenary Assembly, Geneva, 1982, Vol. IX.

11. *Satellite Communications Reference Data Handbook*, Computer Sciences Corp., Falls Church, VA, March 1983, DCA contract DCA100-81-C-0044.

12. "Facility Design for Tropospheric Scatter," U.S. Dept. of Defense, Washington, DC, November 1977, MIL-HDBK-417.

INDEX